W9-CRZ-232

Günther Rüdiger and Rainer Hollerbach

The Magnetic Universe

Geophysical and Astrophysical Dynamo Theory

Günther Rüdiger and Rainer Hollerbach

The Magnetic Universe

Geophysical and Astrophysical Dynamo Theory

WILEY-VCH

WILEY-VCH Verlag GmbH & Co. KGaA

Authors

Günther Rüdiger
Astrophysical Institute Potsdam
gruediger@aip.de

Rainer Hollerbach
Dept. of Mathematics, University of Glasgow
rh@maths.gla.ac.uk

Cover picture
Total radio emission and magnetic field vectors of
M51, obtained with the Very Large Array and the
Effelsberg 100-m telescope (λ=6.2 cm, see Beck
2000). With kind permission of Rainer Beck, Max-
Planck-Institut für Radioastronomie, Bonn.

This book was carefully produced. Nevertheless,
authors, and publisher do not warrant the information contained therein to be free of errors.
Readers are advised to keep in mind that statements, data, illustrations, procedural details or
other items may inadvertently be inaccurate.

Library of Congress Card No.: applied for
British Library Cataloging-in-Publication Data:
A catalogue record for this book is available from
the British Library

Bibliographic information published by
Die Deutsche Bibliothek
Die Deutsche Bibliothek lists this publication in
the Deutsche Nationalbibliografie; detailed bibliographic data is available in the Internet at
<http://dnb.ddb.de>.

© 2004 WILEY-VCH Verlag GmbH & Co. KGaA,
Weinheim

All rights reserved (including those of translation
into other languages). No part of this book may
be reproduced in any form – nor transmitted or
translated into machine language without written
permission from the publishers. Registered
names, trademarks, etc. used in this book, even
when not specifically marked as such, are not to
be considered unprotected by law.

Printed in the Federal Republic of Germany
Printed on acid-free paper

Printing Strauss GmbH, Mörlenbach
Bookbinding Litges & Dopf GmbH ,
Heppenheim
ISBN 3-527-40409-0

Contents

Preface

It is now 85 years since Sir Joseph Larmor first proposed that electromagnetic induction might be the origin of the Sun's magnetic field (Larmor 1919). Today this so-called dynamo effect is believed to generate the magnetic fields of not only the Sun and other stars, but also the Earth and other planets, and even entire galaxies. Indeed, most of the objects in the Universe have associated magnetic fields, and most of these are believed to be due to dynamo action. Quite an impressive record for a paper that is only two pages long, and was written before galaxies other than the Milky Way were even known!

However, despite this impressive list of objects to which Larmor's idea has now been applied, in no case can we say that we fully understand all the details. Enormous progress has undoubtedly been made, particularly with the huge increase in computational resources available in recent decades, but considerable progress remains to be made before we can say that we understand the magnetic fields even just of the Sun or the Earth, let alone some of the more exotic objects to which dynamo theory has been applied.

Our goal in writing this book was therefore to present an overview of these various applications of dynamo theory, and in each case discuss what is known so far, but also what is still unknown. We specifically include both geophysical and astrophysical applications. There is an unfortunate tendency in the literature to regard stellar and planetary magnetic fields as somehow quite distinct. How this state of affairs came about is not clear, although it is most likely simply due to the fact that geophysics and astrophysics are traditionally separate departments. Regardless of its cause, it is certainly regrettable. We believe the two have enough in common that researchers in either field would benefit from a certain familiarity with the other area as well. It is our hope therefore that this book will not only be of interest to workers in both fields, but that they will find new ideas on the 'other side of the fence' to stimulate further developments on their side (and maybe thereby help tear down the fence entirely).

Much of the final writing was done in the 2nd half of 2003. Without the technical support of Mrs. A. Trettin and M. Schultz from the Astrophysical Institute Potsdam it would not have been possible to finish the work in time. We gratefully acknowledge their kind and constant help. Many thanks also go to Axel Brandenburg, Detlef Elstner, and Manfred Schüssler – to name only three of the vast dynamo community – for their indispensable suggestions and never-ending discussions.

Potsdam and Glasgow, 2004

1 Introduction

Magnetism is one of the most pervasive features of the Universe, with planets, stars and entire galaxies all having associated magnetic fields. All of these fields are generated by the motion of electrically conducting fluids, via the so-called dynamo effect. The basics of this effect are almost trivial to explain: moving an electrical conductor through a magnetic field induces an emf (Faraday's law), which generates electric currents (Ohm's law), which have associated magnetic fields (Ampere's law). The hope is then that with the right combination of flows and fields the induced field will reinforce the original field, leading to (exponential) field amplification.

Of course, the details are rather more complicated than that. The basic physical principles may date back to the 19th century, but it was not until the middle of the 20th century that Backus (1958) and Herzenberg (1958) rigorously proved that this process can actually work, that is, that it is possible to find 'the right combination of flows and fields.' And even then their flows were carefully chosen to make the problem mathematically tractable, rather than physically realistic. For most of these magnetized objects mentioned above it is thus only now, at the start of the 21st century, that we are beginning to unravel the details of how their fields are generated.

The purpose of this book is to examine some of this work. We will not discuss the basics of dynamo theory as such; for that we refer to the books by Roberts (1967), Moffatt (1978) and Krause & Rädler (1980), which are still highly relevant today. Instead, we wish to focus on some of the details specific to each particular application, and explore some of the similarities and differences.

For example, what is the mechanism that drives the fluid flow in the first place, and hence ultimately supplies the energy for the field? In planets and stars it turns out to be convection, whereas in accretion disks it is the differential rotation in the underlying Keplerian motion. In galaxies it could be either the differential rotation, or supernova-induced turbulence, or some combination of the two.

Next, what is the mechanism that ultimately equilibrates the field, and at what amplitude? The basic physics is again quite straightforward; what equilibrates the field is the Lorentz force in the momentum equation, which alters the flow, at least just enough to stop it amplifying the field any further. But again, the details are considerably more complicated, and again differ widely between different objects.

Another interesting question to ask concerns the nature of the initial field. In particular, do we need to worry about this at all, or can we always count on some more or less arbitrarily small stray field to start this dynamo process off? And yet again, the answer is very different for different objects. For planets we do not need to consider the initial field, since both the

The Magnetic Universe: Geophysical and Astrophysical Dynamo Theory.
Günther Rüdiger, Rainer Hollerbach
Copyright © 2004 Wiley-VCH Verlag GmbH & Co. KGaA
ISBN: 3-527-40409-0

advective and diffusive timescales are so short compared with the age that any memory of the precise initial conditions is lost very quickly. In contrast, in stars the advective timescale is still short, but the diffusive timescale is long, so so-called fossil fields may play a role in certain aspects of stellar magnetism. And finally, in galaxies even the advective timescale is relatively long compared with the age, so there we do need to consider the initial field.

Accretion disks provide another interesting twist to this question of whether we need to consider the initial condition. The issue here is not whether the dynamo acts on a timescale short or long compared with the age, but whether it can act at all if the field is too weak. In particular, this Keplerian differential rotation by itself cannot act as a dynamo, so something must be perturbing it. It is believed that this perturbation is due to the Lorentz force itself, via the so-called magnetorotational instability. In other words, the dynamo can only operate at finite field strengths, but cannot amplify an infinitesimal seed field. One must therefore consider whether sufficiently strong seed fields are available in these systems.

Accretion disks also illustrate the effect that an object's magnetic field may have on its entire structure and evolution. As we saw above, the magnetic field always affects the flow, and hence the internal structure, in some way, but in accretion disks the effect is 'particularly dramatic.' It turns out that the transport of angular momentum outward – which of course determines the rate at which mass moves inward – is dominated by the Lorentz force. Something as fundamental as the collapse of a gas cloud into a proto-stellar disk and ultimately into a star is thus magnetically controlled. That is, magnetism is not only pervasive throughout the Universe, it is also a crucial ingredient in forming stars, the most common objects found within it.

We hope therefore that this book will be of interest not just to geophysicists and astrophysicists, but to general physicists as well. The general outline is as follows: Chapter 2 presents the theory of planetary dynamos. Chapters 3 and 4 deal with stellar dynamos. We consider only those aspects of stellar hydrodynamics and magnetohydrodynamics that are relevant to the basic dynamo process; see for example Mestel (1999) for other aspects such as magnetic braking. Chapter 5 discusses this magnetorotational instability in Keplerian disks. Chapter 6 considers galaxies, in which the magnetorotational instability may also play a role. Chapter 7, concerning neutron stars, is slightly different from the others. In particular, whereas the other chapters deal with the origin of the particular body's magnetic field, in Chapt. 7 we take the neutron star's initial field as given, and consider the details of its subsequent decay. We consider only the field in the neutron star itself though; see Mestel (1999) for the physics of pulsar magnetospheres. Lastly, Chapt. 8 discusses the magnetorotational instability in cylindrical Couette flow. This geometry is not only particularly amenable to theoretical analysis, it is also the basis of a planned experiment. However, we also point out some of the difficulties one would have to overcome in any real cylinder, which would necessarily be bounded in z.

Where relevant, individual chapters of course refer to one another, to point out the various similarities and differences. However, most chapters can also be read more or less independently of the others. Most chapters also present both numerical as well as analytic/asymptotic results, and as much as possible we try to connect the two, showing how they mutually support each other. Finally, we discuss fields occurring on lengthscales from kilometers to megaparsecs, and ranging from 10^{-20} to 10^{15} G – truly the magnetic Universe.

2 Earth and Planets

2.1 Observational Overview

We begin with a brief overview of the field as it is today, as well as how it has varied in the past. See also Merrill, McElhinny & McFadden (1998) or Dormy, Valet & Courtillot (2000) for considerably more detailed accounts of the observational data, or Hollerbach (2003) for a discussion of the theoretical origin of some of the timescales on which the field varies.

Figure 2.1 shows the Earth's magnetic field as it exists today. The two most prominent features, are (i) that it is predominantly dipolar, and (ii) that this dipole is quite closely aligned with the rotation axis, with a tilt of only $11°$. We would expect a successful geodynamo theory to be able to explain both of these features, as well as others, of course, such as why the field has the particular amplitude that it does.

Turning to the dipole dominance first, we begin by noting that much of this is an artifact of where we have chosen to observe the field, namely at the surface of the Earth. As we will see later, the field is actually created deep within the Earth, in the molten iron core, with the overlying mantle playing no direct role. Because the mantle (consisting of rock) is largely insulating, we can project the field back down to the core-mantle boundary (CMB). All components of the field are amplified when we do this, but the nondipole components are also amplified relative to the dipole, since they drop off faster with increasing radius, and hence increase faster when projected back inward again. Figure 2.1 also shows the resulting field at the CMB, which we note is indeed considerably less dipole dominated.

Figure 2.2 shows the corresponding power spectra, both at the surface and the CMB. The enhancement of the higher harmonics at the CMB is clearly visible. The other important point to note is that whereas the surface spectrum has been plotted to spherical harmonic degree $l = 25$, only the modes up to $l = 12$ have been projected inward to obtain the CMB spectrum. The reason for this is the sharp break observed in the surface spectrum at $l \approx 13$, with the power dropping off quite steeply up to there, but not at all thereafter. The generally accepted interpretation of this phenomenon is that this power in the $l > 12$ modes is due to crustal magnetism. These modes cannot therefore be projected back down to the CMB to obtain the spectrum there. Figure 2.1 (bottom) is thus not the true field at the CMB, but merely a filtered version of it, with all of the smallest scales having been filtered out. That is, the true field could very well exhibit highly localized features like sunspots, but this crustal contamination prevents us from ever observing them.

Turning next to the alignment of the dipole with the rotation axis, the probability that two vectors chosen at random would be aligned to within $11°$ or better is less than 2%. It seems more plausible therefore that this degree of alignment is not a coincidence, but instead reflects

The Magnetic Universe: Geophysical and Astrophysical Dynamo Theory.
Günther Rüdiger, Rainer Hollerbach
Copyright © 2004 Wiley-VCH Verlag GmbH & Co. KGaA
ISBN: 3-527-40409-0

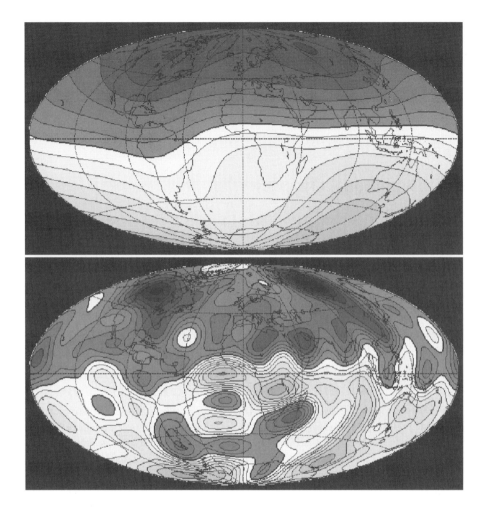

Figure 2.1: The radial component of the Earth's field at the surface (*top*), and projected down to the core-mantle boundary (*bottom*). Courtesy A. Jackson.

some controlling influence of rotation on the geodynamo. And indeed, we will see below that rotation exerts powerful constraints on the field (although it is not immediately obvious why this influence should lead to an alignment of the field with the rotation axis).

2.1.1   Reversals

Figure 2.1 shows the field as it is today. The field is not static, however, varying instead on timescales as short as minutes or even seconds, and as long as tens or even hundreds of millions of years. Of all of these variations, the most dramatic are reversals, in which the entire

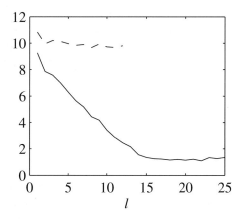

Figure 2.2: Power spectra of the Earth's field at the surface (solid) and the core-mantle boundary (dashed).

field switches polarity. See, for example, Gubbins (1994) or Merrill & McFadden (1999) for reviews devoted specifically to reversals.

Figure 2.3 shows the reversal record for the past 40 million years. The field is seen to reverse on the average every few hundred thousand years, but with considerable variation about that average. These relatively infrequent and irregular reversals of the Earth's field are thus very different from the comparatively regular, and much faster solar cycle.

Unlike the interval between reversals, the time it takes for the reversal itself seems to be a relatively constant 5–10 thousand years. During the reversal, the field is weaker, and considerably more complicated and less dipolar than in Fig. 2.1. Between reversals, however, it is generally similar to today's field, in terms of both field strength, dipole-dominated structure, and alignment with the rotation axis. This last point, of course, provides additional evidence that this alignment is not due to chance, but instead reflects the powerful influence of rotation.

Finally, the average interval between reversals itself varies on timescales of tens and hundreds of millions of years. For example, there were no reversals at all between 83 and 121 million years ago. Because these timescales are so much longer than any of the timescales 'naturally' present in the core, it is generally believed that this very long-term behavior is

Figure 2.3: The reversal sequence for the past 40 million years. Courtesy A. Witt.

of external origin. In particular, the timescale of mantle convection is precisely tens to hundreds of millions of years (e.g. Schubert, Turcotte & Olson 2001), so the thermal boundary conditions that the mantle imposes on the core will also evolve on these timescales. See, for example, Glatzmaier et al. (1999) for a series of numerical simulations in which different thermal boundary conditions did indeed lead to different reversal rates.

2.1.2   Other Time-Variability

As noted above, reversals are only the most dramatic variation in time found in the field. Between reversals the field varies as well, again with a broad range of timescales and amplitudes. Most familiar is the so-called secular variation, in which some of the nondipolar features fluctuate on timescales of decades to centuries. See for example Bloxham, Gubbins & Jackson (1989) or Jackson, Jonkers & Walker (2000) for summaries of the secular variation observed in the historical record. Intermediate between secular variation and reversals are also excursions, in which the field varies by considerably more than the usual secular variation, but does not actually reverse either. Excursions are around ten times more numerous than reversals, but of similar duration.

At the other extreme, the shortest timescales that can be observed within the core are geomagnetic jerks, in which the usual secular variation changes abruptly – and over the whole Earth – within a single year. Around three or four such events have been recorded in the past century (LeHuy et al. 1998). Note also that these events may well occur even faster than the one-year timescale on which they are recorded at the surface; the mantle is not a perfect insulator, and its weak conductivity effectively screens out any variations in the core occurring on timescales faster than a year. (For this reason also the variations in the field occurring on timescales as short as minutes or seconds must be of external origin, i.e. magnetospheric or ionospheric.)

2.2   Basic Equations and Parameters

The Earth's interior consists of a series of concentric spherical shells nested rather like the layers of an onion. The most fundamental division is that between the core and the mantle. The core, consisting mostly of iron, extends from the center out to a radius of 3480 km; the mantle, consisting of rock, extends from there essentially to the Earth's surface at $R = 6370$ km. In fact, the top 30 km or so are sufficiently different in their material properties (brittle rather than plastic, due to the much lower pressures and temperatures) that they are further distinguished from the mantle, and referred to as the crust. However, as important as the distinction between crust and mantle may be for phenomena such as plate tectonics, volcanism, earthquakes, etc. (e.g. Schubert, Turcotte & Olson 2001), the fact that both consist largely of rock, which is a very poor electrical conductor, immediately suggests that we must seek the origin of the Earth's magnetic field elsewhere, namely in the core. From the point of view of geodynamo theory, the mantle and crust are merely 3000 km of 'inconvenience' blocking what we would really like to observe (see Sect. 2.9.1 though).

Turning to the core then, it is further divided into a solid inner core of radius $R_{in} = 1220$ km, and a fluid outer core of radius $R_{out} = 3480$ km. The inner core was first detected seismically in 1936. See for example Gubbins (1997) for a review devoted specifically to the inner core. Further seismic studies show it to be sufficiently rigid to sustain shear waves (although it may actually be a so-called mushy layer right to the center, see, for example, Fearn, Loper & Roberts 1981). In contrast, the outer core is as fluid as water, with a viscosity of around 10^{-2} cm^2/s (Poirier 1994, De Wijs et al. 1998).

Further seismic (and other) studies also indicate that the density of the outer core increases from around 9.9 g/cm^3 at R_{out} to 12.2 g/cm^3 at R_{in}, at which point there is an abrupt jump to 12.8 g/cm^3 in the inner core. This value for the inner core is consistent with the density of around 90% pure iron (at the corresponding pressures and temperatures). The 5% jump across the inner core boundary cannot be explained purely by the phase transition from solid to liquid though; the outer core must contain perhaps 15–20% lighter impurities (with S, Si and O being the most likely candidates, e.g. Alfè et al. 2002).

With this basic structure of the core in place, we can begin to understand the dynamics that ultimately lead to the emergence of the Earth's magnetic field. As the Earth slowly cooles over billions of years, the core gradually solidifies, that is, the inner core grows. (The reason it solidifies from the center, even though it is hottest there, is due to the influence of the pressure on the melting temperature.) As it freezes, most of the impurities get rejected back into the fluid (just as freezing salt water will reject most of the salt, leaving relatively fresh water in the ice). As Braginsky (1963) was the first to point out, there are then two sources of buoyancy at the inner core boundary, namely that due to these light impurities being rejected back into the fluid, and that due to the release of latent heat from the freezing process itself. Additionally, of course, there is the usual source of (negative) buoyancy at the outer core boundary, namely that due to the fluid there losing heat to the mantle and hence becoming denser. It is these various sources of buoyancy that drive the convection that ultimately generates the magnetic field.

Incidentally, note also that we can extrapolate this cooling process backward to estimate when the inner core first formed. Buffett et al. (1992, 1996) considered detailed models of the thermal evolution of the core, and concluded that the inner core started to solidify around two billion years ago, and also that at present thermal and compositional effects are of comparable importance in powering the geodynamo. The precise age of the inner core continues to be debated though; recent estimates vary between one and three billion years (Labrosse & Macouin 2003 and Gubbins et al. 2003, respectively). It is quite interesting then that there is paleomagnetic evidence for the existence of a field as long ago as 3.5 billion years (McElhinny & Senanayake 1980). That is, there was most likely a dynamo even before the inner core formed, and hence before these various buoyancy sources at the inner core boundary became available.

2.2.1 Anelastic and Boussinesq Equations

Having discussed in qualitative terms the dynamics that lead to core convection and ultimately a magnetic field, our next task is to write down the specific equations. The most detailed analysis of these equations, and the various approximations one can make, is by Braginsky & Roberts (1995); here we merely summarize some of their findings. Linearizing the thermodynamics about an adiabatic reference state with density ρ_{a}, the momentum equation they ultimately end up with is

$$\frac{D\boldsymbol{u}}{Dt} + 2\boldsymbol{\Omega} \times \boldsymbol{u} = -\nabla P + C\,\boldsymbol{g}_{\mathrm{a}} + \frac{1}{\mu_0 \rho_{\mathrm{a}}}(\nabla \times \boldsymbol{B}) \times \boldsymbol{B} + \nu \Delta \boldsymbol{u}. \tag{2.1}$$

The so-called co-density C is given by $C = -\alpha_S S - \alpha_\xi \xi$, where S and ξ are the entropy and composition perturbations, respectively, and

$$\alpha_S = -\frac{1}{\rho}\frac{\partial \rho}{\partial S}, \qquad \alpha_\xi = -\frac{1}{\rho}\frac{\partial \rho}{\partial \xi} \tag{2.2}$$

determine how variations in S and ξ translate into relative density variations (this means of course that we also need a suitable equation of state $\rho = \rho(P, S, \xi)$ to determine these coefficients). One other point worth stating explicitly is that the gravity $\boldsymbol{g}_a$ appearing in Eq. (2.1) is that due to the adiabatic reference state only (hence the subscript); Braginsky & Roberts show that the self-gravity induced by the convective density perturbations themselves can be incorporated into the reduced pressure P. This is obviously a considerable simplification, as $\boldsymbol{g}_a$ is then known (varying roughly as $-\boldsymbol{r}$), rather than having to be solved for at every timestep of the other equations.

The continuity equation associated with Eq. (2.1) is $\nabla \cdot (\rho_a \boldsymbol{u}) = 0$, that is, rather than considering the fully compressible continuity equation we have made the anelastic approximation, and thereby filtered out sound waves[1]. The timescale for sound waves to traverse the entire core is around ten minutes, which is so much faster than any of the other dynamics we will be interested in that filtering them out completely is a reasonable approximation. (Note that this is very different from many astrophysical situations, where the Alfvén speed is often comparable with or even greater than the sound speed.) Finally, with the usual advection-diffusion equations for S and ξ, and of course the induction equation for $\boldsymbol{B}$, we have a complete set of equations that we should be able to timestep for S, ξ, $\boldsymbol{u}$ and $\boldsymbol{B}$.

As we will see in the remainder of this chapter, making actual progress with these equations is a formidable undertaking, primarily because some of the nondimensional parameters take on such extreme values. Many models therefore simplify these equations further still, in a variety of ways. For example, even though we saw that compositional and thermal sources of buoyancy are both important, most models neglect compositional effects, and consider thermal convection only. Given how different thermal and compositional convection can be (e.g. Worster 2000), this probably does affect at least the details of the solutions; neglecting compositional effects certainly cannot be rigorously justified. The only 'justification' one can offer is that we cannot even get the details of thermal convection right, so there is little point in worrying about the precise differences between thermal and compositional convection. For example, the compositional diffusivity is several orders of magnitude smaller than the thermal (e.g. Roberts & Glatzmaier 2000), but even the thermal diffusivity is orders of magnitude smaller than anything that any numerical model can cope with. So if both have to be increased to artificially large values, much of the difference between the two effects is also likely to disappear (although there are other differences as well, such as very different boundary conditions).

Another common simplification is to make the Boussinesq approximation, in which density variations are neglected everywhere except in the buoyancy term itself. That is, we replace

---

[1] see Lantz & Fan (1999) for a recent discussion of the anelastic approximation

the adiabatic density profile ρ_a by a constant, ρ_0. The Boussinesq approximation also cannot be rigorously justified (once again, see Braginsky & Roberts 1995). In particular, the variations in ρ_a that are being neglected are orders of magnitude greater than the convective density perturbations that are being included (very much unlike laboratory convection). However, given that the density contrast across the outer core is only $\sim 20\%$ (as we saw above), it seems likely that Boussinesq and anelastic results also will not differ by too much. There certainly do not appear to be any fundamental differences between the two.

We are therefore left with

$$\frac{D\boldsymbol{u}}{Dt} + 2\boldsymbol{\Omega} \times \boldsymbol{u} = -\nabla P - \alpha T \boldsymbol{g} + \frac{1}{\mu_0 \rho_0}(\nabla \times \boldsymbol{B}) \times \boldsymbol{B} + \nu \Delta \boldsymbol{u},$$

$$\frac{\partial \boldsymbol{B}}{\partial t} = \nabla \times (\boldsymbol{u} \times \boldsymbol{B}) + \eta \Delta \boldsymbol{B}, \qquad \left(\frac{\partial}{\partial t} + \boldsymbol{u} \cdot \nabla\right)T = \chi \Delta T, \qquad (2.3)$$

with $\nabla \cdot \boldsymbol{u} = 0$ and $\nabla \cdot \boldsymbol{B}$ as the simplest set of equations still 'reasonably' consistent with the original physics. (Note that when we neglect compositional effects, the entropy S can be replaced by the temperature T, with α then being the usual coefficient of thermal expansion.) These are the equations we will focus on, although in Sect. 2.4.7 we will return briefly to the original anelastic equation.

2.2.2   Nondimensionalization

Having settled on the equations, the next point we want to consider is how to nondimensionalize them, and what that might already tell us about the dynamics (that is, which terms are small or large, etc.). For a lengthscale, the obvious choice is the outer core radius $R_{\text{out}} = 3480$ km (many numerical models actually take $R_{\text{out}} - R_{\text{in}}$, but such minor details need not concern us here). The timescale is not quite so obvious, but a natural choice is the magnetic diffusive timescale R_{out}^2/η. Using the value $\eta \approx 2 \cdot 10^4$ cm^2/s appropriate for molten iron (Poirier 1994), this comes out to around 200,000 yr. (Incidentally, we see therefore that the range of timescales observed in the field varies from much shorter to much longer than this diffusive timescale.)

The fluid flow is then scaled by length/time $= \eta/R_{\text{out}} = O(10^{-6})$ m/s, so the advective and diffusive terms in the induction equation are (formally) comparable. Note though that the actual magnitude of the flow can only emerge from a full solution of the problem, and may turn out to be different from this value. Indeed, if the time evolution of the field at the core-mantle boundary is used to estimate the flow, one obtains magnitudes on the order of $10^{-4} - 10^{-3}$ m/s (Bloxham & Jackson 1991). That is, we would expect $\boldsymbol{u}$ to equilibrate at 10^{2-3} rather than order 1. This value of a few hundred is then also the magnetic Reynolds number $\text{Rm} = u R_{\text{out}}/\eta$ in the core.

The magnetic field is scaled by $(\Omega \rho_0 \mu_0 \eta)^{1/2} \approx 10$ G, which ensures that the Coriolis and Lorentz forces in the momentum equation are formally comparable. This is believed to be the appropriate balance at which the field equilibrates, for reasons that will become clear later. It also compares rather well with the ~ 3-G field observed at the CMB (particularly when we remember that the field deep within the core is likely to be at least somewhat stronger than right at the boundary). But once again, the actual magnitude of the field can only emerge from the complete solution. And as before with the magnitude of $\boldsymbol{u}$ giving us Rm, the magnitude of B (squared in this case) gives us the Elsasser number

$$\Lambda = \frac{B^2}{\Omega \rho_0 \mu_0 \eta}. \tag{2.4}$$

We see therefore that Λ is 0.1 to perhaps 1 in the core.

Finally, the natural scale for the temperature is simply the temperature difference δT across the core. However, there is one very considerable difficulty with this, namely estimating what δT actually is. In particular, the dynamically relevant temperature difference is only what is left over after the adiabatic temperature difference has been subtracted out. This ends up being virtually everything though: of the more than 1000 K difference across the core, the super-adiabatic δT that actually drives convection amounts to a small fraction of 1 K. In other words, δT cannot be estimated by taking the known temperature difference and subtracting out the adiabat; the errors would overwhelm the signal. Instead, δT can only be inferred indirectly by energetic/thermodynamic considerations.

With these scalings, the nondimensionalized Boussinesq equations become

$$\mathrm{Ro}\,\frac{D\boldsymbol{u}}{Dt} + 2\hat{\boldsymbol{e}}_z \times \boldsymbol{u} = -\nabla P + q\,\widehat{\mathrm{Ra}}\,T\,\boldsymbol{r} + (\nabla \times \boldsymbol{B}) \times \boldsymbol{B} + \mathrm{E}\Delta\boldsymbol{u},$$

$$\frac{\partial \boldsymbol{B}}{\partial t} = \nabla \times (\boldsymbol{u} \times \boldsymbol{B}) + \Delta \boldsymbol{B}, \qquad \left(\frac{\partial}{\partial t} + \boldsymbol{u} \cdot \nabla\right)T = q\Delta T. \tag{2.5}$$

The nondimensional parameters appearing in these equations are, first, the (modified) Rayleigh number

$$\widehat{\mathrm{Ra}} = \frac{g_0 \alpha \delta T R_{\mathrm{out}}}{\Omega \chi}, \tag{2.6}$$

where $g_0 = |\boldsymbol{g}(R_{\mathrm{out}})|$ (and by replacing $\boldsymbol{g}$ by $-\boldsymbol{r}$ in Eq. (2.5)$_1$ we are assuming for simplicity that gravity varies linearly with r). Note that this Rayleigh number measures the buoyancy force against the Coriolis force, rather than against the viscous force, as in classical Rayleigh–Benard convection. And once again, we remember that because of these uncertainties in δT, it is not clear just how large $\widehat{\mathrm{Ra}}$ is in the core. See, however, Gubbins (2001) for the latest estimates, and also Kono & Roberts (2001) for how $\widehat{\mathrm{Ra}}$ should even be defined when both thermal and compositional effects are important.

Next we have the Rossby number

$$\mathrm{Ro} = \frac{\eta}{\Omega R_{\mathrm{out}}^2}, \tag{2.7}$$

measuring the ratio of the rotational timescale Ω^{-1} ($=1/2\pi$ day) to the diffusive timescale R_{out}^2/η ($=200{,}000$ yr, as we saw above). That is, $\mathrm{Ro} = O(10^{-9})$. The Ekman number E (measuring viscous to Coriolis forces) and the Roberts number q

$$\mathrm{E} = \frac{\nu}{\Omega R_{\mathrm{out}}^2}, \qquad q = \frac{\chi}{\eta}, \tag{2.8}$$

(the latter measuring the ratio of thermal to magnetic duffusivity) come out to be $O(10^{-15})$ and $O(10^{-6})$, resp.

It is the extreme smallness of these three parameters that then makes the geodynamo equations so difficult. For example, if the advective term is at least as important as the diffusive term in Eq. $(2.5)_2$ (as we saw it is, and indeed must be to have any chance of achieving dynamo action), then in Eq. $(2.5)_3$ the advective term will dominate the diffusive term by many orders of magnitude, leading to extremely small lengthscales in T, which will certainly cause numerical difficulties, if nothing else. See also Christensen, Olson & Glatzmaier (1999) for further difficulties associated with the smallness of q.

These difficulties associated with q are usually 'solved' by invoking turbulent diffusivities, in which case all three diffusivities ν_T, η_T and χ_T will most likely be comparable, yielding $q_T = O(1)$ – which is indeed the range used in virtually all numerical models. However, one has not really solved the problem thereby, merely deferred it to a proper investigation of this small-scale turbulence. See, for example, Braginsky & Meytlis (1990), St. Pierre (1996), Davidson & Siso-Nadal (2002) and Buffett (2003) for models that begin to explore the precise nature of such rotating MHD turbulence.

And finally, even if an appeal to turbulent diffusivities solves (or rather ignores) the difficulties associated with q, those associated with Ro and E remain. In particular, η_T (and hence also ν_T) cannot be increased much beyond 100 m²/s, otherwise the field would simply decay faster than it can be sustained. This means though that even Ro_T and E_T are at most 10^{-7} – which is still several orders of magnitude smaller than most numerical models can cope with. Much of the remainder of this chapter will be devoted to discovering just why small Ro and E should pose such problems.

But first, there is one more general feature of Eqs. (2.5) worth mentioning, namely the associated energy equation. If we add the dot products of Eq. $(2.5)_1$ with u and Eq. $(2.5)_2$ with B, after a little algebra we obtain the global energy balance

$$\frac{\partial}{\partial t} \frac{1}{2} \int \left(|B|^2 + Ro\,|u|^2 \right) dV$$

$$= q\widehat{Ra} \int u_r Tr\,dV - \int \left(|\nabla \times B|^2 + E|\nabla \times u|^2 \right) dV. \tag{2.9}$$

The point we wish to focus on here is not so much the right-hand side (that is, how the energy changes), but rather the left, what the energy is in the first place. In particular, we recognize that if our nondimensionalization is correct, so that u and B do indeed equilibrate at roughly $O(1)$ values, then the magnetic energy will be several orders of magnitude *greater* than the kinetic. And because Ro is *so* small, this remains true even if u equilibrates at $O(10^3)$, as we saw above that it does. This is in sharp contrast to most astrophysical systems, where the magnetic energy is typically orders of magnitude smaller, or at best reaches equipartition.

Of course, if we included the energy stored in the Earth's rotation, we would be back in the astrophysically more familiar situation where the kinetic energy dominates by far. The rotational energy is not available though, since angular momentum must be conserved, so only deviations from solid-body rotation could be converted into magnetic (or other) forms of energy. And here again we see an enormous difference between the Earth and the Sun, for example; whereas in the Sun the differential rotation is a significant fraction of the overall rotation ($\sim 28\%$), in the Earth it is almost infinitesimal ($< 0.01\%$).

2.3   Magnetoconvection

Rotating, magnetic convection is a complicated process. Following Chandrasekhar (1961), let us therefore begin with classical Rayleigh–Benard convection, and first consider how rotation and magnetism separately alter the dynamics. Then we will explore how they act together, and finally what implications that might have for planetary dynamos, where the magnetic field is created by the convection itself, rather than being externally imposed.

Consider an infinite plane layer, heated from below and cooled from above. Additionally, there is an overall rotation $\Omega\hat{e}_z$, and an externally imposed magnetic field $B_0\hat{e}_z$. Linearizing about this basic state, the perturbation equations become

$$\frac{\partial \boldsymbol{u}}{\partial t} + 2\mathrm{E}^{-1}\hat{e}_z \times \boldsymbol{u} = -\nabla P + \Delta \boldsymbol{u} + \mathrm{Ra}\,\mathrm{Pr}^{-1}\,T\,\hat{e}_z + \mathrm{Ha}^2\,\mathrm{Pm}^{-1}\,(\nabla \times \boldsymbol{b}) \times \hat{e}_z$$

$$\frac{\partial \boldsymbol{b}}{\partial t} = \nabla \times (\boldsymbol{u} \times \hat{e}_z) + \mathrm{Pm}^{-1}\Delta \boldsymbol{b}, \qquad\qquad \frac{\partial T}{\partial t} - \boldsymbol{u} \cdot \hat{e}_z = \mathrm{Pr}^{-1}\Delta T, \qquad (2.10)$$

where length has been nondimensionalized by the layer thickness d, time by d^2/ν, $\boldsymbol{u}$ by ν/d, $\boldsymbol{b}$ by the imposed field B_0, and T by the imposed temperature difference δT. The nondimensional parameters are then the usual two Prandtl numbers $\mathrm{Pr} = \nu/\chi$ and $\mathrm{Pm} = \nu/\eta$, the Rayleigh number

$$\mathrm{Ra} = \frac{g\alpha\delta T d^3}{\nu\chi}, \qquad\qquad\qquad\qquad\qquad\qquad\qquad\qquad (2.11)$$

measuring the thermal forcing, the (inverse) Ekman number

$$\mathrm{E}^{-1} = \frac{\Omega d^2}{\nu}, \qquad\qquad\qquad\qquad\qquad\qquad\qquad\qquad\qquad (2.12)$$

measuring the rotation, and finally the Hartmann number

$$\mathrm{Ha} = \frac{B_0 d}{\sqrt{\mu_0 \rho \nu \eta}}, \qquad\qquad\qquad\qquad\qquad\qquad\qquad\qquad (2.13)$$

measuring the imposed magnetic field. Note also that the details of the nondimensionalization here – and hence the nondimensional parameters that arise – are different from those in Sect. 2.2.2. The reason for this is that here we want to start with classical Rayleigh–Benard convection, and only then add in rotation and magnetism, and study their effects. We must therefore also start with the classical nondimensionalization, so, for example, the usual Rayleigh number measuring buoyancy against viscosity, rather than against the Coriolis force, as in Eq. (2.6). Later on we will 'translate' the insight gained here into the geophysically more relevant parameters introduced in Sect. 2.2.2.

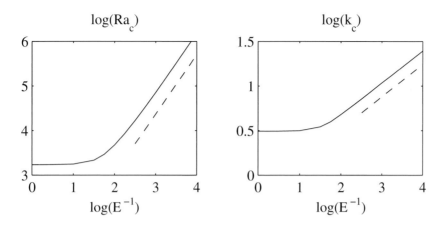

Figure 2.4: The influence of rotation without magnetism. *Left:* $\mathrm{Ra_c}$ as a function of E^{-1}. *Right:* k_c as a function of E^{-1}. The dashed lines have slopes 4/3 and 1/3, respectively, and indicate the scalings in the asymptotic limit.

Taking all quantities in Eq. (2.10) proportional to $\exp(\sigma t + ik_x x + ik_y y)$, we end up with the five equations

$$\sigma T = u_z + \mathrm{Pr}^{-1}\Delta T,$$

$$\sigma \Delta u_z = -2\mathrm{E}^{-1}\omega_z' + \nabla^4 u_z - \mathrm{Ra}\,\mathrm{Pr}^{-1} k^2 T + \mathrm{Ha}^2\,\mathrm{Pm}^{-1}\,\Delta b_z',$$

$$\sigma \omega_z = 2\mathrm{E}^{-1}u_z' + \Delta\omega_z + \mathrm{Ha}^2\,\mathrm{Pm}^{-1}j_z',$$

$$\sigma b_z = u_z' + \mathrm{Pm}^{-1}\Delta b_z,$$

$$\sigma j_z = \omega_z' + \mathrm{Pm}^{-1}\Delta j_z, \tag{2.14}$$

where u_z and b_z are the z-components of $\boldsymbol{u}$ and $\boldsymbol{b}$, ω_z and j_z the z-components of $\nabla \times \boldsymbol{u}$ and $\nabla \times \boldsymbol{b}$, the primes denote differentiation with respect to z, and $k^2 = k_x^2 + k_y^2$. Together with the boundary conditions

$$T = 0, \qquad u_z = u_z' = 0, \qquad \omega_z = 0, \qquad b_z = \pm b_z'/k, \qquad j_z = 0, \tag{2.15}$$

at $z = \pm d/2$, corresponding to rigid boundaries and electrically insulating exteriors, this system forms a well-defined eigenvalue problem that can be solved (numerically) for σ for any set of values for k, Ra, E^{-1} and Ha. Just as in Rayleigh–Benard convection, we are interested in the particular values $\mathrm{Ra_c}$ (and corresponding k_c) for which we first obtain exponentially growing solutions, that is, modes with $\Re(\sigma) > 0$. In the absence of rotation and magnetism, this critical Rayleigh number for the onset of convection is 1708, with associated wave number $k_c = 3.12$. We would like to discover then what effect nonzero E^{-1} and Ha have on this value, that is, whether rotation and magnetism help or hinder the onset of convection, and most importantly, how they interact with one another.

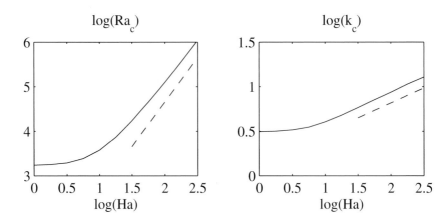

Figure 2.5: The influence of magnetism without rotation. *Left*: Ra_c as a function of Ha. *Right*: k_c as a function of Ha. The dashed lines have slopes 2 and 1/3, respectively, and indicate the scalings in the asymptotic limit.

2.3.1   Rotation or Magnetism Alone

Figure 2.4 (left) shows Ra_c as a function of E^{-1}, when $Ha = 0$. We note that it increases monotonically, ultimately scaling as $E^{-4/3}$ in the rapidly rotating limit. Rotation therefore suppresses convection. To see why, we turn to Eq. (2.14)$_3$, and note that for increasingly rapid rotation it becomes increasingly difficult to balance the term $2E^{-1}u'_z$ against any of the others: the magnetic term is out, because we are taking $Ha = 0$ here; the inertial term is also out, because these modes turn out to be steady, so $\sigma = 0$. If it were not for the viscous term, we would therefore have $u'_z = 0$ – which is of course just the familiar Taylor–Proudman theorem. Together with the boundary conditions, this would imply $u_z = 0$ though, eliminating the possibility of convective overturning. For convection to occur we must therefore break this Taylor–Proudman result, and as we just saw, the only way to achieve that is to balance the Coriolis term $2E^{-1}u'_z$ against the viscous term $\Delta\omega_z$. This in turn implies that the convection must occur on very short horizontal lengthscales, since only then can the viscous term compete with this very large factor E^{-1} in the Coriolis term. Indeed, we see in Fig. 2.4 (right) that k_c also increases monotonically, ultimately scaling as $E^{-1/3}$. Convection on ever shorter horizontal lengthscales is increasingly inefficient though, thereby explaining why Ra_c increases.

Figure 2.5 shows Ra_c and k_c as functions of Ha, when $E^{-1} = 0$. Both again increase monotonically, with Ra_c scaling as Ha^2 in the strongly magnetic limit, and k_c scaling as $Ha^{1/3}$. The reason why Ra_c increases is therefore just as before, because the convection is again being forced to occur on ever shorter horizontal lengthscales. This in turn is also easy to understand; the magnetic field tends to suppress all motion perpendicular to it, forcing the flow into tall, thin convection cells. More mathematically, the difficulty this time is in balancing the term $Ha^2\,Pm^{-1}\,\Delta b'_z$ in Eq. (2.14)$_2$. If b'_z were zero though, Eq. (2.14)$_4$ would again yield the unacceptable result $u'_z = 0$.

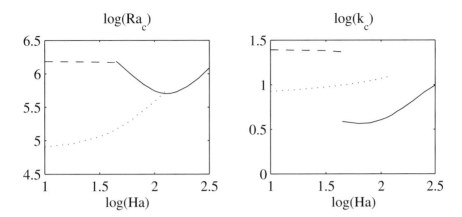

Figure 2.6: The effect of rotation and magnetism together. *Left*: Ra_c as a function of Ha. *Right*: k_c as a function of Ha. The dashed and solid lines denote the two different modes of convection discussed in the text. The dotted lines will be discussed in Sect. 2.3.4.

2.3.2   Rotation and Magnetism Together

We see therefore that acting alone, rotation and magnetism each suppress convection. When both act together though, the results could well be quite different. In particular, we note that then we can balance the Coriolis term $2E^{-1}u'_z$ against the magnetic term $Ha^2\,Pm^{-1}j'_z$ in Eq. (2.14)$_3$, and similarly in Eq. (2.14)$_2$. That is, the mechanisms that forced the convection to adopt very short horizontal lengthscales in either of the previous two cases do not apply here. If convection can occur with $k_c = O(1)$ though, Ra_c should also be much less than in either of the previous two cases.

Figure 2.6 shows Ra_c and k_c as functions of Ha, when $E^{-1} = 10^4$, and validates this argument. We see that initially (the dashed line) increasing Ha has almost no effect, with the rapid rotation continuing to suppress the convection. However, once Ha reaches a critical value, a transition takes place to a completely different mode of convection (the solid line), which occurs with $k_c = O(1)$, and correspondingly much lower Ra_c, exactly as suggested above. Doing the asymptotic analysis (Chandrasekhar 1961), one finds that this transition takes place when $Ha = O(E^{-1/3})$. And once on this second branch, the minimum occurs when $Ha = O(E^{-1/2})$, at which point Ra_c is also $O(E^{-1})$ (so the Coriolis, buoyancy and magnetic terms in Eq. (2.14)$_2$ are all comparable).

To summarize then, we have seen that while rotation and magnetism separately suppress convection, adding a magnetic field to a rotating system can facilitate convection again, reducing Ra_c from $O(E^{-4/3})$ for $Ha < O(E^{-1/3})$ down to $O(E^{-1})$ for $Ha = O(E^{-1/2})$. In the next section we will then (i) translate these results back into the geophysically more relevant parameters, and (ii) try to understand what implications they might have for planetary dynamos.

2.3.3   Weak versus Strong Fields

Doing the translation first, we note that the Ekman number is the same here and in Sect. 2.2.2, whereas the Rayleigh numbers are related by $\widehat{\mathrm{Ra}} = \mathrm{E}\,\mathrm{Ra}$. The Hartmann number is similarly related to the Elsasser number by $\Lambda = \mathrm{E}\,\mathrm{Ha}^2$. We therefore have that $\widehat{\mathrm{Ra}}_c = O(\mathrm{E}^{-1/3})$ for $\Lambda < O(\mathrm{E}^{1/3})$, and $\widehat{\mathrm{Ra}}_c = O(1)$ for $\Lambda = O(1)$ (having these last two quantities independent of E is, of course, what makes the nondimensionalization in Sect. 2.2.2 particularly convenient).

To assess what these results might imply for the geodynamo, we must consider the differences between our idealized Rayleigh–Benard problem and the real Earth. Most obviously, in the Earth we have a spherical shell rather than an infinite plane layer. This certainly makes the analysis considerably more complicated, and indeed adds various subtleties not present before. However, the main results are unchanged. Roberts (1968) and Busse (1970) considered rotating, nonmagnetic convection in spherical shells, and found that just as in the plane layer, it does not occur until $\widehat{\mathrm{Ra}} = O(\mathrm{E}^{-1/3})$. See also Jones, Soward & Mussa (2000) for the final(?) word on this problem. Similarly, Eltayeb & Kumar (1977), Fearn (1979) and Jones, Mussa & Worland (2003) considered rotating, magnetic convection, and found that there too the main results are as above.

Far more fundamental than this geometrical difference is the origin of the magnetic field; in this idealized problem it is externally imposed, whereas in the real Earth it is internally generated. That is, in the analysis above we could adjust Ha at will, but in the Earth we cannot adjust Λ; the amplitude of the field can only emerge as part of the full solution. Needless to say, this makes the problem considerably more difficult. Nevertheless, let us at least speculate about some of the implications that these results might have for internally generated rather than externally imposed fields.

In particular, imagine taking the Earth's core, and gradually increasing the Rayleigh number from zero. What sort of a sequence of bifurcations would we obtain? For $\widehat{\mathrm{Ra}} = 0$ we would clearly have $\boldsymbol{u} = 0$, and hence also $\boldsymbol{B} = 0$. The initial onset of convection therefore would be nonmagnetic, and would thus occur when $\widehat{\mathrm{Ra}} = O(\mathrm{E}^{-1/3})$. Increasing $\widehat{\mathrm{Ra}}$ further, the convection would presumably become more and more vigorous, until eventually a second critical value is reached where the flow acts as a dynamo. Immediately beyond this value, the field would most likely equilibrate as some very small value, but increasing $\widehat{\mathrm{Ra}}$ further still, both $\boldsymbol{u}$ and hence also $\boldsymbol{B}$ would presumably equilibrate at ever larger values.

In slowly rotating systems, this would presumably be all there is to it; the greater $\widehat{\mathrm{Ra}}$ is, the greater $\boldsymbol{u}$ and eventually $\boldsymbol{B}$ are, and that is it. If the system is rotating sufficiently rapidly though, the above analysis suggests that something quite dramatic could happen. Roberts (1978) conjectured that once the field exceeds $\Lambda = O(\mathrm{E}^{1/3})$, it would begin to facilitate the convection. A more vigorous flow would then yield a stronger field, which would further increase the flow, and so on. The resulting runaway growth would cease only when the field reaches $\Lambda = O(1)$, and the whole pattern of convection has switched from $O(\mathrm{E}^{1/3})$ to $O(1)$ lengthscales. Then once the system has switched to this new mode of convection, according to the results above it should also be possible to reduce $\widehat{\mathrm{Ra}}$ back down to some $O(1)$ value, and still maintain both the flow as well as the field. That is, the magnetic field facilitates convection to such an extent that one can have not only convection, but dynamo action, at a Rayleigh number lower than that for the initial onset of nonmagnetic convection. Indeed,

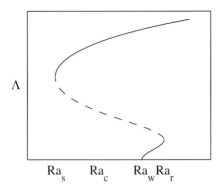

Λ

Ra_s Ra_c Ra_w Ra_r

Figure 2.7:   The sequence of bifurcations discussed in the text. The initial onset of nonmagnetic convection is denoted by $\widehat{\mathrm{Ra}}_c$, the onset of the weak-field regime by $\widehat{\mathrm{Ra}}_w$. The runaway growth occurs at $\widehat{\mathrm{Ra}}_r$. Once on the strong field branch, one can reduce $\widehat{\mathrm{Ra}}$ back down to $\widehat{\mathrm{Ra}}_s$ and still maintain both convection and dynamo action.

Malkus (1959) suggests that the Earth generates its field precisely in order to facilitate the convection, and that the $\Lambda = O(1)$ amplitude of the field is precisely that amplitude that most facilitates it.

As plausible as the above scenario might be, is there any compelling evidence that it is actually true, and if so, how small must the Ekman number be before distinct weak and strong-field regimes exist? Childress & Soward (1972), Soward (1974) and Fautrelle & Childress (1982) considered the infinite plane layer version of this problem, and concluded that there is indeed a point beyond which the weak-field regime ceases to exist. They were not able to prove the existence of a strong-field regime though, since the multiscale asymptotic methods that work for the weak-field do not work for the strong field. St. Pierre (1993) solved this problem numerically, and demonstrated that a strong-field regime does exist, and is subcritical, at $\mathrm{E} = 10^{-5}$. To date though no one has proven the existence of a subcritical, strong-field dynamo in the proper spherical shell geometry. Establishing that such solutions exist, and how small E must be before they exist, is one of the major issues facing geodynamo theory today.

Assuming that subcritical strong-field solutions do exist, what might be the geophysical implications? As we will see in the next section, the strong-field regime is particularly delicate, with small variations in B capable of inducing very large variations in u, which in turn act back on B, and so on. That is, where the weak-field regime was vulnerable to this runaway growth, the strong-field regime could suffer from runaway collapse. Where such a collapse would lead to depends on how large $\widehat{\mathrm{Ra}}$ is. If it is larger than where the runaway growth of the weak-field regime occurs, then even if one occasionally collapsed to the weak-field regime, one would just bounce right back. See for example Zhang & Gubbins (2000), who suggest that excursions may be caused by such temporary transitions.

If, however, $\widehat{\mathrm{Ra}}$ is less than the initial onset of nonmagnetic convection, the system could undergo a so-called *dynamo catastrophe*, in which both the dynamo and the convection suddenly switch off completely. If that happened, there would be no way of 'bouncing back'; the field would be gone forever (unless one could somehow increase the Rayleigh number again). Gubbins (2001) suggests that $\widehat{\mathrm{Ra}}$ is sufficiently large that this cannot happen in the Earth. It could conceivably have happened in other planets though, or could also happen in the Earth at some point in the future, when the core has cooled further, and $\widehat{\mathrm{Ra}}$ is smaller. The possibility of such a dynamo catastrophe is certainly a major concern in numerical simulations, where $\widehat{\mathrm{Ra}}$ cannot be too large, to avoid excessively fine structures appearing in the solution.

2.3.4   Oscillatory Convection Modes

Perceptive readers will notice that we have not mentioned either of the two Prandtl numbers Pr or Pm in any of the above discussion. The reason for this is that all of the convection modes considered so far turn out to be steady, so right at $\mathrm{Ra_c}$ we have not only $\Re(\sigma) = 0$, but $\Im(\sigma) = 0$ as well. If $\sigma = 0$ though, Eq. (2.14) can be rescaled to eliminate Pr and Pm entirely; simply define $\tilde{T} = \mathrm{Pr}^{-1}T$, $\tilde{b}_z = \mathrm{Pm}^{-1}b_z$ and $\tilde{j}_z = \mathrm{Pm}^{-1}j_z$. All of the $\mathrm{Ra_c}$ curves considered so far are therefore valid for any Prandtl numbers. However, it turns out that for sufficiently small Pr and/or Pm (and both are indeed small for liquid iron, with Pr around 0.1, and $\mathrm{Pm} \approx 10^{-6}$), oscillatory modes set in at lower Rayleigh numbers than these steady modes. The dotted line in Fig. 2.6 (left) shows $\mathrm{Ra_c}$ for one of these oscillatory modes. We see that now convection sets in so far below the previous results that the destabilizing influence of the magnetic field has disappeared completely; $\mathrm{Ra_c}$ is a monotonically increasing function of Ha, with the only other effect being to switch from oscillatory to steady convection once $\Lambda = O(1)$.

It might appear then that the existence of these oscillatory convection modes completely invalidates our entire previous discussion of weak versus strong-field regimes, since that was based specifically on the destabilizing influence of the field once $\Lambda > O(\mathrm{E}^{1/3})$. One must remember though that what ultimately matters is not just the initial onset of convection, but also how vigorous and efficient it is. These oscillatory modes turn out not to be very efficient, precisely because they oscillate far too rapidly (on the rotational timescale, as they must if inertia is to break the Taylor–Proudman theorem). See, for example, Zhang (1994) for an asymptotic analysis of these modes in a sphere, or Tilgner & Busse (1997) for numerical solutions demonstrating that the efficiency is indeed low in the weakly supercritical regime, and only increases in the much more strongly supercritical regime. See also Nakagawa (1959) and Aurnou & Olson (2001) for laboratory experiments in rotating magnetoconvection, using mercury and gallium, respectively. All of the phenomena discussed in this section were observed in one or other of these experiments, including transitions from less efficient oscillatory convection in the weakly supercritical regime to more efficient steady convection in the strongly supercritical regime. It seems likely, therefore, that the existence of these oscillatory convection modes will complicate, but not completely disrupt our weak versus strong-field bifurcation diagram in Fig. 2.7.

Finally, we should mention the work of Busse (2002b), who considered the onset of (nonmagnetic) convection in rapidly rotating systems, when both thermal and compositional buoyancy sources are included. These, of course, have very different diffusivities, hence different Prandtl numbers. Busse showed that in this case one can again obtain convection modes with lower $\mathrm{Ra_c}$ than if the Prandtl numbers were the same. The geophysical significance of these modes is not yet known though.

2.4   Taylor's Constraint

In the previous section we discovered that not all convectively driven dynamos are alike; unless the rotation is sufficiently rapid to obtain distinct weak and strong-field regimes, a given model is not even qualitatively in the right parameter range. In this section (and the next as well) we will see why increasingly rapid rotation is unfortunately also increasingly difficult.

2.4.1 Taylor's Original Analysis

We begin with the Navier-Stokes equation $(2.5)_1$. Taylor (1963) then argued that since Ro and E are so small, one should be able to set them identically to zero, thereby obtaining the so-called magnetostrophic balance

$$2\hat{e}_z \times \boldsymbol{u} = -\nabla P + \boldsymbol{F}_B + (\nabla \times \boldsymbol{B}) \times \boldsymbol{B}. \tag{2.16}$$

Taking the curl, and using also $\nabla \cdot \boldsymbol{u} = 0$, we get

$$-2 \frac{\partial}{\partial z} \boldsymbol{u} = \nabla \times \left[\boldsymbol{F}_B + (\nabla \times \boldsymbol{B}) \times \boldsymbol{B} \right], \tag{2.17}$$

so one might suppose that the solution is simply $\boldsymbol{u} = \boldsymbol{U}_t + \boldsymbol{U}_M$, where the thermal and magnetic winds are given by

$$\boldsymbol{U}_t = -\frac{1}{2} \int \nabla \times \boldsymbol{F}_B \, dz, \qquad \boldsymbol{U}_M = -\frac{1}{2} \int \nabla \times \left[(\nabla \times \boldsymbol{B}) \times \boldsymbol{B} \right] dz, \tag{2.18}$$

and we are temporarily ignoring the issue of constants of integration, and the boundary conditions used to determine them.

This analysis turns out to be oversimplified though; in general Eq. (2.16) simply has no solution at all. To see why, consider its ϕ-component

$$2u_s = -\frac{1}{s} \frac{\partial P}{\partial \phi} + \left[(\nabla \times \boldsymbol{B}) \times \boldsymbol{B} \right]_\phi, \tag{2.19}$$

where the subscripts denote the indicated components, and (z, s, ϕ) are *cylindrical* coordinates. The first point to note is that by considering the ϕ-component we have eliminated the purely radial buoyancy force $\boldsymbol{F}_B$. We next eliminate the pressure gradient by integrating over the so-called geostrophic contours $C(s)$ consisting of cylinders parallel to the axis of rotation (see Fig. 2.8 below). These contours are axisymmetric, so integrating $\partial P/\partial \phi$ once around in ϕ yields zero.

At this stage we are therefore left with

$$\int_{C(s)} 2u_s \, dS = \int_{C(s)} \left[(\nabla \times \boldsymbol{B}) \times \boldsymbol{B} \right]_\phi dS. \tag{2.20}$$

Now, what is $\int u_s \, dS$? Physically, it is just the net flow through the cylinder $C(s)$. And since we are taking the fluid to be incompressible, that net flow must vanish; the fluid cannot pile up either inside or outside $C(s)$. We are therefore left with just

$$\int_{C(s)} \left[(\nabla \times \boldsymbol{B}) \times \boldsymbol{B} \right]_\phi dS = 0, \tag{2.21}$$

stating that the integrated Lorentz torque must vanish, and on *each* such cylinder $C(s)$, if Eq. (2.16) is to have a solution at all.

Furthermore, even if Eq. (2.21) is satisfied, so that Eq. (2.16) has a solution, there is the additional complication that the solution is then not unique, since one can add to it an arbitrary geostrophic flow $U_g(s)\,\hat{e}_\phi$ (physically this amounts to each cylinder $C(s)$ undergoing solid-body rotation). To see why, we need only note that U_g trivially satisfies the Navier-Stokes equation (2.17), $\nabla \cdot \boldsymbol{u} = 0$, and also the no normal flow boundary condition (having dropped the viscous term, we can of course no longer impose no slip, but only no normal flow).

We see therefore that Eq. (2.16) either has no solution, or else an infinite number of solutions, depending on whether or not Taylor's constraint (2.21) is satisfied. In fact, the two problems of satisfying Eq. (2.21) and determining U_g are linked, as Taylor also showed. In particular, suppose the field satisfies Taylor's constraint at some initial time t_0. As the field then evolves (according to the induction equation), how do we ensure that it will continue to satisfy Eq. (2.21)? Clearly, what we need is not only Eq. (2.21), but also its time derivative

$$\frac{\mathrm{d}}{\mathrm{d}t} \int_{C(s)} \left[(\nabla \times \boldsymbol{B}) \times \boldsymbol{B} \right]_\phi \mathrm{d}S = 0. \tag{2.22}$$

This turns out to determine U_g in the following way: First replace Eq. (2.22) by

$$\int_{C(s)} \left[(\nabla \times \frac{\partial \boldsymbol{B}}{\partial t}) \times \boldsymbol{B} + (\nabla \times \boldsymbol{B}) \times \frac{\partial \boldsymbol{B}}{\partial t} \right]_\phi \mathrm{d}S = 0. \tag{2.23}$$

Then we simply remember that the induction equation gives us

$$\frac{\partial \boldsymbol{B}}{\partial t} = \nabla \times \left((\boldsymbol{U}_t + \boldsymbol{U}_M + U_g(s)\,\hat{e}_\phi) \times \boldsymbol{B} \right) + \Delta \boldsymbol{B}, \tag{2.24}$$

where $\boldsymbol{U}_t$ and $\boldsymbol{U}_M$ are known (according to Eq. (2.18)), and only U_g is unknown. Inserting Eq. (2.24) into Eq. (2.23) therefore gives us an equation in which the only unknown is U_g. Just counting the derivatives on U_g, we can already see that the general form of this equation will be

$$A_2(s) \frac{\mathrm{d}^2}{\mathrm{d}s^2} U_g + A_1(s) \frac{\mathrm{d}}{\mathrm{d}s} U_g + A_0(s)\,U_g = B(s), \tag{2.25}$$

where the coefficients $A_n(s)$ and $B(s)$ involve integrals over the geostrophic contours $C(s)$ of various combinations of the known quantities $\boldsymbol{B}$, $\boldsymbol{U}_t$ and $\boldsymbol{U}_M$. Subject to suitable boundary conditions at $s = 0$ and 1 (we will consider various complications introduced by the inner core and its associated tangent cylinder later), one might then hope to invert Eq. (2.25) for U_g.

Taylor's original idea therefore was that the field would exactly satisfy Eq. (2.21), and would evolve according to Eq. (2.24), with the geostrophic flow determined at each instant by Eq. (2.25), thereby ensuring that Eq. (2.21) continues to be satisfied. However, as elegant as this prescription undoubtedly is, no one has ever succeeded in following it. There are a number of reasons for this. One difficulty concerns the distinction between weak- and strong-field regimes introduced in the previous section. If we now set $\mathrm{E} = 0$, the weak-field regime disappears off to $\widehat{\mathrm{Ra}} \to \infty$. The remaining strong-field regime is therefore especially vulnerable to the dynamo catastrophe mentioned before. In the next section we will also consider further difficulties that result when one attempts to set $\mathrm{E} = 0$, and that may also play a role in this lack of success in following Taylor's prescription.

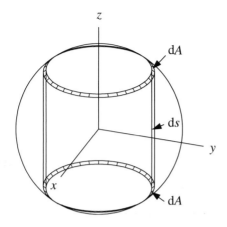

Figure 2.8: The shells used in the derivation of the torque balance Eq. (2.31). In the limit $ds \rightarrow 0$ these shells become the geostrophic contours $C(s)$. Note also the absence of an inner core; we will consider it and its associated tangent cylinder in Sect. 2.6.

2.4.2 Relaxation of Ro = E = 0

Let us return to the original Navier-Stokes equation, and consider how Taylor's development is modified if we do not attempt to set Ro and E identically equal to zero. After all, Taylor's constraint only arose because we made that rather drastic step, so relaxing it will also relax Taylor's constraint in some way. So what we want to consider here is precisely how Taylor's constraint is modified, and whether this new prescription can be more successfully implemented.

Returning to the geostrophic contours $C(s)$, it is convenient at this stage to consider a shell of finite thickness ds, as indicated in Fig. 2.8. Later we will simply let $ds \rightarrow 0$. So, let us consider the torque balance on such a shell when we restore inertia and viscosity. As always, the general balance is just

$$I\frac{d\Omega}{dt} = \Gamma, \tag{2.26}$$

where Ω is the shell's angular velocity, I its moment of inertia, and Γ the sum of all the torques acting on it. More specifically then, $\Omega = U_g/s$, where U_g is the same geostrophic flow as before (we are interested in the torque balance on the whole shell, after all, so only the z-independent part of its rotation is relevant here). The moment of inertia is similarly straightforward, yielding $I = \text{Ro} \, 4\pi \, (1 - s^2)^{1/2} \, s^3 \, ds$. The torque balance (2.26) therefore becomes

$$\text{Ro} \, 4\pi \, (1 - s^2)^{1/2} \, s^2 \, ds \, \frac{dU_g}{dt} = \Gamma. \tag{2.27}$$

Next, what is Γ, that is, what are the various torques acting on this shell? The magnetic torque is of course much the same as before, namely just

$$\Gamma_{\text{M}} = \int \left[(\nabla \times \boldsymbol{B}) \times \boldsymbol{B} \right]_{\phi} s \, dV, \tag{2.28}$$

where the extra factor of s comes about because we are now explicitly considering the torque rather than the ϕ-component of the force, and the dV rather than dS reflects the finite thickness of the shell ds, and hence a volume rather than surface integral.

So finally, the only other contribution we need to include is the viscous torque. Because E is so small, we will not consider viscosity in the interior, but only in the top and bottom Ekman boundary layers. In these layers, the viscous drag per unit area is then $E(-U_g)/\delta$, where δ is the thickness of the layer, and $(-U_g)$ the jump in the zonal flow across it, from zero at $r = 1$ to U_g at $r = 1 - \delta$. The viscous torque is thus $(-E\,U_g/\delta)s\,dA$, where dA is the area indicated in Fig. 2.8, and is related by

$$dA = 2 \cdot 2\pi s \frac{ds}{(1 - s^2)^{1/2}} \tag{2.29}$$

to the infinitesimal thickness ds of the shell. Finally, we just apply the standard result (e.g. Greenspan 1968) that the thickness of the Ekman layer on a spherical boundary is $\delta = E^{1/2}/(1 - s^2)^{1/4}$ to obtain

$$\Gamma_\nu = -E^{1/2} \frac{4\pi s^2 \, ds}{(1 - s^2)^{1/4}} \, U_g. \tag{2.30}$$

Putting it all together, and letting $ds \to 0$, our torque balance thus becomes

$$\text{Ro}\, 4\pi \, (1 - s^2)^{1/2} \, s^2 \, \frac{\partial U_g}{\partial t}$$

$$= \int_{C(s)} \left[(\nabla \times \boldsymbol{B}) \times \boldsymbol{B} \right]_\phi s \, dS - E^{1/2} \frac{4\pi s^2}{(1 - s^2)^{1/4}} \, U_g. \tag{2.31}$$

This therefore is the generalization of Taylor's constraint to include inertial and viscous effects. In terms of the physics, we see that the two versions are very similar; both are simply torque balances applied to the geostrophic contours $C(s)$. From a mathematical point of view, however, the two are very different; whereas Taylor's original constraint is a solvability condition that must be satisfied exactly, in Eq. (2.31) the magnetic torque must be small (because Ro and $E^{1/2}$ are small) but not necessarily identically zero. Furthermore, whereas in Taylor's prescription the geostrophic flow is determined in this very roundabout manner, Eq. (2.25), in this new prescription it is determined directly by Eq. (2.31) itself. For example, if we set Ro $= 0$ again (which is indeed the most commonly implemented version of Eq. (2.31), and hence the one we will focus attention on), we have simply

$$U_g = E^{-1/2} \frac{(1 - s^2)^{1/4}}{4\pi s} \int_{C(s)} \left[(\nabla \times \boldsymbol{B}) \times \boldsymbol{B} \right]_\phi dS. \tag{2.32}$$

2.4.3　Taylor States versus Ekman States

Given how different these two prescriptions, Eqs. (2.32) and (2.25) are, depending on whether one does or does not include viscosity, it seems appropriate to begin by showing that the viscous version can nevertheless recover Taylor's inviscid solutions. We expand $\boldsymbol{B}$ as

$$\boldsymbol{B} = \boldsymbol{B}_0 + E^{1/2}\boldsymbol{B}_1 + E\boldsymbol{B}_2 + \dots, \tag{2.33}$$

where $\boldsymbol{B}_0$ must satisfy Taylor's constraint, Eq. (2.21), identically (we will see in a moment how this is enforced precisely by Taylor's original prescription). According to Eq. (2.32), the geostrophic flow is then given by

$$U_{\mathrm{g}} = (U_{01} + U_{10}) + \mathrm{E}^{1/2}(U_{02} + U_{11} + U_{20}) + \dots, \tag{2.34}$$

where

$$U_{ij} = \frac{(1 - s^2)^{1/4}}{4\pi s} \int_{C(s)} \left[(\nabla \times \boldsymbol{B}_i) \times \boldsymbol{B}_j \right]_\phi \mathrm{d}S. \tag{2.35}$$

Similarly, the magnetic wind, Eq. (2.18), becomes

$$\boldsymbol{U}_{\mathrm{M}} = \boldsymbol{U}_{00} + \mathrm{E}^{1/2}(\boldsymbol{U}_{01} + \boldsymbol{U}_{10}) + \dots, \tag{2.36}$$

where

$$\boldsymbol{U}_{ij} = -\frac{1}{2} \int \nabla \times \left[(\nabla \times \boldsymbol{B}_i) \times \boldsymbol{B}_j \right] \mathrm{d}z. \tag{2.37}$$

Explicitly separated out order by order, the induction equation therefore becomes

$$\frac{\partial}{\partial t} \boldsymbol{B}_0 = \Delta \boldsymbol{B}_0 + \nabla \times (\boldsymbol{U}_{\mathrm{t}} \times \boldsymbol{B}_0) +$$

$$+ \nabla \times (\boldsymbol{U}_{00} \times \boldsymbol{B}_0) + \nabla \times \left[(U_{01} + U_{10})\hat{\boldsymbol{e}}_\phi \times \boldsymbol{B}_0 \right],$$

$$\frac{\partial}{\partial t} \boldsymbol{B}_1 = \Delta \boldsymbol{B}_1 + \nabla \times (\boldsymbol{U}_{\mathrm{t}} \times \boldsymbol{B}_1) +$$

$$+ \nabla \times (\boldsymbol{U}_{00} \times \boldsymbol{B}_1) + \nabla \times \left[(U_{01} + U_{10}) \times \boldsymbol{B}_0 \right] +$$

$$+ \nabla \times \left[(U_{01} + U_{10})\hat{\boldsymbol{e}}_\phi \times \boldsymbol{B}_1 \right] + \nabla \times \left[(U_{02} + U_{11} + U_{20})\hat{\boldsymbol{e}}_\phi \times \boldsymbol{B}_0 \right], \tag{2.38}$$

and so on at ever higher order.

Now, how might one solve this system? Noting that $\boldsymbol{B}_1$ is not yet determined in Eq. (2.38)$_1$, the combination $(U_{01} + U_{10})$ is unknown, so we may relabel it as some new quantity $\hat{U}_1$, say, to obtain

$$\frac{\partial}{\partial t} \boldsymbol{B}_0 = \Delta \boldsymbol{B}_0 + \nabla \times (\boldsymbol{U}_{\mathrm{t}} \times \boldsymbol{B}_0) + \nabla \times (\boldsymbol{U}_{00} \times \boldsymbol{B}_0) + \nabla \times (\hat{U}_1 \hat{\boldsymbol{e}}_\phi \times \boldsymbol{B}_0). \tag{2.39}$$

In solving this equation, we choose $\hat{U}_1$ such that $\boldsymbol{B}_0$ continues to satisfy Taylor's constraint (2.21). This solution for $\boldsymbol{B}_0$ is thus precisely Taylor's original prescription.

Next, we note that $\boldsymbol{B}_2$ is not yet determined in Eq. (2.38)$_2$, so the combination $(U_{02} + U_{11} + U_{20})$ is unknown, so we relabel it as some new quantity $\hat{U}_2$. In solving Eq. (2.38)$_2$ for $\boldsymbol{B}_1$, then choosing $\hat{U}_2$ such that $\boldsymbol{B}_0$ and $\boldsymbol{B}_1$ together also satisfy $U_{01} + U_{10} = \hat{U}_1$, where we remember that $\hat{U}_1$ has indeed already been determined at the previous order. Continuing in this fashion, we can (in principle at least) solve this system of equations to arbitrary order. And as we just saw, the first step, solving for $\boldsymbol{B}_0$, is precisely Taylor's original prescription. We see therefore that the viscous version of Taylor's constraint (2.32) does indeed allow us to recover Taylor's solutions, and also obtain the higher-order viscous corrections.

The real significance of Eq. (2.32), however, is that it is more general than Taylor's prescription, and allows other solutions as well. In particular, we can also expand B as

$$B = E^{1/4} B_0 + E^{3/4} B_1 + E^{5/4} B_2 + \ldots, \tag{2.40}$$

where now B_0 does not necessarily satisfy Taylor's constraint. As a result, the geostrophic flow is now given by

$$U_g = U_{00} + E^{1/2}(U_{01} + U_{10}) + E(U_{02} + U_{11} + U_{20}) + \ldots, \tag{2.41}$$

and the magnetic wind by

$$U_M = E^{1/2} U_{00} + E(U_{01} + U_{10}) + \ldots, \tag{2.42}$$

where U_{ij} and U_{ij} are defined as before. Again separated out order by order, the induction equation now becomes

$$\frac{\partial}{\partial t} B_0 = \Delta B_0 + \nabla \times (U_t \times B_0) + \nabla \times (U_{00}\hat{e}_\phi \times B_0),$$

$$\frac{\partial}{\partial t} B_1 = \Delta B_1 + \nabla \times (U_t \times B_1) + \nabla \times (U_{00}\hat{e}_\phi \times B_1) +$$

$$+ \nabla \times \left[(U_{01} + U_{10})\hat{e}_\phi \times B_0\right] + \nabla \times (U_{00} \times B_0). \tag{2.43}$$

We see therefore that these states, known as Ekman states, are very different from the previous Taylor states. Not only are the leading-order amplitudes different, $O(E^{1/4})$ versus $O(1)$, the dynamical balances are also different. In the Ekman state B_0 is equilibrated by the geostrophic flow U_{00}, with the magnetic wind U_{00} only entering at second order. In contrast, in the Taylor state B_0 is equilibrated by the magnetic wind U_{00}, with the geostrophic flow $\hat{U}_1$ merely enforcing Taylor's constraint (which we note is homogeneous, and hence does not determine the amplitude of B).

As different as they are, Ekman and Taylor states still do not exhaust the possibilities present in Eq. (2.32). Hollerbach (1997) showed that there is also an intermediate state for which

$$B = E^{1/8} B_0 + E^{3/8} B_1 + E^{5/8} B_2 + \ldots. \tag{2.44}$$

This state is nongeneric, however, and so will not be considered further here. Finally, if one allows for boundary layers scalings different from the $E^{1/2}$ Ekman layers implicit in Eq. (2.32), one can obtain yet further solutions, such as Braginsky's model-Z (e.g. Braginsky & Roberts 1987). These are again nongeneric though, and so will also not be considered further.

2.4.4   From Ekman States to Taylor States

Given this plethora of possible solutions, how can we know which one applies in any given situation? The answer is not to impose any of the above scalings (as following Taylor's prescription would do). Instead, apply Eq. (2.32) directly, and allow the solutions themselves to sort out which particular scaling to follow.

To see in detail how this works, let us start with the mean-field induction equation

$$\frac{\partial \boldsymbol{B}}{\partial t} = \Delta \boldsymbol{B} + \nabla \times (\alpha \boldsymbol{B}) + \nabla \times (\boldsymbol{u} \times \boldsymbol{B}), \tag{2.45}$$

where $\boldsymbol{u} = \boldsymbol{U}_{\mathrm{M}} + U_{\mathrm{g}} \hat{\boldsymbol{e}}_\phi$, with $\boldsymbol{U}_{\mathrm{M}}$ and U_{g} given by Eqs. (2.18) and (2.32). (By excluding the thermal wind $\boldsymbol{U}_{\mathrm{t}}$ here, we are restricting attention to α^2-dynamos. We will consider $\alpha\Omega$-dynamos below.) Now imagine gradually increasing the amplitude of α. Eventually one will reach the kinematic eigenvalue α_{c}, beyond which the linearized equation would yield exponentially growing rather than decaying solutions. The question then is, what equilibrates these solutions in the supercritical regime, and at what amplitude? This is turn will decide whether we have an Ekman state, a Taylor state, or something else.

What equilibrates the solutions is, of course, the nonlinear feedback via $\boldsymbol{u}$. For $O(1)$ supercriticality one would therefore expect the solutions to equilibrate when $\boldsymbol{u} = O(1)$. So how large must $\boldsymbol{B}$ be before $\boldsymbol{u} = O(1)$? In particular, we saw above that the Ekman state and the Taylor state have very different scalings for $\boldsymbol{B}$, but both have $U_{\mathrm{g}} = O(1)$. So which one is it to be? The key point to note is that because this linear, kinematic eigensolution has no knowledge of Taylor's constraint, in general it will not satisfy it. According to the above analysis, this means that initially at least, just beyond α_{c}, the solution can only be an Ekman state, in which the geostrophic flow equilibrates the solution, with the magnetic wind having no effect at leading order. We begin therefore by considering this equilibration via the geostrophic flow.

If $\boldsymbol{B}$, and thus also $\boldsymbol{u}$, are axisymmetric, we can decompose them as

$$\boldsymbol{B} = \boldsymbol{B}_{\mathrm{t}} + \boldsymbol{B}_{\mathrm{p}} = B\hat{\boldsymbol{e}}_\phi + \nabla \times (A\hat{\boldsymbol{e}}_\phi),$$

$$\boldsymbol{u} = \boldsymbol{u}_{\mathrm{t}} + \boldsymbol{u}_{\mathrm{p}} = v\hat{\boldsymbol{e}}_\phi + \nabla \times (\psi\hat{\boldsymbol{e}}_\phi), \tag{2.46}$$

where U_{g} contributes to v only, but $\boldsymbol{U}_{\mathrm{M}}$ to both v and ψ. Incidentally, note also that because Taylor's constraint is inherently axisymmetric, mean-field models are ideally suited for studying it. Separated out into these poloidal and toroidal components, the induction equation becomes

$$\frac{\partial A}{\partial t} = D^2 A + \alpha B + \hat{\boldsymbol{e}}_\phi \cdot (\boldsymbol{u}_{\mathrm{p}} \times \boldsymbol{B}_{\mathrm{p}}),$$

$$\frac{\partial B}{\partial t} = D^2 B + \hat{\boldsymbol{e}}_\phi \cdot \nabla \times (\alpha \boldsymbol{B}_{\mathrm{p}}) + \hat{\boldsymbol{e}}_\phi \cdot \nabla \times (\boldsymbol{u}_{\mathrm{t}} \times \boldsymbol{B}_{\mathrm{p}} + \boldsymbol{u}_{\mathrm{p}} \times \boldsymbol{B}_{\mathrm{t}}), \tag{2.47}$$

where $D^2 = \Delta - 1/s^2$. After a little algebra, Eq. (2.32) similarly yields

$$U_{\mathrm{g}} = -\mathrm{E}^{-1/2} \frac{(1 - s^2)^{1/4}}{2s^2} \frac{\mathrm{d}}{\mathrm{d}s}\left[s^2 T\right], \tag{2.48}$$

where the Taylor integral

$$T = \int_{-z_{\mathrm{T}}}^{+z_{\mathrm{T}}} B \frac{\partial A}{\partial z} \, \mathrm{d}z, \tag{2.49}$$

and $z_{\mathrm{T}} = (1 - s^2)^{1/2}$. Since we are assuming that initially at least the solution is equilibrated entirely by the geostrophic flow, the precise form of $\boldsymbol{U}_{\mathrm{M}}$ is not important here.

So, can this geostrophic flow equilibrate the field (and if so, how could we ever get any-thing other than an Ekman state)? To answer these questions, we need to work out the effect of U_g on the field. According to (2.47), the only effect of U_g is on the toroidal field, via the term $\hat{\boldsymbol{e}}_\phi \cdot \nabla \times (\boldsymbol{u}_t \times \boldsymbol{B}_p)$. After a little algebra, this yields

$$\frac{\partial B}{\partial t} = -s\frac{\mathrm{d}}{\mathrm{d}s}\left(\frac{U_g}{s}\right)\frac{\partial A}{\partial z} \qquad \Longrightarrow \qquad \frac{1}{2}\frac{\partial}{\partial t}B^2 = -s\frac{\mathrm{d}}{\mathrm{d}s}\left(\frac{U_g}{s}\right)B\frac{\partial A}{\partial z}. \qquad (2.50)$$

Inserting (2.48) for U_g and integrating over the sphere, after a little more algebra one finally obtains the result

$$\frac{1}{2}\frac{\partial}{\partial t}\int B^2\,\mathrm{d}V = -\mathrm{E}^{-1/2}\pi\int\limits_0^1\frac{(1-s^2)^{1/4}}{s^3}\left(\frac{\mathrm{d}}{\mathrm{d}s}(s^2T)\right)^2\mathrm{d}s. \qquad (2.51)$$

The effect of the geostrophic flow on the magnetic energy is thus negative definite, vanishing only when $T = 0$. That is, as long as Taylor's constraint is not satisfied, U_g will indeed equilibrate the field – and of course at the $O(\mathrm{E}^{1/4})$ Ekman state scaling.

Now let us consider what might happen as we further increase the amplitude of α. Malkus & Proctor (1975) conjectured that the solutions – which now do know about Taylor's con-straint, because they are being equilibrated by U_g, which involves T – might evolve in such a way that they tend to satisfy it more and more closely. The reason one might expect such behavior is essentially a 'competition' between different field structures satisfying Taylor's constraint more or less closely. According to Eq. (2.51), those structures that come closest to satisfying $T = 0$ will be least affected by the geostrophic flow, and can therefore grow the most before ultimately being equilibrated. They should therefore win out over structures more affected by U_g. In the increasingly supercritical regime, as more and more structure becomes available, it is then indeed plausible that the field might tend to satisfy Taylor's constraint more and more closely.

Malkus & Proctor therefore conjectured that one would eventually reach a second crit-ical value α_T where Taylor's constraint is satisfied exactly. Once that occurs though, the geostrophic flow is no longer capable of equilibrating the field. It thus grows beyond the $O(\mathrm{E}^{1/4})$ Ekman scaling, until it reaches the $O(1)$ Taylor scaling, at which point the mag-netic wind equilibrates it. Figure 2.9 (left) shows this hypothesized transition from the Ekman regime to the Taylor regime.

Turning to the results then, the first model to demonstrate the existence of this second critical value α_T was the plane-layer model of Soward & Jones (1983). They only included U_g though, not $\boldsymbol{U}_M$, so were only able to show that there exists an α_T beyond which U_g alone is no longer capable of equilibrating the field, but not whether $\boldsymbol{U}_M$ will then equilibrate it, and at what amplitude. The first model to include both U_g and $\boldsymbol{U}_M$, and hence obtain the full transition from the Ekman regime to the Taylor regime, was by Hollerbach & Ierley (1991), working in a full sphere, as presented here.

Incidentally, Taylor's constraint is, in general, quite different – and indeed far more com-plicated – in a plane layer than in a sphere. The geostrophic flow is then two-dimensional, rather than merely one-dimensional as in a sphere. That is, in a sphere U_g only has one com-ponent (ϕ), and only depends on one coordinate (s). In contrast, in a plane layer any flow of the form $\nabla \times [\Phi(x, y)\,\hat{\boldsymbol{e}}_z]$ satisfies all three of the requirements we demanded of U_g (namely

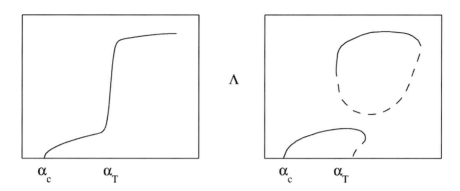

Figure 2.9: *Left*: The transition from the Ekman state to the Taylor state originally conjectured by Malkus & Proctor (1975). *Right*: An alternative transition, as found by Soward & Jones (1983) or Hollerbach & Ierley (1991).

that it be independent of z, have zero divergence, and satisfy the no normal flow boundary conditions). This planar U_g therefore has both x and y components, and also depends on both x and y. Not surprisingly, the associated solvability condition, namely Taylor's constraint, is then also considerably more complicated. As valuable as they undoubtedly are, plane-layer dynamos such as those of St. Pierre (1993) or Jones & Roberts (2000) are thus also potentially quite different from the spherical dynamos we are really interested in. In the mean-field model of Soward & Jones, however, this distinction does not arise, since they constrained their solutions to be independent of y. Their geostrophic flow, and hence also Taylor's constraint, is thus much the same as in a sphere after all.

Another interesting feature discovered by both Soward & Jones and Hollerbach & Ierley is shown in Fig. 2.9 (right). In particular, we note that the Taylor state that exists for $\alpha > \alpha_T$ is disconnected from the Ekman state that exists for $\alpha > \alpha_c$, very much unlike the original conjecture of Malkus & Proctor. Which half of Fig. 2.9 the bifurcation diagram looks like depends on the detailed spatial structure of α. Almost all choices of α do seem to yield some type of transition from the Ekman to the Taylor regime though.

At this point it is perhaps also worth comparing and contrasting this distinction between Ekman and Taylor states here with that between the weak and strong-field regimes in Sect. 2.3.3; the bifurcation diagrams in Figs. 2.9 and 2.7 do after all look rather similar. Nevertheless, the issues involved are quite different, and should not be confused. In particular, the distinction between weak and strong fields came about because of the effect of the field on the pattern of convection, which is completely neglected in mean-field models. And similarly, we did not mention Taylor's constraint at all in our discussion of weak versus strong fields. As tempting as it may be, we cannot therefore necessarily identify the weak-field regime with an Ekman state, and the strong-field regime with a Taylor state. So which of these various regimes is truly relevant to the real geodynamo then?

Let us begin with the distinction between Ekman and Taylor states discussed here. The crucial point to note is that mean-field theory is in a sense inconsistent, in taking the same (more or less arbitrarily prescribed) spatial structure for α, when we know, precisely from our discussion of weak versus strong fields, that really the pattern of convection, and hence its parameterisation into α, would be quite different in the two regimes. That is, even if these two states do exist in the real geodynamo, the bifurcation sequence cannot be as presented in Fig. 2.9. These mean-field models are thus simplified models that allow us to explore some of the other dynamics not previously captured by the distinction between weak and strong-field regimes (which we believe to be the real bifurcation sequence).

So what precisely does this distinction between Ekman and Taylor states tell us about the weak versus strong-field regimes? Well, in the strong-field regime B is order one, so presumably it must be in a Taylor state. In contrast, in the weak-field regime the Lorentz force is less than the viscous force, so B need not satisfy Taylor's constraint (which after all arises only if we neglect the viscous force compared with the Lorentz force). So in this sense we can perhaps say that 'weak field equals Ekman state' and 'strong field equals Taylor state' after all, so long as we also remember that the considerations that led us to distinguish weak from strong, and Ekman from Taylor, are very different, as just discussed.

2.4.5   Torsional Oscillations

Although we speak of Taylor's constraint as being satisfied in the Taylor regime, and not satisfied in the Ekman regime, it is not quite correct to say that the former is characterized by $T = 0$, and the latter by $T \neq 0$. In fact, $T = O(E^{1/2})$ in both regimes. The difference is how this comes about. In the Ekman regime it is trivially accomplished by the $O(E^{1/4})$ scaling of the field itself; the Taylor integral, being quadratic in B, then obviously scales as $E^{1/2}$. In the Taylor regime it is rather less trivially accomplished by having sufficient internal cancellation that the integral scales as $E^{1/2}$ even though the field scales as $O(1)$. (To see why this cancellation occurs to precisely $O(E^{1/2})$, simply insert the expansion (2.33) into Taylor's constraint (2.21) and use the fact that B_0 by itself satisfies Eq. (2.21) identically.)

The need for this increasingly high degree of internal cancellation in the integrated Lorentz torque then demonstrates just how delicate the Taylor state is. Imagine a solution evolving along in time in such a way that this cancellation suddenly breaks down. According to Eq. (2.51), if the cancellation breaks down so completely that $T = O(1)$ rather than $O(E^{1/2})$, that will induce an unsustainably large drain on the magnetic energy. The field must therefore evolve back toward the proper Taylor state balance – and on an extremely rapid timescale – or else it will necessarily collapse to an Ekman state. Remembering the above identification that 'Ekman state equals weak-field regime', we see therefore how a breakdown of Taylor's constraint could trigger this dynamo catastrophe discussed in Sect. 2.3.3; according to Eq. (2.32), only a very subtle change in B is required to induce an enormous geostrophic flow, which then reacts back on B to produce further changes, and all too quickly the whole dynamo could shut off.

As noted above, the alternative (and considerably more desirable) possibility is that the field simply evolves back toward a Taylor state on a very rapid timescale. Formally, this timescale could be as fast as $O(E^{1/2})$, since U_g could be large as $O(E^{-1/2})$, so the timescale on which such a flow would advect everything else is $O(E^{1/2})$. In fact, Taylor's constraint should never break down so completely that $T = O(1)$ (according to Eq. (2.51) $T = O(E^{1/4})$ is the most that is energetically sustainable), so in practice U_g will not be quite as large, and hence its advective timescale not quite as fast.

Nevertheless, we see that even if it does not lead to a dynamo catastrophe, any breakdown of Taylor's constraint will necessarily lead to the emergence of rather short timescales. Indeed, these timescales are sufficiently short that we probably should not neglect inertia after all, as we did in Eq. (2.32), but instead return to the more general Eq. (2.31). We recognize then that another possibility for dealing with breakdowns of Taylor's constraint is that U_g remains $O(1)$, but instead oscillates on $O(\mathrm{Ro})$ timescales – which once again are extremely short though. Either way then, any breakdown of Taylor's constraint will indeed induce very short timescales in the dynamics associated with the geostrophic flow.

Flows of this type, in which the integrated Lorentz torque is balanced by inertia, with the result that each individual shell $C(s)$ oscillates in essentially solid-body rotation, are known as torsional oscillations, and have been observed in the core (Zatman & Bloxham 1997, 1999). Indeed, Bloxham, Zatman & Dumberry (2002) show that the geomagnetic jerks mentioned in Sect. 2.1.2 are probably caused by torsional oscillations, which would certainly explain the very short timescale of these events. As noted above, the timescale of these events in the core may well be even shorter than the one-year timescale at which they are observed at the surface; in terms of the dynamics of torsional oscillations we now understand how such events could indeed occur on timescales considerably shorter than a year. On this view then, the Earth's field is always close to a Taylor state, but small deviations are continually exciting torsional oscillations, of various periods and amplitudes, and occasionally a somewhat larger deviation from Taylor's constraint causes a jerk, on a timescale of a year or even less.

2.4.6 $\alpha\Omega$-Dynamos

This idea of the field evolving along in time, satisfying Taylor's constraint more or less closely, also leads quite naturally to a discussion of Taylor's constraint in $\alpha\Omega$-dynamos. In particular, the α^2-dynamos we considered above are typically steady. At any given amplitude of α, they are therefore unambiguously in one state or the other. In contrast, $\alpha\Omega$-dynamos in spheres or shells are usually oscillatory. This leads to the unpleasant possibility that they could be in different states at different parts of the cycle, making any overall classification almost impossible. This is precisely what was found by Barenghi & Jones (1991) and also Hollerbach, Barenghi & Jones (1992). Both found clear evidence for the existence of this second critical value α_T, beyond which the geostrophic flow alone is no longer capable of equilibrating the solutions. Including the magnetic wind then did equilibrate the solutions again, and at more or less the $O(1)$ amplitude one would expect for a Taylor state. The details of the temporal evolution, however, continued to depend on E, suggesting that the solution is indeed oscillating between the Taylor and Ekman states. Similarly, Hollerbach (1997) suggested that excursions might be caused by the field temporarily dropping from a Taylor state to this $O(\mathrm{E}^{1/8})$ intermediate state mentioned above. See also Zhang & Gubbins (2000), who suggest a particular mechanism as to why it might temporarily switch from a Taylor state to something else. Given that the Earth's field does evolve in time, and how easily Taylor's constraint can break down, the idea that it occasionally (if only temporarily, to avoid the dynamo catastrophe) switches from one state to another certainly seems more plausible than that it should remain in a Taylor state forever.

2.4.7   Taylor's Constraint in the Anelastic Approximation

The last point to note regarding Taylor's constraint is whether it must be modified if we make the anelastic rather than the Boussinesq approximation. In particular, if we think back to Eq. (2.16), we note that $\nabla \cdot \boldsymbol{u} = 0$ was an essential ingredient in the derivation of Taylor's constraint and all the subsequent analysis. If we therefore make the anelastic approximation $\nabla \cdot (\rho_a \boldsymbol{u}) = 0$ instead, to what extent is this analysis still valid? The equivalent of Eq. (2.19) is then

$$2\rho_a u_s = -\frac{1}{s}\frac{\partial P}{\partial \phi} + \left[(\nabla \times \boldsymbol{B}) \times \boldsymbol{B}\right]_\phi, \tag{2.52}$$

and $\nabla \cdot (\rho_a \boldsymbol{u}) = 0$ now yields $\int \rho_a u_s \, \mathrm{d}S = 0$. We see therefore that we obtain exactly the same result, Eq. (2.21), as before. All of the dynamics associated with Taylor's constraint are thus the same in the anelastic as in the Boussinesq approximation (although to our knowledge no one has ever developed a model along the lines of Soward & Jones (1983) or Hollerbach & Ierley (1991) using the anelastic approximation).

2.5   Hydromagnetic Waves

In the previous section we saw some of the difficulties that result from the extreme smallness of inertia and viscosity in the momentum equation. This smallness turns out to generate not only the global difficulties associated with Taylor's constraint, but local difficulties associated with wave motions as well. To see how these come about, it is perhaps convenient to revert to the dimensional equations

$$\frac{\partial \boldsymbol{B}}{\partial t} = \nabla \times (\boldsymbol{u} \times \boldsymbol{B}) + \eta \Delta \boldsymbol{B},$$

$$\frac{D\boldsymbol{u}}{Dt} + 2\boldsymbol{\Omega} \times \boldsymbol{u} = -\nabla P + \nu \Delta \boldsymbol{u} + \frac{1}{\mu_0 \rho_0}(\nabla \times \boldsymbol{B}) \times \boldsymbol{B}, \tag{2.53}$$

where we are neglecting the buoyancy force for convenience. Even without the coupling to T these equations already turn out to generate a surprising variety of wave motions. This, incidentally, is also the reason for considering the dimensional rather than nondimensional equations; since the resulting waves occur on a broad range of timescales, it is best not to bias the analysis by nondimensionalizing on any one particular timescale.

Linearizing these equations about the basic state $\boldsymbol{B} = \boldsymbol{B}_0$, $\boldsymbol{u} = 0$, and looking for perturbations proportional to $\exp[\mathrm{i}(\boldsymbol{k} \cdot \boldsymbol{r} - \omega t)]$, after a certain amount of algebra one obtains the dispersion relation

$$(\omega + \mathrm{i}\eta k^2)(\omega - \omega_\mathrm{C} + \mathrm{i}\nu k^2) = \omega_\mathrm{A}^2, \tag{2.54}$$

where

$$\omega_\mathrm{A} = \pm\frac{\boldsymbol{k} \cdot \boldsymbol{B}_0}{\sqrt{\rho_0 \mu_0}}, \qquad \omega_\mathrm{C} = \pm 2\frac{\boldsymbol{k} \cdot \boldsymbol{\Omega}}{k} \tag{2.55}$$

are the dispersion relations of pure Alfvén waves (in magnetic but nonrotating systems) and inertial oscillations (in rotating but nonmagnetic systems), respectively. What we would like to discover therefore is whether these waves still exist when both magnetism and rotation are important – or alternatively what other waves may exist in that case.

We begin by noting that ω_C and ω_A already define two very different timescales. The timescale associated with inertial oscillations is, of course, just the daily rotational timescale. In contrast, if we take ~ 50 G as a typical field strength in the core, the timescale for Alfvén waves to cross the core comes out on the order of decades. That is, $\omega_A \ll \omega_C$ – unless of course k and Ω are almost perpendicular, in which case ω_C can be arbitrarily small. As we shall see, this considerably extends the range of timescales one can obtain from Eq. (2.54).

We can solve Eq. (2.54) easily enough, yielding

$$\omega = \frac{1}{2}\left[\omega_C - i(\eta + \nu)k^2 \pm \sqrt{\left(\omega_C + i(\eta - \nu)k^2\right)^2 + 4\omega_A^2}\right]. \tag{2.56}$$

In order to make sense of this result, we need to simplify it further. To do this, we note that the viscous and magnetic diffusive timescales are far longer than this Alfvén timescale even. Neglecting quantities quadratic in the diffusive terms, we therefore have

$$\omega \approx \frac{1}{2}\left[\omega_C - i(\eta + \nu)k^2 \pm \sqrt{\omega_C^2 + 4\omega_A^2 + 2i\omega_C(\eta - \nu)k^2}\right]. \tag{2.57}$$

Next, we saw that in general $\omega_C \gg \omega_A$, so we can Taylor-expand the square root as

$$\omega \approx \frac{1}{2}\left[\omega_C - i(\eta + \nu)k^2 \pm \omega_C\left(1 + 2\frac{\omega_A^2}{\omega_C^2} + \frac{i(\eta - \nu)k^2}{\omega_C}\right)\right] \tag{2.58}$$

to obtain

$$\omega_+ = \omega_C\left(1 + \frac{\omega_A^2}{\omega_C^2}\right) - i\nu k^2, \qquad \omega_- = -\frac{\omega_A^2}{\omega_C} - i\eta k^2. \tag{2.59}$$

Remembering that ω_C is a completely nonmagnetic inertial oscillation, we recognize that ω_+ is an inertial oscillation that has been modified slightly by the presence of the magnetic field. The modification is indeed only slight though, since $\omega_A/\omega_C \ll 1$. And not surprisingly then, this essentially nonmagnetic mode is damped by viscosity rather than magnetic diffusivity.

The second mode, ω_-, is very different, and has no analog in either rotating, nonmagnetic or nonrotating, magnetic systems. Inserting the above values that the inertial timescale is one day, and the Alfvén timescale a few years, we find that the timescale associated with this mode is around 10^4 to 10^5 years, that is, the same as the magnetic diffusive timescale, which incidentally is also the damping mechanism of this mode. Note also how both ω_C and ω_A combine to yield this timescale, emphasizing once again that both rotation and magnetism are fundamental to the existence of these so-called slow magnetohydrodynamic waves.

So far we have seen that Eq. (2.54) supports these two types of waves $\omega_\pm$ on these two very different timescales of a day and $\sim 10^5$ years. We recall though that these two were derived on the assumption that $\omega_C \gg \omega_A$, which need not be the case if k is almost perpendicular to Ω. To see what happens in this case, let us return to Eq. (2.57) and simply insert $\omega_C = 0$, yielding

$$\omega = \pm\omega_A - i(\eta + \nu)k^2/2. \tag{2.60}$$

That is, even though rotation is ordinarily an important ingredient in the dynamics of the core, if $k \cdot \Omega = 0$ we recover classical Alfvén waves just as in a nonrotating system. We see therefore that our original dispersion relation, Eq. (2.54), supports at least three timescales, namely the decadal Alfvén timescale in addition to the above two. In fact, since the solutions of Eq. (2.54) must depend continuously on ω_C, we realize that as $|\omega_C|$ varies from 0 to its maximum value 2Ω, we will necessarily obtain everything in between as well. That is, Eq. (2.54) supports waves covering the entire range of timescales from $\sim 10^5$ years to 1 day.

The existence of these very short timescales is extremely undesirable from a numerical point of view, since the timestep will then also have to be very small, so an enormous number of them will be needed to cover even one magnetic diffusion time. One would therefore like to filter out some of these short timescales. The obvious thing to try is to neglect inertia, which is known to eliminate inertial oscillations in nonmagnetic systems anyway. The dispersion relation one then obtains is $(\omega + i\eta k^2)(-\omega_C + i\nu k^2) = \omega_A^2$, so now there is the single solution

$$\omega = -\frac{\omega_A^2}{\omega_C - i\nu k^2} - i\eta k^2. \tag{2.61}$$

Comparing this result with the previous ones, it looks like we have indeed eliminated the inertial oscillations Eq. (2.59)$_1$, and even the Alfvén waves Eq. (2.60), with only the equivalent of the slow waves Eq. (2.59)$_2$ remaining. Unfortunately, this is an illusion. While Eq. (2.61) and Eq. (2.59)$_2$ may look almost identical, there is also one very important difference between them. In particular, while Eq. (2.59)$_2$ is only valid for $\omega_C > \omega_A$, and can therefore never yield timescales shorter than decadal, Eq. (2.61) is valid for all ω_C, and can therefore yield almost arbitrarily short timescales (particularly if one attempts to neglect viscosity as well). That is, we may have succeeded in reducing the number of allowed waves, but if anything the timescale problem is even worse. This inability to filter out any of the short timescales was first pointed out by Walker, Barenghi & Jones (1998). See also Chapt. 10 of Moffatt (1978) for a discussion of waves in the core.

2.6   The Inner Core

We have already noted above how the existence of the inner core is crucial to the dynamo, in terms of allowing this compositional convection to take place. In this section we want to consider various other aspects of the inner core, and some of its effects on the flow and field in the outer core. The most important point to note here is the existence of the so-called tangent cylinder, the cylinder parallel to the axis of rotation and just touching the inner core. That is, in terms of these geostrophic contours $C(s)$ introduced in the context of Taylor's constraint, the tangent cylinder, denoted by $\mathcal{C}$, is simply the particular cylinder $C(R_{in})$. With this identification $\mathcal{C} = C(R_{in})$, we can also immediately see at least one reason why this tangent cylinder is important: the contours over which Taylor's constraint is integrated change abruptly across $\mathcal{C}$, with one single integral outside $\mathcal{C}$ but two separate integrals above and below the inner core inside $\mathcal{C}$. The significance of this for the geodynamo is not immediately obvious, but Jault (1996) has suggested that it might then be more difficult to satisfy Taylor's constraint inside $\mathcal{C}$, since this cancellation discussed above must now occur separately in the two integrals. And, of course, the torque balance Eq. (2.31) must be similarly modified, not only to take into account things like the viscous torque in the Ekman layers on the inner core, but more fundamentally that inside $\mathcal{C}$ there are now also two separate geostrophic flows above and below the inner core.

2.6.1   Stewartson Layers on $\mathcal{C}$

The importance of the tangent cylinder is not confined to Taylor's constraint, however, but arises in nonmagnetic problems as well. For example, suppose we take a rapidly rotating spherical shell, and additionally impose a slight differential rotation on the inner sphere. (We will see below that such a differential rotation of the inner core may indeed exist in the Earth.) The tangent cylinder then divides the flow into two distinct regions, with fluid outside $\mathcal{C}$ co-rotating with the outer sphere, but fluid inside $\mathcal{C}$ rotating at a rate roughly intermediate between the inner and outer spheres. The reason the flow adjusts itself in this peculiar fashion is due to the Taylor–Proudman theorem, stating that in rapidly rotating systems the flow will tend to align itself such that $\partial\boldsymbol{u}/\partial z \approx 0$. With this result, this abrupt change across $\mathcal{C}$ follows quite naturally: For fluid columns outside $\mathcal{C}$ the boundary conditions are $\Omega = \Omega_{\mathrm{out}}$ at both ends, so having $\Omega = \Omega_{\mathrm{out}}$ everywhere along the column will indeed satisfy both the boundary conditions and the Taylor–Proudman theorem. In contrast, for fluid columns inside $\mathcal{C}$ the boundary conditions are still $\Omega = \Omega_{\mathrm{out}}$ at one end, but $\Omega = \Omega_{\mathrm{in}}$ at the other. It is therefore not possible to satisfy the Taylor–Proudman theorem everywhere along the column. Instead, it is satisfied in the interior by having $\Omega \approx (\Omega_{\mathrm{in}} + \Omega_{\mathrm{out}})/2$ (in fact, the precise weightings used in this average vary with s), with all of the required z-dependence concentrated in the Ekman layers at the boundaries.

The detailed structure of the shear layer that resolves the resulting jump in angular velocity across $\mathcal{C}$ was deduced by Stewartson (1966), and shown to consist of an inner thickness $\mathrm{E}^{1/3}$ right on $\mathcal{C}$, and outer thicknesses $\mathrm{E}^{2/7}$ just inside $\mathcal{C}$ and $\mathrm{E}^{1/4}$ just outside. The Stewartson layer was also reproduced numerically by Hollerbach (1994a) down to $\mathrm{E} = 10^{-5}$, and by Dormy, Cardin & Jault (1998) down to $\mathrm{E} = 10^{-7}$. Even these values are not small enough to clearly distinguish all the different scalings, but the results are certainly consistent with Stewartson's asymptotics. Figure 2.10 shows the solution at $\mathrm{E} = 10^{-3}$, 10^{-4} and 10^{-5}.

Having obtained this basic Stewartson layer, the next question to ask is what effect a magnetic field would have on it. Given the stiffness imparted by the magnetic tension, it seems likely that the Lorentz force would suppress the shear (at least if the field has a component perpendicular to the layer). Both Hollerbach (1994a) and Dormy, Cardin & Jault (1998) also considered this magnetic problem, and showed that this is indeed the case; imposing an order one (as measured by the Elsasser number) dipole field almost completely suppresses the layer again, with all the adjustment then occurring in the inner and outer Ekman layers (actually called Ekman-Hartmann layers in this rotating and magnetic problem). Kleeorin et al. (1997) considered the asymptotics of this problem, and showed that the suppression of the Stewartson layer begins when the Elsasser number is as small as $\mathrm{E}^{1/3}$. In particular, there is almost certainly no Stewartson layer in the Earth's core after all, even if the inner core should turn out to be differentially rotating.

2.6.2   Nonaxisymmetric Shear Layers on $\mathcal{C}$

Both Taylor's constraint as well as the Stewartson layer are axisymmetric phenomena. It turns out that there are nonaxisymmetric structures associated with $\mathcal{C}$ as well, which if anything are even more severe. To see how these arise, we return to the magnetostrophic balance Eq. (2.16)

$$2\hat{\boldsymbol{e}}_z \times \boldsymbol{u} = -\nabla P + \boldsymbol{F}, \tag{2.62}$$

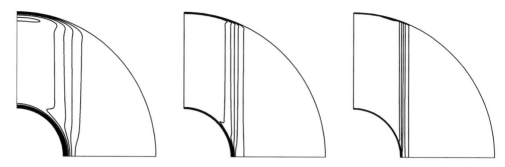

Figure 2.10: The angular velocity in the Stewartson layer problem. From left to right $E = 10^{-3}, 10^{-4}$ and 10^{-5}.

where $\boldsymbol{F}$ now denotes both the buoyancy and Lorentz forces. As before, the solution ought to be just

$$\boldsymbol{u} = -\frac{1}{2} \int^{z} \nabla \times \boldsymbol{F} \, \mathrm{d}z' + \hat{\boldsymbol{u}}(s, \phi). \tag{2.63}$$

Now, however, let us consider in detail how (or whether) this constant of integration $\hat{\boldsymbol{u}}(s, \phi)$ can be determined, and whether the resulting solutions make sense.

We begin by noting that since Eq. (2.63) is linear in $\boldsymbol{u}$, we can restrict attention to single $\exp(im\phi)$ azimuthal modes at a time. There is then a considerable difference between axisymmetric ($m = 0$) versus nonaxisymmetric ($m \neq 0$) solutions. In particular, we will see in a moment how the no normal flow boundary conditions (again, the only conditions we can impose in the absence of viscosity) determine the z and s components of $\hat{\boldsymbol{u}}$. The incompressibility condition $\nabla \cdot \boldsymbol{u} = 0$ then yields

$$\frac{1}{s} \frac{\partial}{\partial s} \left(s \hat{u}_s \right) + \frac{im}{s} \hat{u}_\phi = \frac{1}{2} \nabla \cdot \int^{z} \nabla \times \boldsymbol{F} \, \mathrm{d}z'. \tag{2.64}$$

So if $m = 0$, $\hat{u}_\phi$ continues to be undetermined, and we have instead the constraint

$$\frac{1}{s} \frac{\partial}{\partial s} \left(s \hat{u}_s \right) = \frac{1}{2} \nabla \cdot \int^{z} \nabla \times \boldsymbol{F} \, \mathrm{d}z'. \tag{2.65}$$

That is, $\boldsymbol{F}$ must be such that $\hat{u}_s$, which we remember has already been determined by the boundary conditions, also just happens to satisfy Eq. (2.65). If it does not there is no solution at all, whereas if it does, there are infinitely many solutions, since $\hat{u}_\phi$ is still arbitrary. We recognize that this is precisely the situation we encountered before with regard to Taylor's constraint, and indeed Eq. (2.65) is Taylor's constraint, just expressed in a less intuitive form.

In contrast, for $m \neq 0$ Eq. (2.64) determines $\hat{u}_\phi$, and there is also no constraint for $\hat{u}_s$ to satisfy. That is, for nonaxisymmetric modes this approach Eq. (2.63) seems to work, yielding a unique solution for any $\boldsymbol{F}$. The difficulty that can then lead to shear layers on $\mathcal{C}$ is that these solutions do not necessarily make sense. To see why, let us consider in more detail how the boundary conditions determine $\hat{u}_z$ and $\hat{u}_s$. At the outer boundary no normal flow yields

$$z u_z + s u_s = 0 \quad \text{at} \quad z = \sqrt{R_{\mathrm{out}}^2 - s^2}, \tag{2.66}$$

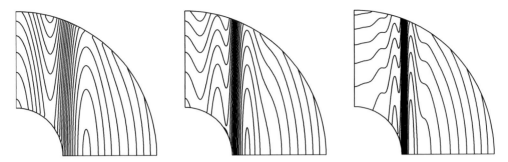

Figure 2.11: Contours of u_s in one of these nonaxisymmetric shear layers on $\mathcal{C}$. From left to right $E = 10^{-4}$, 10^{-5} and 10^{-6}. (Hollerbach 1994b).

and this applies for all $s \leq R_{\text{out}}$. Similarly, at the inner boundary we have

$$zu_z + su_s = 0 \qquad \text{at} \quad z = \sqrt{R_{\text{in}}^2 - s^2}, \tag{2.67}$$

but this only applies for $s \leq R_{\text{in}}$. For $R_{\text{in}} \leq s \leq R_{\text{out}}$, we take instead the symmetry condition

$$u_z = 0 \qquad \text{at} \quad z = 0, \tag{2.68}$$

appropriate for solutions having u_z antisymmetric and u_s and u_ϕ symmetric about the equator. (See, for example, Gubbins & Zhang (1993) for a discussion of the symmetry classes into which the geodynamo equations and solutions can be split.)

We see therefore that both inside and outside $\mathcal{C}$ we have two conditions on two different linear combinations of u_z and u_s, so we can indeed solve them for $\hat{u}_z$ and $\hat{u}_s$. The reason this procedure nevertheless goes wrong is that the conditions (2.67,2.68) do not join smoothly at $s = R_{\text{in}}$, where Eq. (2.67) becomes $u_s = 0$ rather than $u_z = 0$. If the boundary conditions being imposed change discontinuously across $\mathcal{C}$ though, one must expect that the resulting values for $\hat{u}_z$ and $\hat{u}_s$ will too, and according to Eq. (2.63) these discontinuities will be transmitted undiminished along the entire tangent cylinder. That is, Eq. (2.63) may indeed have a unique solution, but all three components of that solution, including u_s, are, in general, discontinuous across $\mathcal{C}$. (In contrast, in the Stewartson layer problem only u_ϕ is discontinuous in the $E \to 0$ limit.) The existence of this discontinuity in the inviscid solution was first pointed out by Hollerbach & Proctor (1993). Figure 2.11 shows how the inclusion of viscosity smooths out the discontinuity.

Hollerbach & Proctor went on to show that this discontinuity could be avoided if $\boldsymbol{F}$ satisfies a certain integral constraint on $\mathcal{C}$. In particular, the discontinuity will be eliminated if u_z and u_s both happen to be zero at $s = R_{\text{in}}$, $z = 0$, since then Eqs. (2.67,2.68) will match smoothly after all. At $s = R_{\text{in}}$ we must therefore have that

$$u_z = -\frac{1}{2} \int_0^z (\nabla \times \boldsymbol{F})_z \, dz', \qquad u_s = -\frac{1}{2} \int_0^z (\nabla \times \boldsymbol{F})_s \, dz'. \tag{2.69}$$

We also still need to satisfy the outer boundary condition (2.66) though, yielding the constraint

$$z_{\mathrm{T}} \int\limits_0^{z_{\mathrm{T}}} (\nabla \times \boldsymbol{F})_z \, \mathrm{d}z' + R_{\mathrm{in}} \int\limits_0^{z_{\mathrm{T}}} (\nabla \times \boldsymbol{F})_s \, \mathrm{d}z' = 0, \qquad (2.70)$$

where $z_{\mathrm{T}} = (R_{\mathrm{out}}^2 - R_{\mathrm{in}}^2)^{1/2}$. That is, if $\boldsymbol{F}$ satisfies Eq. (2.70) the inviscid solutions will match smoothly across $\mathcal{C}$, eliminating the need for the viscous shear layer that otherwise resolves the discontinuity (as in Fig. 2.11).

Finally, as before in the Stewartson layer problem, one might suspect that a sufficiently strong magnetic field would tend to suppress these shear layers as well. That this is indeed the case was verified numerically by Hollerbach (1994b), and asymptotically by Soward & Hollerbach (2000). However, unlike in the Stewartson layer problem, this time there is also a price to be paid, namely that $\boldsymbol{F}$ must adjust itself to satisfy this constraint (2.70). Whether this constraint has an impact on the dynamics of the field comparable with Taylor's constraint is not known though. (Also very much unlike Taylor's constraint, Eq. (2.70) does not appear to have any simple physical interpretation.)

2.6.3   Finite Conductivity of the Inner Core

We note that the existence of the tangent cylinder is a purely geometrical consequence of working in a spherical shell, when the Taylor–Proudman theorem dictates that cylindrical co-ordinates are really more natural. Insofar as these dynamics associated with $\mathcal{C}$ are concerned, the precise material properties of the inner core are thus irrelevant; it must merely exist. Many early geodynamo models therefore took the inner core to be either insulating or perfectly conducting, which in both cases means that one need not solve for the magnetic field in it, but can simply impose appropriate boundary conditions on the field in the outer core. As we saw above though, the composition of the inner core is only slightly different from that of the outer core, so really its conductivity should also be comparable with that of the outer core. As soon as one introduces a finitely conducting inner core, one must also solve for the field in it. In the process one introduces a new timescale, namely the magnetic diffusive timescale $R_{\mathrm{in}}^2/\eta =$ 24,000 yr. There are two important points to note then: (i) this timescale is much longer than some of the advective timescales found in the outer core, and (ii) since the inner core is solid (at least on these timescales), the field in the inner core can change *only* on this diffusive timescale. That is, whereas the field in the outer core can change on timescales as short as a year or even shorter (as we saw with geomagnetic jerks), the field in the inner core can only change on timescales of thousands of years. Since the field must be continuous across the inner core boundary, this suggests that the inner core could have a stabilizing influence, damping out some of the most rapid fluctuations in the outer core. This potentially stabiliz-ing role of the inner core was first noted by Hollerbach & Jones (1993), who suggested that without it the field would reverse far more frequently than it does. See also Gubbins (1999), who suggested that the distinction between reversals and mere excursions is precisely whether a given fluctuation in the outer core lasts long enough to reverse the field in the inner core as well.

That this stabilizing mechanism can indeed work was verified by the numerical model of Glatzmaier & Roberts (1995), who found that the field was much more stable with a finitely conducting inner core than with an insulating one. However, one should be careful not to ascribe too much influence to the inner core. For example, Wicht (2002) finds a regime in which it has little influence. In particular, at only 4% of the volume of the total core, it cannot possibly 'control' the outer core. That is, if the outer core is sufficiently 'determined' to achieve a reversal, it will do so, and there is nothing the inner core can do to stop it. Or, of course, if the field in the outer core is stable anyway, then obviously the inner core has no effect either. If, however, there are substantial short-term fluctuations in the outer core that come close to achieving reversals, then the inner core may have a considerable effect, as found by Glatzmaier & Roberts. Whether the real Earth is more like the Glatzmaier-Roberts model, or more like that of Wicht, is an open question.

2.6.4 Rotation of the Inner Core

The last point we wish to address is whether the inner core rotates relative to the mantle. Let us begin by simply estimating how quickly one might expect it to rotate. In particular, it has been known since the 18th century that some of the nondipolar parts of the field tend to drift westward at rates of around 0.1–0.2° per year. If we interpret this as due to advection of the field (e.g. Jault, Gire & LeMouel 1988), this gives us an estimate of the differential rotation just beneath the core-mantle boundary. Extrapolating from there all the way down to the inner core boundary is, of course, quite a leap, but certainly suggests that we could expect the inner core to rotate at least that fast, and perhaps considerably faster.

And indeed, the numerical model of Glatzmaier & Roberts (1996) yielded inner core rotation rates of around 2–3° per year (eastward though, so slightly faster than the mantle). In ~30 yr therefore, the inner core will have rotated a substantial fraction of one complete revolution. This observation motivated various seismologists to re-examine old data sets to see whether such a rotation could be found. The first results were in excellent agreement with the Glatzmaier-Roberts prediction, with Song & Richards (1996) getting 1°/yr, and Su, Dziewonsky & Jeanloz (1996) even 3°/yr. Subsequent analyses, however, (using different data sets and methods), have yielded ever lower rates, with Creager (1997) getting at most 0.3°/yr and Vidale, Dodge & Earle (2000) 0.15°/yr. Finally, Laske & Masters (1999) and Souriau, Garcia & Poupinet (2003) find that the rate is at most 0.2°/yr, but could also be zero.

Given this observation above that the westward drift at the CMB is already as much as 0.2°/yr, it is really quite astonishing that the rotation rate of the inner core should be at most that much. One possible explanation was provided by Buffett (1996, 1997), who suggested that gravitational anomalies in the mantle could first of all induce corresponding anomalies in the inner core, and then lock it into a fixed longitude with respect to the mantle. In this scenario there could therefore be a considerable differential rotation somewhere in the middle of the core, but right at the inner core boundary this additional coupling brings it to zero again. Following this suggestion, Buffett & Glatzmaier (2000) incorporated this effect into a numerical model, and found that it was indeed sufficient to almost completely suppress the previously observed rotation. See also Aurnou & Olson (2000) for a particularly illuminating analytical model illustrating the competing effects of the gravitational and electromagnetic torques on the inner core, with the electromagnetic torque trying to overcome this gravitational locking effect (and occasionally succeeding). Incidentally, note also that the inner core's finite conductivity plays an important role here as well, since without it there could be no electromagnetic torque on the inner core (e.g. Gubbins 1981).

Finally, we note very briefly that for seismologists to be able to say anything at all about the inner core's rotation rate, it must have some type of structure that allows one to define longitude. That is, it cannot be perfectly axisymmetric, let alone radially symmetric. Discussing the possible origins of the lack of axisymmetry would take us too far from our main topic here, but there are a number of obvious reasons why the inner core should lack radial symmetry. For example, it is quite likely that it will grow by different rates in the polar and equatorial regions (since the details of the convection are likely to be different). Since it must maintain hydrostatic equilibrium though, this implies a slow readjustment via viscous creep. The electromagnetic stresses acting on the inner core are also likely to depend on latitude, and could also induce viscous creep. See, for example, Karato (1993, 1999), Bergman (1997), or Buffett & Wenk (2001) for discussions of some of these effects, and the seismic anisotropy they induce in the inner core.

2.7   Numerical Simulations

We have already mentioned several numerical simulations in previous sections. In this section we want to discuss some more of them, and consider what we have learned, but also what the limitations still are. See also Dormy, Valet & Courtillot (2000) or Jones (2000) for very thorough discussions, or Busse (2002a), Glatzmaier (2002), and Kono & Roberts (2002) for recent reviews by some of the leading computationalists themselves. See also Glatzmaier (1984), Kuang & Bloxham (1999), Tilgner (1999) or Hollerbach (2000) for detailed descriptions of some of the numerical algorithms used, and Christensen et al. (2001) for a numerical benchmark between various codes. Note in particular that none of these simulations make use of any of the asymptotic expansions of the momentum equations that we have seen in the context of Taylor's constraint, for example. As successful as these expansions were in that context, no one has ever succeeded in extending them from a full sphere to a spherical shell, and from mean-field to fully three-dimensional models. Instead, the momentum equation is solved directly, and the Rossby and Ekman numbers are simply reduced as far as possible. As we have seen though over and over again, reducing these parameters (particularly E) is enormously difficult, and most models are limited to the range $E \geq 10^{-6}$ or so. Indeed, achieving even that often requires the use of so-called hyperviscosities, in which the Ekman number is increased with spherical harmonic degree l. That is, $E = O(10^{-6})$ applies only to the lowest modes; the highest ones are damped several orders of magnitude more strongly. Using hyperviscosities is known to distort the dynamics in important ways (Zhang & Jones 1997; Grote, Busse & Tilgner 2000a), but without it one would be forced to damp all modes more strongly.

The first to achieve $E = O(10^{-6})$ (with hyperviscosities) were Glatzmaier & Roberts (1995). As noted above, one of their findings was that a finitely conducting inner core seemed to stabilize the field. Even so, their field was probably still not stable enough, reversing after only one or two magnetic diffusive timescales. They were unfortunately not able to run long enough to obtain reversal statistics, which would have allowed a better comparison with the real reversal record. Longer runs, long enough to obtain multiple reversals, were done by Glatzmaier et al. (1999), who studied the influence of the thermal boundary conditions imposed at the core-mantle boundary, and found that in this way the mantle could indeed control the overall reversal frequency. However, even these runs were not done for the hundreds of diffusive timescales that one would really like. At these values of E integrating for even one diffusive timescale is quite a challenge.

Another particularly interesting set of runs was done by Roberts & Glatzmaier (2001), who looked at the effect of varying inner core sizes. Rather surprisingly, they found that a larger inner core tended to yield a less stable field, so exactly the opposite of what one might have expected if the inner core is supposedly stabilizing the field. A similar result had previously also been found by Morrison & Fearn (2000), although their model was not fully three-dimensional. The resolution to this particular paradox is presumably that changing the inner core also changes the size of the region inside the tangent cylinder, and that the behavior of the field inside and outside $\mathcal{C}$ are often very different. So in these two models the dynamics inside $\mathcal{C}$ presumably favor less stable fields, and this outweighs the potentially stabilizing role of the inner core itself. This is not necessarily the case though; Sakuraba & Kono (1999) and Bloxham (2000a) also consider different inner core sizes, and find that it has relatively little effect.

Following Glatzmaier & Roberts, the next to achieve $E = O(10^{-6})$ (again with hyperviscosities) were Kuang & Bloxham (1997, 1999), see also Kuang (1999). One interesting aspect of this work is the effect that viscous coupling can have. In particular, in the real Earth inertial effects should be more important that viscous effects (E being several orders of magnitude smaller than Ro). In the numerical models, however, reducing Ro is much easier than reducing E (indeed, one can set Ro $\equiv 0$). Kuang & Bloxham therefore argued that one should apply stress-free rather than no-slip boundary conditions, to minimize the effect of the artificially large viscosity. They found that this has a surprisingly large effect, with stress-free conditions tending to concentrate the field outside $\mathcal{C}$, but no-slip conditions tending to concentrate it inside $\mathcal{C}$.

As noted above, Bloxham (2000a) subsequently used this model to explore the effect of different inner core sizes. Like Glatzmaier et al. (1999), he also considered the influence of different thermal boundary conditions. However, unlike their study of the resulting reversal statistics, he focused on the much simpler question of whether the field continues to be dominated by an axial dipole, with all other harmonics averaging to zero on timescales of a few thousand years. As simple as it sounds, this is quite an important issue, as this so-called geocentric axial dipole hypothesis is widely used in interpreting paleomagnetic data (for example, in reconstructing plate tectonic motions). He then found that imposing different inhomogeneities at the core-mantle boundary could indeed lead to different field structures. Bloxham (2000b, 2002) further considered the influence of variations in heat flux at the CMB, this time with a view to understanding aspects of the present-day field. In particular, he found that some type of inhomogeneities were needed to match the amplitude of the secular variation, and may also play a role in other features, such as the existence of persistent flux bundles. See also Kono, Sakuraba & Ishida (2000), and Christensen & Olson (2003).

Finally, the last set of simulations we want to discuss are those of Grote, Busse & Tilgner (1999, 2000b), see also Busse (2000, 2002a). Unlike some of the previously mentioned results, the emphasis here was less on trying to reproduce the real Earth as closely as possible, but rather on exploring more general properties of the field generation process, such as whether dipolar fields are always preferred over quadrupolar fields. (See, for example, Gubbins & Zhang (1993) for a discussion of the various symmetries allowed by the geodynamo equations.) Busse et al. then found that there are indeed regions in parameter space where the field is quadrupolar rather than dipolar. Something as basic as the strong dipole dominance of the Earth's field is thus not fully understood. Perhaps even more fascinating, they also found examples of so-called 'hemispherical' dynamos, in which the dipole and quadrupole components interact in such a way that the field is almost completely confined to just one hemisphere. These solutions presumably do not apply to the Earth, but again, it is not entirely clear why not.

To summarize then, numerical geodynamo modeling has progressed to the point where surprisingly realistic models exist, models that reproduce not only the dominant axial dipole, but also secondary features like the secular variation. On the other hand, in a slightly different region in parameter space one can also find solutions radically different from the present-day Earth. And given that most of the relevant parameters are still many orders of magnitude removed from real Earth values, it is far too soon to say that the geodynamo problem has been solved. Indeed, even at Ekman numbers as small as 10^{-6}, it is not clear that something as basic as the distinction between weak and strong fields has been reached; certainly none of these models has conclusively demonstrated it.

2.8   Magnetic Instabilities

All of the models discussed in the previous section equilibrate at order one Elsasser numbers, as we would expect based on the discussion in Sect. 2.3.3 (although once again, none of them exhibit a clear distinction between weak and strong-field regimes). It is nevertheless of interest to consider the detailed mechanism whereby the field equilibrates (in a sense, the specific numerical constant multiplying the basic order-one scaling). Magnetic instabilities are one possible equilibration mechanism. See also the review by Fearn (1998).

Suppose we take the momentum and induction equations (but for simplicity neglect the temperature equation), and linearize them about some large-scale field $\boldsymbol{B}$,

$$\mathrm{Ro}\, \frac{\partial \boldsymbol{u}}{\partial t} + 2\hat{\boldsymbol{e}}_z \times \boldsymbol{u} = -\nabla P + (\nabla \times \boldsymbol{B}) \times \boldsymbol{b} + (\nabla \times \boldsymbol{b}) \times \boldsymbol{B} + \mathrm{E}\Delta\boldsymbol{u},$$

$$\frac{\partial \boldsymbol{b}}{\partial t} = \nabla \times (\boldsymbol{u} \times \boldsymbol{B}) + \Delta\boldsymbol{b}. \tag{2.71}$$

The idea behind magnetic instabilities is to ask how large must $\boldsymbol{B}$ become (for some given spatial structure) before this system will spawn some type of (typically small-scale) instability. Now, in this linearized system these instabilities would simply grow indefinitely, without any feedback on $\boldsymbol{B}$. In the nonlinear regime though, one would have a continual transfer of energy from the large-scale field $\boldsymbol{B}$ to the smaller-scale $\boldsymbol{b}$. The suggestion then is that it is this energy drain that equilibrates the large-scale field. That is, we might expect the amplitude that emerges from the full system of equations to be close to the critical amplitude that emerges from these linearized equations. And because these linearized equations are so much simpler than the original set, one can explore the full range of possibilities in far more detail. For example, if $\boldsymbol{B}$ is taken to be axisymmetric (as it invariably is, and as befits a 'large-scale' field), the different nonaxisymmetric instability modes decouple, thereby reducing the problem from 3D to 2D. As a result of simplifications like this, the Ekman numbers that can be achieved in magnetic instability studies are typically far smaller than in the full simulations, and the true asymptotic limit can often be reached.

The earliest work on magnetic instabilities was by Malkus (1967) and Acheson (1972), who considered ideal instabilities of fields of the form $\boldsymbol{B} = B(s)\hat{\boldsymbol{e}}_\phi$, and by Fearn (1984), who considered resistive instabilities. More general – and more realistic – fields of the form $\boldsymbol{B} = B(s,z)\hat{\boldsymbol{e}}_\phi$ were considered by Fearn & Proctor (1983), Fearn & Weiglhofer (1991), and Zhang & Fearn (1993), for example. These results suggest that the large-scale field within the core cannot be much greater than 50 G – in reasonable agreement, incidentally, with the results of the full simulations.

There are a number of directions in which one can extend these basic studies. For example, Fearn et al. (1997) included a large-scale differential rotation of the form $\boldsymbol{U} = u(s)\hat{\boldsymbol{e}}_\phi$ (that is, like the geostrophic flow that $\boldsymbol{B}$ itself would induce), and found that it could be either stabilizing or destabilizing. We note then that not only $\boldsymbol{B}$, but $\boldsymbol{b}$ also induces a geostrophic flow. This flow induced by the instability itself is one of the mechanisms that ultimately equilibrate it, so the fact that its effect can be either stabilizing or destabilizing may be quite important, since this determines whether the bifurcation is supercritical or subcritical. Indeed, Hutcheson & Fearn (1996) have suggested that subcritical magnetic instabilities may cause the entire field to switch to a completely different state. That is, events like reversals or excursions, for example, may be triggered by magnetic instabilities.

Finally, one might ask why we did not consider magnetic instabilities in Sect. 2.3. That is, how can we be sure that the modes we obtained there really are convective rather than magnetic instabilities? The energy equation associated with Eq. (2.71) is

$$\frac{\partial}{\partial t} \frac{1}{2} \int \left(|\boldsymbol{b}|^2 + \mathrm{Ro}\, |\boldsymbol{u}|^2 \right) \mathrm{d}V$$

$$= \int \boldsymbol{u} \cdot (\boldsymbol{J} \times \boldsymbol{b})\, \mathrm{d}V - \int \left(|\boldsymbol{\nabla} \times \boldsymbol{b}|^2 + \mathrm{E} |\boldsymbol{\nabla} \times \boldsymbol{u}|^2 \right) \mathrm{d}V, \tag{2.72}$$

where $\boldsymbol{J} = \boldsymbol{\nabla} \times \boldsymbol{B}$. That is, magnetic instabilities are really instabilities of the $\boldsymbol{J}$ rather than the field $\boldsymbol{B}$ – and so the potential fields considered in Sect. 2.3 cannot exhibit magnetic instabilities at any field strength.

It is nevertheless of interest to ask what might happen if we include the temperature equation in our system (2.71) after all, thereby yielding a system that allows both convective and magnetic instabilities. This problem was considered by Zhang (1995), who showed that one can have a smooth transition from one type of instability to another. In particular, he showed that unlike the solid line in Fig. 2.6 (left), where $\widehat{\mathrm{Ra}}_\mathrm{c}$ decreases up to some $O(1)$ Elsasser number, and increases again thereafter, if $\boldsymbol{B}$ is not a potential field, so that magnetic instabilities may occur, $\widehat{\mathrm{Ra}}_\mathrm{c}$ simply decreases monotonically, and eventually even becomes negative. The physical interpretation of this phenomenon is that in the region where $\widehat{\mathrm{Ra}}_\mathrm{c} > 0$, the instability is primarily convective, and is therefore driving the dynamo. In contrast, in the region where $\widehat{\mathrm{Ra}}_\mathrm{c} < 0$, the instability is primarily magnetic, and is therefore dissipating the field again. On this view we would thus expect the field to equilibrate at precisely that Elsasser number where $\widehat{\mathrm{Ra}}_\mathrm{c} = 0$. Note though that one obtains a smooth transition from convective to magnetic instabilities only for $\mathrm{q} = O(1)$; see Fearn & Proctor (1983) or Zhang & Jones (1996) for the behavior at small q. Nevertheless, the overall conclusion remains that if the field is too strong, magnetic instabilities will most likely dissipate it again.

2.9   Other Planets

The Earth is not the only planet in the solar system to have a magnetic field; most of the planets (and larger moons) either have a field, or at least show evidence of having had one in the past. We conclude therefore with a tour through the solar system, noting which bodies have magnetic fields and which do not, and whether we can understand these results in terms of the same theoretical framework presented here. See also Connerney (1993), Russell (1993), Ness (1994), and Stevenson (2003) for summaries of the observational data.

Starting closest to home, the Moon has no large-scale field at present. It does have weak small-scale fields, of order 10^{-4} G. It appears also that rocks between 3.2 and 3.9 Gyr old are particularly strongly magnetized (Runcorn 1994). One suggestion therefore is that the Moon did have a dynamo back then, but that it switched off some 3 Gyr ago. See, for example, Stegman et al. (2003) for the most recent lunar dynamo model. As to why it switched off, there is a ready explanation for that too; since the Moon is so much smaller than the Earth, its thermal evolution is also much quicker, so its core almost completely solidified long ago, thereby necessarily switching off any dynamo that may have existed before. (In another few billion years, the Earth's core too will freeze completely, and the geodynamo will also switch off.)

2.9.1   Mercury, Venus and Mars

Turning next to Mercury, things already get considerably more interesting. Although Mercury was only visited once, by Mariner 10 in 1974/75, that was sufficient to establish that it has a global field, with a dipole moment of $2 - 5 \cdot 10^{19}$ A m^2. Or, extrapolating back down to the surface, we obtain field strengths of around $3 \cdot 10^{-3}$ G. Of course, what we are really interested in is not the field strength as such, but rather the Elsasser number. Extrapolating not just down to the surface, but on down to the core (whose radius can be estimated by other means) poses no problems. Mercury's rotation rate is also known, namely one revolution every 59 days. The density ρ and magnetic diffusivity η are not known, but are unlikely to be radically different from the corresponding values for the Earth. Inserting the numbers, we therefore end up with an Elsasser number of around 10^{-4}.

What to make of this value then? It seems too small to correspond to the strong-field regime, but could well be in the weak-field regime. We note in particular that the Ekman number for Mercury is around 10^{-12} (assuming again the same viscosity as in the Earth's core), so $\Lambda \approx E^{1/3}$ could fit the weak-field regime quite nicely. However, the difficulty then is to understand why the Earth managed the transition from the weak to the strong-field regime, but Mercury did not. In particular, one cannot simply postulate that $\widehat{Ra}$ is less in Mercury, since that is more likely to shut the dynamo off entirely (via this dynamo catastrophe) than to switch it to the weak-field regime. Another possibility is that the growth of an inner core has progressed much further in Mercury than in the Earth – it is considerably smaller, after all, hence ought to evolve faster. And as we saw above, the relative size of the inner core can have a significant influence on the field, in a number of important ways. We conclude therefore that Mercury's field could conceivably be explained by our theoretical framework here, but there are certainly a lot of details that still need pinning down.

Moving out, we next encounter Venus, which was visited by numerous probes, both Soviet and American. The most recent was Pioneer Venus, which orbited the planet between 1978 and 1992. It found no evidence of an internally generated magnetic field; any field that does exist could have a dipole moment of at most $4 \cdot 10^{18}$ Am^2. Again converted to an Elsasser number, this yields $\Lambda < 10^{-7}$. This is so much smaller than the weak-field regime even that we conclude that Venus does indeed not have a dynamo at all. Note also that while Venus might be slowly rotating in comparison with the Earth (requiring 243 days/revolution), in terms of an Ekman number it is still rotating extremely rapidly, with $E \approx 10^{-12}$. The distinction between weak and strong-field regimes should therefore apply to it just as much as to the Earth or Mercury.

So, why does Venus not have a dynamo? In particular, in terms of size it is the planet most similar to the Earth, with a radius of 6050 km versus 6370, and an estimated core radius of ~ 3000 km versus 3480. Why should the Earth therefore not only have a dynamo, but a strong field dynamo, whereas Venus has none at all? One suggestion, by Stevenson (1983), is that Venus has not yet nucleated an inner core, and that thermal convection by itself is not sufficient to drive a dynamo in this case. The reason why Venus might not yet have an inner core, even though it is slightly smaller than the Earth, and hence ought to evolve quicker, is – paradoxically – precisely because it is smaller. In particular, its smaller size means the central pressure is also less, and the effect of this on the freezing temperature could be sufficient that its core is still fluid all the way through.

Another point to bear in mind is that the pattern of mantle convection is very different in Venus than in the Earth. Unlike the Earth, Venus has no plate tectonics; instead, the entire surface is effectively one single plate (e.g. Schubert, Turcotte & Olson, 2001). This rigid lid then acts as an insulating blanket covering the convection below, with the result that even though Venus is smaller than the Earth, it may actually cool slower (which would also be consistent with it not having an inner core yet). We realize therefore that (i) there is not necessarily a one-to-one relationship between a planet's size and the rate at which it cools, and (ii) in order to understand a planet's core one must ultimately understand its mantle as well. That is, up to now we have viewed the mantle merely as an obstacle that prevents us from seeing the core we are really interested in. We realize now though that the rate at which the core cools, and hence the vigor of its convection, is ultimately controlled by processes in the mantle.

Just like Venus, Mars also has no large-scale field, with the results from the Mars Global Surveyor probe placing an upper bound of $2 \cdot 10^{18}$ A m^2 on the dipole moment (corresponding to $\Lambda < 10^{-8}$). Like the Moon though, it has small-scale fields, indeed surprisingly strong ones, occasionally even comparable with the Earth's surface field. See also Acuña et al. (2001) for summaries of the data from the MGS mission. We conclude therefore that Mars also had a dynamo once, but that it switched off just like the Moon's did (and the reason the small-scale fields left over today are so much stronger than on the Moon has more to do with the details of the minerals that got magnetized than the underlying dynamo).

The only difficulty with this explanation is that Mars is bigger than Mercury (although their cores are believed to be comparable), so it ought to evolve slower, so why has it already switched off whereas Mercury has not? One suggestion, by Nimmo & Stevenson (2000), again illustrates the role that the mantle may play in ultimately controlling the core. In particular, they suggest that the pattern of mantle convection in Mars switched from Earth-like plate tectonics to a Venus-like rigid lid. Because this is so much less efficient at removing heat, the mantle temporarily stops cooling (indeed, it may even heat up slightly, due to the radioactive trace elements contained in it), so the core stops cooling too, causing the dynamo to switch off. See also Schubert, Russell & Moore (2000) and Stevenson (2001) for elaborations on this idea, including details of when this is believed to have happened.

As plausible as it may be, this idea raises even more questions than it answers. For one, what about Venus? Did it too once have plate tectonics, and hence perhaps a dynamo? Unfortunately, Venus is above the Curie temperature even at the surface, so any evidence of such an ancient dynamo is lost forever. Next, what about Mercury? It has no plate tectonics either, so why does it have a dynamo whereas Venus and Mars do not? One possibility might be that when eventually the core does start cooling again, it cools sufficiently quickly that the dynamo switches back on. And because Mercury is smallest, one might expect this process to happen first in it. So might Venus and/or Mars have dynamos at some point in the future? In the case of Mars, its dynamo has been switched off for so long that it seems rather unlikely that it would suddenly switch back on again, but Venus is certainly a strong candidate for a future dynamo.

To summarize our tour of the terrestrial planets then, we conclude that while none of them obviously contradicts our theoretical framework of strong versus weak versus no dynamo, there is far more to understanding their magnetism than just this framework. In particular, we recognize that ultimately one cannot understand a planet's magnetism without understanding its entire thermal evolution, and that this evolution is not necessarily a one-way process, with the mantle and core always cooling, and smaller planets always cooling faster than larger ones.

2.9.2 Jupiter's Moons

Jupiter and its moons have been extensively studied, most recently by the Galileo mission, which found evidence of internally generated magnetic fields in Ganymede and possibly Io, but not in Europa and Callisto. See also Showman & Malhotra (1999) and Russell (2000) for summaries of the Galileo results, and their most likely interpretations.

Considering Ganymede first, its dipole moment is around 10^{20} A m^2, so slightly larger than Mercury's. Given that Ganymede is also slightly larger than Mercury, this is perhaps not too surprising. However, its core is believed to be considerably smaller than Mercury's (by a factor of perhaps 2 in radius). The field within the core must therefore be considerably stronger than in Mercury. The result is that Ganymede's Elsasser number is around 0.1, so strong rather than weak.

Turning next to Io, its dipole moment too is slightly larger than Mercury's, which is somewhat surprising, considering that in size it is closer to the Moon. So why has it not long since solidified all the way through? The answer is that while Jupiter and its moons may look like a miniature solar system, there are also important differences. In particular, Jupiter's moons are in much tighter orbits than Mercury even, and even allowing for the fact that Jupiter is much smaller than the Sun. This means that tidal effects are far more important in the Jovian system than in the solar system. That is, unlike the Moon and the planets, the Galilean satellites (and especially Io, the innermost one) do not just gradually cool, but are instead continuously being reheated by tidal friction. Indeed, far from being cold and dead like the Moon, Io is the most volcanically active body in the solar system.

However, before concluding that this extra source of heat has generated a true dynamo, there is an additional complication we need to point out, again caused by the fact that Io orbits so close to Jupiter. In particular, whereas the background solar magnetic field is very small in comparison with the Earth's field, the background Jovian field is almost comparable with Io's field. It is thus conceivable that Io is not really acting as a dynamo at all, but is merely amplifying this externally imposed field through magnetoconvection. See, for example, Sarson et al. (1997) for some numerical calculations exploring this issue, and Walker & Hollerbach (1999) for a study of how the adjustment to Taylor's constraint is modified in the presence of a background field.

Finally, what about Europa and Callisto? They may not have internally generated fields, but both do show a so-called induction response, caused by moving a conductor through a magnetic field. The field they are moving through is clear, of course, namely this same background field as in Io, but what is the conductor? Perhaps somewhat surprisingly, it is believed not to be a metallic core. Callisto is believed not to have a core at all, and Europa's is too small (the response is such that the conductor must be quite close to the surface). Instead, in both cases the conductor is believed to be an ocean, a layer of salty water covering the entire moon (and covered by ice of course, as the surface temperatures are well below 0°C). See, for example, Kivelson et al. (2000).

2.9.3 Jupiter and Saturn

As noted above, Jupiter not only has a magnetic field, this field is strong enough to make itself felt as far out as Io at least. The origin of this field is believed to be a convection-driven dynamo, just as in the Earth. Unlike the bodies considered up to now though, the electrically conducting fluid in this case is not molten iron, but rather metallic hydrogen. Fortunately, the conductivity can still be estimated, and indeed comes out to be rather similar to that of molten iron (Nellis 2000). We can therefore again convert the observed dipole moment to an Elsasser number, which comes out at around 1. Jupiter is therefore also in the strong-field regime.

There are, nevertheless, also a number of differences between Jupiter and the terrestrial planets, besides just this difference between molten iron versus metallic hydrogen. First, whereas the density contrast across the Earth's outer core was no more than 20%, in Jupiter it varies by perhaps a factor of 4 across the metallic-hydrogen zone. In modelling Jupiter one should therefore use the anelastic rather than the Boussinesq approximation. Secondly, the core-mantle boundary in the Earth represents a very abrupt change in terms of both the mechanical as well as the electromagnetic properties, going from fluid to solid, and from conducting to almost completely insulating. In contrast, in Jupiter there is a much more gradual transition from metallic to molecular hydrogen, and, of course, no rigid boundary anywhere. What effect this might have on the details of the dynamo is not known, but should probably be taken into account as well if one really wished to model Jupiter's magnetic field.

Saturn was last visited by Voyager 2 in 1981, so we have less data on it than on Jupiter. Nevertheless, it too has a magnetic field, again believed to originate in a zone of metallic hydrogen. And again, it is a strong-field dynamo, with an Elsasser number of around 0.1. One particularly interesting feature about Saturn's field is how astonishingly closely it is aligned with the rotation axis. As we saw above, the Earth's field is aligned to within 11°, which we agreed is already too close to be just chance. (Mercury, Jupiter, Ganymede and Io are similarly closely aligned, making chance alignment even less plausible.) In contrast, Saturn's field is aligned to within 0.1°. Indeed, the entire field, not just the dipole, appears to be almost completely axisymmetric. We know from Cowling's theorem that it cannot be axisymmetric in the region where it is actually being generated. So why is the external field so strongly axisymmetric? Stevenson (1982) suggests it has to do with the fact that the metallic hydrogen region is buried so deep within the planet (because Saturn is considerably less massive than Jupiter, one must go much deeper to reach the metallization pressure). He then suggests that differential rotation in the overlying, weakly conducting layers wipes out the nonaxisymmetric components of the field. However, Love (2000) shows by way of some sample calculations that differential rotation need not have this effect.

2.9.4   Uranus and Neptune

Uranus and Neptune were each visited only once, by Voyager 2 in 1986 and 1989, respectively. Both have magnetic fields, with dipole moments of order 10^{24} A m^2 for both. Unlike all the other planets, their fields are not aligned with their rotation axes, being tilted by 59° and 47°, respectively. Also, whereas for all the other planets the dipole is rather closely centered on the planet, for these two it is offset by 0.33 and 0.55 planetary radii, respectively. See, for example, Holme & Bloxham (1996), who show that these results are indeed robust, and not just artifacts of the sparseness of the data.

What to make of these very unexpected results? If only one planet had been anomalous, one would be tempted to say it just happens to be in the middle of a reversal, but two planets reversing at exactly the same time seems rather implausible. It seems more likely therefore that they are indeed operating in a fundamentally different mode than all the other planets. For example, we saw above that in certain regions of parameter space quadrupolar or even hemispherical dynamos may be preferred (Grote, Busse & Tilgner 1999, 2000b; see also Grote & Busse 2000), which could certainly explain these very large offsets.

And leaving aside their peculiar spatial structure, are these fields weak or strong? To answer this we again need to know the magnetic diffusivity (and density) of the conducting fluid. What is that though? In particular, Uranus and Neptune are both too small for metallic hydrogen to form, and any iron cores that might exist deep within are certainly far too small. So what is the conducting fluid? As implausible as it may seem at first, the answer (the only remaining possibility) is a mixture of water, ammonia, methane, and various other constituents, the diffusivity of which has been estimated at around 100 m^2/s (Nellis et al. 1997). (For comparison, we remember that $\eta \approx 2$ m^2/s for molten iron, and $O(10)$ for metallic hydrogen.) The resulting Elsasser numbers are then perhaps $O(10^{-4})$ (although extrapolating the fields of Uranus and Neptune back down into this conducting region is more uncertain than before, because they are less strongly dipole dominated). We conclude therefore that Uranus and Neptune are probably in the weak-field regime.

However, given all these various uncertainties, and how different they are from all the other planets, we cannot really say that we understand either of them. Indeed, that more or less sums up our overview of planetary dynamos in general: none of them obviously contradicts the general theoretical framework laid out in the rest of this chapter (for example, none of them has an Elsasser number greater than one), and here and there we can even make reasonably plausible arguments as to why a given planet does or does not have a field, but we cannot say that we definitely understand any of them, not even the Earth.

3 Differential Rotation Theory

The majority of stellar-activity phenomena are magnetic in origin. Differential rotation, meridional flow and stellar winds, however, can also be understood in the context of mean-field hydrodynamics in stellar convection zones. In stellar convection zones the Schwarzschild criterion ($dS/dr < 0$), where S is the specific entropy, is fulfilled. Their temperature stratification is so steep that possible velocity fluctuations grow exponentially and the entire zone becomes turbulent. Due to the density stratification the turbulence fields are themselves stratified with the radial direction as the preferred direction. If such a turbulence field is subject to an overall rotation the stellar convection zone must lead to the formation of large-scale structure. Wasiutynski (1946), Biermann (1951) and Kippenhahn (1963) were the first to find that differential rotation and meridional flow might be direct consequences of the existence of rotating anisotropic turbulence. Details of the long history of this concept are presented by Rüdiger (1989, Chapt. 2).

We should keep in mind that the convection zones in stars of different mass are very different. In Fig. 3.1 the Sun is shown with its outer convection zone with about 30% extension in radius (only a few % in mass) together with two sorts of cooler stars and two sorts of warmer main sequence (MS) stars. The outer convection zones in cooler stars become deeper and deeper until for M stars the convection zone reaches down to the center. On the other hand, for A stars the outer convection zone starts to become very thin, but an inner zone becomes convectively unstable. This inner convection zone for B stars reaches considerable dimensions.

It thus becomes clear that the level of stellar activity should differ strongly from spectral type to spectral type. There is, however, the striking fact that the linear depth of the outer convection zones, at 200,000 km, does not vary too much along the MS. We shall see later how important the total thickness of a convection zone is, for e.g., the formation of differential surface rotation or the cycletime of an oscillating dynamo.

3.1 The Solar Rotation

Differential rotation is explained here as turbulence-induced with only a small magnetic contribution. It is certainly unrealistic to expect a solution of the complicated problem of the solar dynamo without an understanding of the mean-field hydrodynamics. There is indeed no hope for the stellar dynamo concept if the internal stellar rotation law cannot be predicted or observed (by asteroseismology). Fortunately, the helioseismological inversions yield a detailed

The Magnetic Universe: Geophysical and Astrophysical Dynamo Theory.
Günther Rüdiger, Rainer Hollerbach
Copyright © 2004 Wiley-VCH Verlag GmbH & Co. KGaA
ISBN: 3-527-40409-0

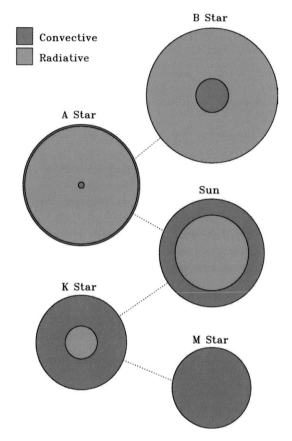

Figure 3.1: The locations of convection zones ($dS/dr < 0$) and radiation zones ($dS/dr > 0$) in MS stars of various spectral types. The radii are not to scale. Courtesy P. Charbonneau.

portrait of the internal solar rotation (see Fig. 3.2). The main empirical features of the solar differential rotation law are

- an equatorial acceleration of about 30% at the surface,
- a strong polar subrotation and weak equatorial superrotation,
- a reduced equator-pole difference in Ω at the lower convection-zone boundary,
- a rigid and slow rotation of the radiative core at least down to 0.2 $R_\odot$ (Couvidat, García & Turck-Chièze 2003)[1],
- a clear negative Ω-gradient in the outermost (supergranulation) layer[2]

(see also Fig. 3.3). A characteristic Taylor–Proudman structure in the equatorial region and a characteristic disk-like structure in the polar region are indicated by these results (Fig. 3.4). In midlatitudes the isolines are almost radial (on average). So the Ω-isolines are far from the Taylor–Proudman geometry that normally results for fast rotation. This fact was called the 'Taylor number puzzle' and forms the main challenge for theoretical explanation.

---

[1] an increase of about 15% at higher latitudes below $x = 0.4$ is reported by Li & Wilson (1998)

[2] young spots are rotating faster by about 4% than the solar surface plasma

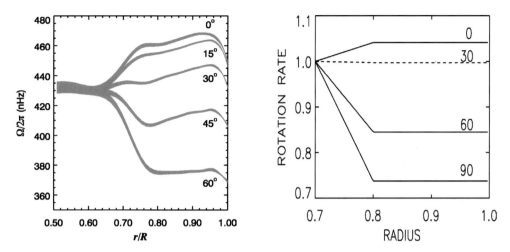

Figure 3.2: *Left*: The solar rotation law in the convection zone as determined by helioseismology (National Solar Observatory). *Right*: The rotation profile (3.8) with $\omega_0(x) = 1$ and $\omega_2(0.7) = 0$, $\omega_2(1) = -0.04$ is rather close to the observations. Note that the observations suggest $\mathrm{d}\omega_0/\mathrm{d}r \gtrsim 0$. See Paternò (1991).

It might easily be true that the overall surface rotation during the times of the Maunder minimum differed (slightly) from its present-day profile. Ribes & Nesme-Ribes (1993) found a rotation rate slower by about 2% at the equator and by about 6% at midlatitudes than at the present time. The differential rotation was thus *stronger* than today. The more magnetic the Sun, the faster and more rigidly its surface rotates (Fig. 3.5).

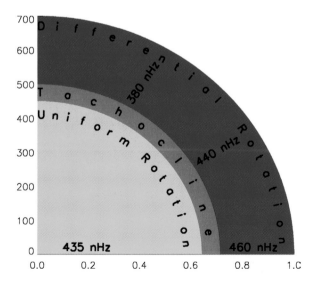

Figure 3.3: The schematic structure of the Sun with respect to the rotation regimes. The tachocline is the transition region between the regime of strong latitudinal differential rotation in the convection zone and the rigid rotation of the radiative interior. The unit of the vertical axes is $1\mathrm{Mm} = 1000$ km.

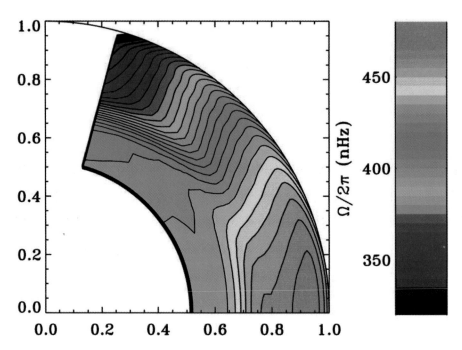

Figure 3.4: Isolines of the angular velocity in the outer part of the Sun after Gilman & Howe (2003). Note that no informations exist for the polar regions. The isolines are radial only in the higher midlatitudes.

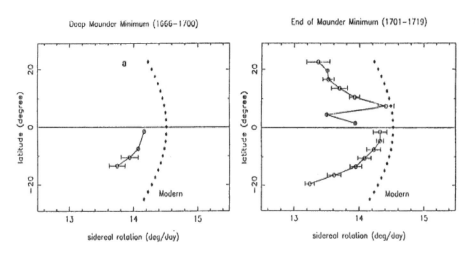

Figure 3.5: The rotation law in the Maunder minimum (*left*) and at its end (*right*). The equatorial acceleration is stronger than in modern times but the equatorial rotation is slower (Ribes & Nesme-Ribes 1993). Balthasar, Vázquez & Wöhl (1986) could not find similar results for a *regular* minimum.

3.1.1 Torsional Oscillations

Schrijver & Zwaan (2000), Stix (2002) and Thompson et al. (2003) presented detailed historical and data-based overviews of all phenomena concerning the temporal variations of the solar rotation law. Besides the small equatorial minimum of the rotation profile, the torsional oscillations are of particular importance. At any epoch, some latitudinal bands at the surface are rotating faster and some are rotating slower than the average. At a fixed latitude there is an oscillation of fast and slow rotation with an 11-year period (see Fig. 3.6). The entire cycle is repeated approximately every 22 years. The whole pattern migrates at about 2 m/s toward the equator. Two fast and two slow belts exist in each hemisphere. The most rapidly rotating zone is always on the equatorial side of the activity belt. The amplitude of the linear velocity is about 6 m/s.

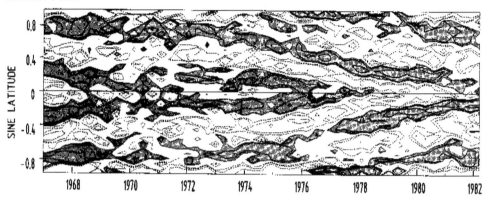

Figure 3.6: Torsional oscillations derived from Doppler shift measurements. The flow pattern drifts in 22 yr from pole to equator, but at a given location the oscillation time is only 11 yr (Howard & LaBonte 1980, 1983).

Howe, Komm & Hill (2002) showed by helioseismological inversions that the migrating bands do not exist only at the surface, but extend downward to 60,000 km, i.e. 30% of the convection zone (Fig. 3.7, left).

An important question in the dynamo theory framework is the possible existence of a polar branch migrating poleward. Using data between 1996 and 2001 Schou (2001) and Vorontsov et al. (2002) report the existence of such a zonal band during the rising phase of cycle 23. According to Vorontsov et al. the high-latitude acceleration around 60° seems to reach deep into the solar convection zone. This is also true for a decelerating flow between both the reported accelerating flows.

For the bottom of the convection zone Howe et al. (2001) reported another oscillatory phenomenon. At the equator and up to latitudes of 60° there is a rather coherent oscillation of the rotation rate with a 1.3-year 'period' (between 1995 and 2000, see Fig. 3.7, right). The same value has been found in sunspot data by Krivova & Solanki (2002). Ternullo (2004) reports the equatorward drift of the sunspot zone as a "sequence of alternating high speed equatorward phases and stationary or even retrograde (poleward) ones". The length of the sequences is about 400–500 days. Also the quasi-two-year oscillations of the surface value of the poloidal magnetic field reported by Benevolenskaya (1995) may have a very similar origin.

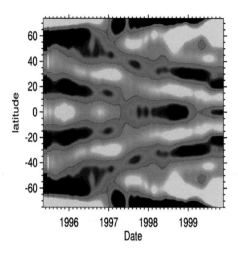
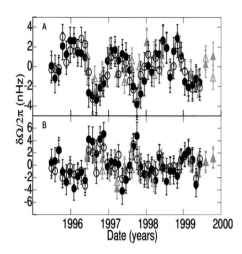

Figure 3.7: *Left*: Deviations from the mean rotation profile as a function of date and latitudes close to the solar surface. Bands of faster (red) and slower (blue) rotation migrate toward the equator in 11 years. Note the existence of a separate branch at midlatitudes. *Right*: The 1.3-year fluctuation deep in the convection zone ($x = 0.72$) of the angular velocity at the equator (*top*) and at midlatitudes (*bottom*). Data from GONG and MDI. Courtesy R. Howe.

3.1.2 Meridional Flow

By analyzing the statistics of sunspot groups observed at Greenwich, Tuominen (1941, 1961) found that there is a mean equatorward motion at low latitudes and a poleward motion at higher latitudes. The order of magnitude, however, was not more than a few m/s. Ward (1973) favored a mean meridional velocity of less than 1 m/s. In a recent paper Vršnak et al. (2003) present a similar result. According to Fig. 3.8 the equatorward motion (of order 5 m/s) is restricted to latitudes less than $10°$, while the poleward motion only reaches up to $40°$. According to this result a new, third cell has been found in the area north of the butterfly diagram where the flow goes toward the equator. If correct, this finding will have enormous consequences for the theory of solar dynamo.

An important step forward was the helioseismological finding of Braun & Fan (1998) of a poleward meridional flow of order 10 m/s. The situation, however, becomes more complicated still for the meridional flows derived by Haber et al. (2002). The flow in the surface layers is given for $-45°$ for 1997–2001, with an amplitude of about 20–30 m/s. The northern hemisphere is similar, but only for 1997. In 1999–2001 the poleward flow only exists in the outermost layer. Beneath this surface layer the flow in the northern hemisphere is toward the equator. If this is true then we have to accept a much higher degree of randomness even for the large-scale flows. A more smooth behavior of the meridional flow in the supergranulation layer has been derived by Zhao & Kosovichev (2004, see Fig. 3.9).

The question whether there is an equatorward backflow deep in the convection zone has been considered by Hathaway et al. (2003). From the sunspot data since 1874 they found an anticorrelation between the drift rate of the center of the butterfly diagram and the cycle length. The faster the drift of the butterfly diagram the shorter are the cycles (Fig. 3.10). With

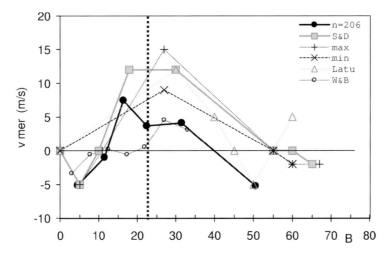

Figure 3.8: The meridional flow at the solar surface vs. the solar latitude. It consists of three cells. The middle cell flows poleward while the smaller ones are directed toward the equator. For details about the other given flow profiles see Vršnak et al. (2003). The dotted vertical line marks the mean latitude of the sunspots in the considered period.

such statistics an amplitude of 1.2 m/s for the meridional backflow velocity at the bottom of the convection zone toward the equator has been derived.

3.1.3   Ward's Correlation

We turn next to the correlation between longitudinal and latitudinal fluctuations, i.e. $Q_{\theta\phi} = \langle u'_\theta u'_\phi \rangle$. The latter is also known as the 'horizontal Reynolds stress'. In order to introduce a suitable normalization we have to keep in mind that $Q_{\theta\phi}$

- vanishes unless the star rotates, hence it must vanish for $\Omega \to 0$,
- is an axial expression that changes sign if $\boldsymbol{\Omega} \to -\boldsymbol{\Omega}$, hence it must be odd in Ω,
- must vanish at the equator if, as we assume, the turbulence pattern has equatorial symmetry,
- must vanish at the poles if the mean quantities are assumed to be axisymmetric.

It thus seems reasonable to introduce a function $w(\theta)$ defined by

$$Q_{\theta\phi} = \nu_\mathrm{T}\Omega_\odot \cos\theta \sin^2\theta\, w(\theta). \tag{3.1}$$

The product $\nu_\mathrm{T}\Omega_\odot$ is the scalar quantity with the correct dimensions. Ward (1965) was the first to present clear evidence for a *positive* correlation, namely

$$Q_{\theta\phi} \simeq 2 \cdot 10^7 \ \mathrm{cm}^2/\mathrm{s}^2. \tag{3.2}$$

He considered the proper motion of sunspot groups, the faster of which tend to move toward the equator. This result indicates that even at the surface of the Sun there must be more than a simple Boussinesq stress-strain relation, $Q_{\theta\phi} = -\nu_\mathrm{T} \sin\theta \partial\Omega/\partial\theta$. Gilman & Howard (1984) extracted the positions of individual spots and groups of spots from a 62-year period

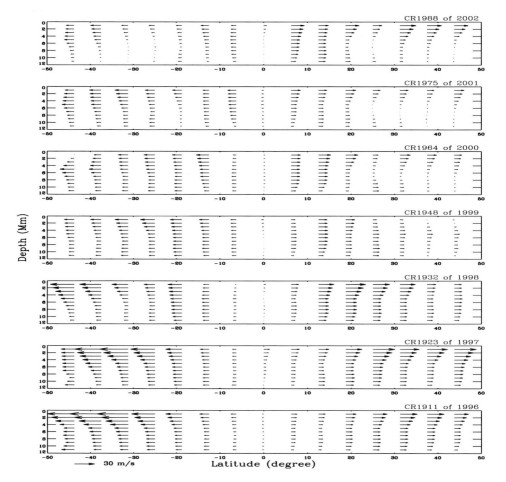

Figure 3.9: The helioseismological results of Zhao & Kosovichev (2004) for the meridional flow in the solar surface layer between 1996 (*bottom*) and 2002 (*top*) for both the hemispheres. The flow is always polewards but it appears to exist in the surface (∼supergranulation) layer only.

and obtained a cross-correlation with the same sign and of similar magnitude as Ward. The result has also been confirmed by Balthasar, Vázquez & Wöhl (1986). However, for small solitary spots, which should especially be almost passive tracers, the result is considerably less (see also Ribes 1986). For old sunspots Nesme-Ribes et al. (1993) even find a (rather small) negative horizontal Reynolds stress.

A recent statistical analysis of bright coronal points by Vršnak et al. (2003) leads to clearer results. The order of the correlation is 10^7 cm^2/s^2, and indeed, it becomes very small at the equator (Fig. 3.11). The main contribution to the positive correlation comes from the youngest structures. If the whole sample of observations is averaged the total cross-correlation is very small. For comparison, Fig. 3.11 also shows the results of Ward and Gilman & Howard.

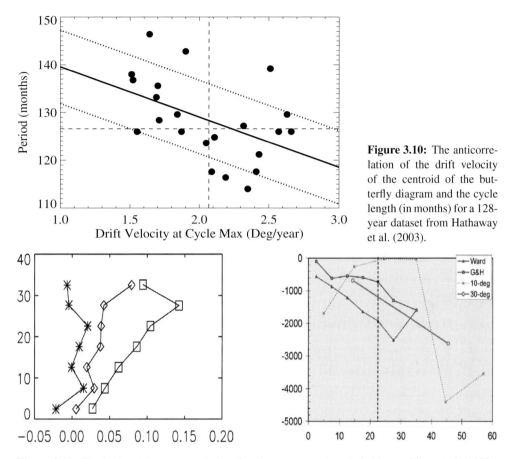

Figure 3.10: The anticorrelation of the drift velocity of the centroid of the butterfly diagram and the cycle length (in months) for a 128-year dataset from Hathaway et al. (2003).

Figure 3.11: The horizontal cross-correlation (3.1) is never negative. *Left*: Nesme-Ribes et al. (1993), note the zero-result for the Meudon data. The vertical axis is the latitude. *Right*: Results of Vršnak et al. (2003) in comparison to those of Ward (1965) and Gilman & Howard (1984). Change the sign at the horizontal axis for comparison with (3.1), note that the latitude is now the horizontal axis.

3.1.4   Stellar Observations

In the search for stellar surface differential rotation, chromospheric activity has been monitored for more than two decades. Surprisingly enough, there is not yet a very clear picture. For example, the rotation pattern of the solar-type star HD 114710 might easily be reversed compared with that of the Sun – under the assumption that the spot migration is toward the equator (Donahue & Baliunas 1992). The same is true for the single K1V star HD 10476 (Donahue 1996). Oláh, Jurcsik & Strassmeier (2003) also report an antisolar rotation law for the spectroscopic binary UZ Lib with $\delta\Omega \simeq -0.003$ day^{-1}. The clear majority of rotation laws, however, lead to the characteristic equatorial acceleration known from the Sun. In order to compare the various observational results a relation $\delta\Omega^{\mathrm{lat}}/\Omega \propto \Omega^{-n'}$ or

$$\delta\Omega^{\mathrm{lat}} \propto \Omega^{n''} \tag{3.3}$$

Table 3.1: The available exponents in the relation (3.3) from different datasets.

	Hall	Henry et al.	Donahue et al.	Reiners/Schmitt	Messina/Guinan
n'	0.85	0.76	0.30	0.34	0.42
$n'' = 1 - n'$	0.15	0.24	0.70	0.66	0.58

with $n'' = 1 - n'$ is introduced. For *positive* values n' of order unity the equator-pole difference $\delta\Omega$ basically does not depend on the global rotation rate Ω. Note that the shearing action
leading to the toroidal fields in the dynamo theory depends on $\delta\Omega$ rather than on $\delta\Omega/\Omega$.

Photometry and also CaII observations led to the first findings regarding the coefficient n'
(Table 3.1). The bulk of the data is summarized by Hall (1991). Henry et al. (1995), Donahue,
Saar & Baliunas (1996), Messina & Guinan (2003) and Reiners & Schmitt (2003a). According
to these observations the equator-pole difference of the rotation rate increases (slightly) with
the rotation rate or – which amounts to the same – decreases with the rotation period[3]. On
the other hand, the single stars with well-known rotational characteristics show more or less
the same equator-pole difference for slow *and* fast rotation (Fig. 3.12). Recent observations
of AB Dor (Donati & CollierCameron 1997) and PZ Tel (Barnes et al. 2000) seem to confirm
this surprising result, where in all cases the value 0.06 day^{-1} is approached. As described
by CollierCameron (2002) these findings resulted from the tracking of starspots over several
stellar rotations by Doppler imaging to find frequency differences for features at different
latitudes. The Doppler-imaging technique for rotating stars and all the available Doppler
images of cool stars have been reviewed by Strassmeier (2002).

The rotation laws are not permanent in time. A weak (strong) time dependence of the
equator-pole difference of the surface rotation rate of AB Dor (LQ Hya) has been reported by
Donati, CollierCameron & Petit (2003) on time scales of a few years. Such time scales are
not incompatible with the possible photometry-cycle of 7 years listed by Saar & Brandenburg
(1999) for LQ Hya[4] but Donati et al. (2003) do *not* find any clear indication for a magnetic-
cycle shorter than (say) 20 years.

Many of the stars with observed antisolar rotation laws are RS CVn binaries. There are,
however, also well-established examples of RS CVn stars with solar-type rotation laws, e.g.
RS CVn itself (Rodonò, Lanza & Catalano 1995) and λ And (Henry et al. 1995). See also
the very detailed review of all the current data by Strassmeier (2004). At first sight one would
expect that obviously the function H (Eq. (3.21) below) might also have negative values, yet
we never encounter this situation. On the other hand, if the meridional flow is very strong,
then the solution (3.11) appears with $\Omega = \mathrm{const.}$ along the streamlines of the meridional flow.
An antisolar rotation low is then the immediate consequence (as $s(\mathrm{pole}) < s(\mathrm{equator})$, see
Rüdiger 1989).

---

[3] for rapidly rotating F-stars the differential rotation appears to become less (Reiners & Schmitt 2003b)

[4] for AB Dor a cycletime of 5.3 years has been announced by Amado et al. (2001)

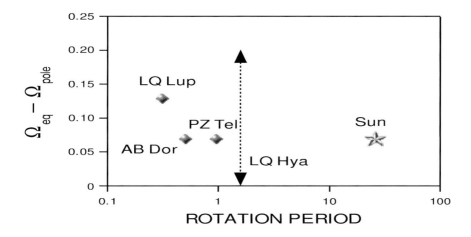

Figure 3.12: $\delta\Omega^{\mathrm{lat}}$ (in day^{-1}) vs. the rotation period (in days). Only for a few young, rapidly rotating stars do we have detailed information about the rotation law at the surface. AB Dor, PZ Tel and LQ Hya are K0 dwarfs, LQ Lup= RXJ1508.6-4423 is a post-TTS (Donati et al. 2000). For LQ Hya a very strong temporal variation of the equator-pole difference has been seen by Donati et al. (2003).

3.2 Angular Momentum Transport in Convection Zones

The theory of nonuniform rotation is mainly the theory of angular momentum conservation. Including meridional flow, Reynolds stress and Lorentz force it reads

$$\frac{\partial}{\partial t}(\rho s^2 \Omega) + \nabla \cdot \left\{ \rho s^2 \Omega \boldsymbol{u}^{\mathrm{m}} + \rho s \langle u'_\phi \boldsymbol{u}' \rangle - \frac{s}{\mu_0} \left(\bar{B}_\phi \bar{\boldsymbol{B}} + \langle B'_\phi \boldsymbol{B}' \rangle \right) \right\} = 0, \qquad (3.4)$$

where Ω is the angular velocity and $s = r \sin\theta$ the distance to the axis, $\bar{\boldsymbol{B}}$ and $\bar{\boldsymbol{u}}$ are the ensemble averages of the magnetic field and fluid velocity, and $\boldsymbol{B}'$ and $\boldsymbol{u}'$ are their fluctuating parts. This equation has often been used to analyze the effect of the meridional flow $\boldsymbol{u}^{\mathrm{m}}$. The Lorentz force is ignored and the Reynolds stress is parameterized by the Boussinesq relation

$$Q_{ij} \equiv \langle u'_i(\boldsymbol{x}, t) u'_j(\boldsymbol{x}, t) \rangle = \cdots - \nu_{\mathrm{T}} (\bar{u}_{i,j} + \bar{u}_{j,i}). \qquad (3.5)$$

The coefficient ν_{T} is called the eddy viscosity. Its determination, for conditions in which the large-scale flow and a magnetic field are influential, is one of the key problems of turbulence theory. The resulting equation describes the angular momentum transport in accordance with

$$\nabla \cdot \left(\rho s^2 \Omega \boldsymbol{u}^{\mathrm{m}} - \rho s^2 \nu_{\mathrm{T}} \nabla \Omega \right) = 0. \qquad (3.6)$$

If $\nabla \cdot (\rho \boldsymbol{u}^{\mathrm{m}}) = 0$ a stream function A with $\rho \boldsymbol{u}^{\mathrm{m}} = \nabla \times (A\hat{\boldsymbol{e}}_\phi/s)$ can be introduced ($\hat{\boldsymbol{e}}_\phi$ being the unit vector in the ϕ-direction). Since then $\boldsymbol{u}^{\mathrm{m}} \cdot \nabla A = 0$, the streamlines are $A = $ const.

In spherical coordinates the components of the meridional flow are

$$u_r^{\mathrm{m}} = \frac{1}{x^2 \sin\theta \rho} \frac{\partial A}{\partial \theta}, \qquad u_\theta^{\mathrm{m}} = -\frac{1}{x \sin\theta \rho} \frac{\partial A}{\partial x}, \qquad (3.7)$$

where $x = r/R$ is the fractional radius. Since the radial drift does not cross the boundaries of the convection zone at $x = 1$ and $x = x_{\mathrm{in}}$ (say), we have $A(x_{\mathrm{in}}) = A(1) = 0$. The stream function may be given by the Legendre function expansion $A = \sum A_n(x) P_n^1 \sin\theta$ with n even. For the angular velocity the series expansion

$$\Omega = \Omega_\odot \sum_{n=1,3,\ldots} \omega_{n-1}(x) \frac{P_n^1}{\sin\theta} \simeq \Omega_\odot \left(\omega_0(x) + \frac{3\omega_2(x)}{2} \left(5\cos^2\theta - 1\right) + \ldots \right) \quad (3.8)$$

is used, so that at leading order the surface equator-pole difference is

$$\frac{\delta\Omega^{\mathrm{lat}}}{\Omega_\odot} = \frac{\Omega_{\mathrm{eq}} - \Omega_{\mathrm{pole}}}{\Omega_\odot} \simeq -\frac{15}{2} \omega_2(1). \qquad (3.9)$$

From Eq. (3.6) one obtains a simple differential equation for ω_2 and A_2 resulting in the relation

$$\Omega_{\mathrm{eq}} - \Omega_{\mathrm{pole}} \sim \int_{x_{\mathrm{in}}}^{1} \xi^3 A_2(\xi) \mathrm{d}\xi, \qquad (3.10)$$

where $x_{\mathrm{in}} = r_{\mathrm{in}}/R$, with r_{in} the inner radius of the convection zone (see Rüdiger 1989). The equator is thus accelerated if $A_2 > 0$, and decelerated if $A_2 < 0$. Positive A_2 describes a meridional flow pattern with clockwise streaming, as seen in the first quadrant of the meridional cross section of the convection zone in Fig. 3.13. Equation (3.10) gives the most basic and most elementary discovery of the early theory (starting with Zöllner 1881). It shows how a meridional flow as shown in Fig. 3.13 can produce a rotation law with accelerated equator. Kippenhahn (1963) demonstrated how a radial rotation law with positive $\mathrm{d}\Omega/\mathrm{d}r$ leads to a clockwise meridional flow of the given geometry. Obviously the flow gains energy at AB from the centrifugal force. The net energy is positive (and the flow can be maintained) if the outer rotation rate exceeds the inner rotation rate. However, as shown by Köhler (1970), the resulting isolines of the angular velocity always end up being nearly parallel to the rotation axis (Taylor–Proudman theorem).

For very rapid flow the rotation law is more complex. Viscosity is then unable to balance the Coriolis force, and the angular velocity must adjust itself to satisfy

$$\rho \boldsymbol{u}^{\mathrm{m}} \cdot \nabla(s^2 \Omega) = 0. \qquad (3.11)$$

This relation requires the angular momentum to be constant on streamlines, i.e. $s^2 \Omega = f(A)$. Characteristic of this limit are therefore large values of Ω near the polar axis! The appearance of such a polar vortex is independent of the sense of the meridional circulation (Köhler 1969). It is, however, true that for counterclockwise flow patterns (the bottom drift is toward the equator) the polar acceleration ('vortex') requires much lower Reynolds numbers of the meridional flow rather than clockwise flows. In the case given in Sect. 5.2 in Rüdiger (1989)

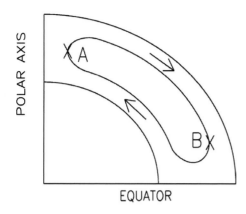

POLAR AXIS

EQUATOR

Figure 3.13: The simplest form of meridional circulation in the convection zone. The flow gains energy along A to B from the centrifugal force and loses energy from B to A. A clockwise flow can thus exist for positive $d\Omega/dr$.

only surface drifts of 2 m/s were enough for the generation of polar acceleration. The main condition for this phenomenon (see Sect. 3.1.4) is that the meridional flow is prescribed from some reasons and is not the result of the internal mean-field hydrodynamics.

There is another important property of the Reynolds equation,

$$\rho\frac{D\boldsymbol{u}}{Dt} = -\nabla\bar{P} - \nabla\cdot\rho Q - \rho\nabla\psi \tag{3.12}$$

with ψ as the gravitational potential. This gives for the stationary solution the relation

$$\mathcal{D}(\boldsymbol{u}^{\mathrm{m}}) = s\frac{d\Omega^2}{dz} + \frac{1}{\bar{\rho}^2}(\nabla\bar{P}\times\nabla\bar{\rho})_\phi, \tag{3.13}$$

in which the left-hand side represents the damping of meridional flow by eddy viscosity[5]. Two sources of flow obviously exist: differential rotation and deformation. The flow resulting from nonparallel ∇P and $\nabla\rho$ is called the 'barocline' or the thermal wind.

When Eq. (3.13) is written in dimensionless units, the first term on the RHS acquires the Taylor number

$$\mathrm{Ta} = \left(\frac{2\Omega R^2}{\nu_{\mathrm{T}}}\right)^2. \tag{3.14}$$

If this number is very large Eqs. (3.11) and (3.13) are simultaneously satisfied by

$$\boldsymbol{u}^{\mathrm{m}} = 0\,, \qquad\qquad \Omega = \Omega(s). \tag{3.15}$$

Cylindrical isorotation contours are thus the immediate consequence of large Coriolis force and centrifugal forces[6]. The Taylor number for the Sun, Ta $\sim 10^7$, is large enough to predict the Ω-contours as cylindrical. Several simulations also led to this result (Kippenhahn 1963, Köhler 1970, Gilman 1977, Brandenburg et al. 1991, Miesch et al. 2000, Brun & Toomre 2002). The helioseismic observations, however, revealed that such isorotation contours only

---

[5] $\mathcal{D}$ is a simple but longwinded operator of 3rd order

[6] the relation leading to the Taylor–Proudman theorem for $\nabla\cdot\boldsymbol{u} = 0$ is $\nabla\times(\boldsymbol{\Omega}\times\boldsymbol{u}) \equiv -\Omega d\boldsymbol{u}/dz = 0$

exist in the equatorial region (Fig. 3.4). We call the problem of resolving this striking contradiction the 'Taylor number puzzle'.

If only the contribution to the angular momentum balance due to turbulence is considered, Eq. (3.6) changes to

$$\frac{\partial}{\partial x_i}(\bar{\rho}sQ_{i\phi}) = 0. \tag{3.16}$$

The one-point correlation tensor Q_{ij} may be divided into several parts, i.e.

$$Q_{ij} = \cdots + Q_{ij}^\nu + Q_{ij}^\Lambda \tag{3.17}$$

with Q^ν as the diffusive part $Q_{ij}^\nu = -\mathcal{N}_{ijkl}\bar{u}_{k,l}$, where the viscosity tensor $\mathcal{N}$ provides the proportionality factors in the stress-strain relation. The remaining part, Q^Λ, does not include any spatial derivatives. It represents any nondiffusive transport of angular momentum by the turbulence (the 'Λ-effect'). Equations (3.16) and (3.17) imply that the differential rotation is maintained by a balance between the damping and driving by the turbulent flow.

According to Eq. (3.16) the turbulence-generated contribution to the angular momentum transport is described by the components $Q_{r\phi}$ and $Q_{\theta\phi}$ of the one-point correlation tensor. This tensor is a symmetric one by definition. Its cross-correlations $Q_{r\phi}$ and $Q_{\theta\phi}$ can only exist if the turbulence rotates. In this case one can formulate

$$Q_{ij}^\Lambda = \Lambda_{ijk}\Omega_k \tag{3.18}$$

with the Λ-tensor symmetric in i and j. The latter property excludes relations such as $\Lambda_{ijk} \propto \epsilon_{ijk}$ though the Λ-tensor must contain the ϵ-tensor in order to be a pseudotensor. However, if an additional preferred direction g exists it is indeed possible to construct a tensor, i.e.

$$\Lambda_{ijk} = A(\epsilon_{ikp}g_j + \epsilon_{jkp}g_i)g_p \tag{3.19}$$

that fulfills all the required conditions. The tensor is symmetric in i and j, it is a pseudotensor and it is even in the vector g, i.e. it is invariant against the transformation $g \to -g$. Another possibility is $\Lambda_{ijk} = B(g \cdot \Omega)(\epsilon_{ikp}\Omega_j + \epsilon_{jkp}\Omega_i)g_p$ that again is invariant for $g \to -g$ but is of second order in Ω so that it does not exist for slow rotation. The tensors describe the interaction of a global rotation and anisotropic turbulence. One finds for the off-diagonal components of Q_{ij} the expressions

$$Q_{r\phi}^\Lambda \equiv \Lambda_V \Omega \sin\theta = (A + B\Omega^2 \cos^2\theta)\sin\theta,$$
$$Q_{\theta\phi}^\Lambda \equiv \Lambda_H \Omega \cos\theta = -B\Omega^2 \sin^2\theta\cos\theta. \tag{3.20}$$

If they are normalized with the eddy viscosity then

$$Q_{r\phi}^\Lambda = \nu_T \Omega V \sin\theta = \nu_T\Omega(V^{(0)} + V^{(1)}\sin^2\theta + \dots)\sin\theta,$$
$$Q_{\theta\phi}^\Lambda = \nu_T \Omega H \cos\theta = \nu_T\Omega(H^{(1)} + H^{(2)}\sin^2\theta + \dots)\sin^2\theta\cos\theta. \tag{3.21}$$

The first treatments (Wasiutynski 1946, Biermann 1951, Kippenhahn 1963, Busse 1970) were restricted to the case of slow rotation. Only the radial Λ-effect survives in this limit while $Q_{\theta\phi}^\Lambda = 0$. The Sun is not, however, a slow rotator. The Coriolis number, $\Omega^* = 2\tau_{\text{corr}}\Omega$, is larger than unity for the solar convection (Durney & Latour 1978). The same is true for all single, cool MS stars.

When the rotation is not slow, a horizontal flux $Q_{\theta\phi}^\Lambda$ also exists. Pidatella et al. (1986) developed models to compare the efficiency of meridional flow and Reynolds stresses in generating the differential rotation. They found the Λ-effect is a more powerful generator. Many other indications confirm this finding (Durney & Spruit 1979, Kitchatinov 1986, Rüdiger & Tuominen 1990, Canuto, Minotti & Schilling 1994).

For slow rotation the zonal components of the one-point correlation tensor are simply

$$Q_{r\phi} = -\nu_\mathrm{T} r \sin\theta \frac{\partial\Omega}{\partial r} + \nu_\mathrm{T} V^{(0)} \sin\theta\,\Omega, \qquad Q_{\theta\phi} = -\nu_\mathrm{T} \sin\theta \frac{\partial\Omega}{\partial\theta}. \qquad (3.22)$$

Insertion of (3.22) into Eq. (3.16) leads to the differential equation

$$\frac{1}{\nu_\mathrm{T}} \frac{\partial}{\partial x} \left\{ \rho x^3 \left(\nu_\mathrm{T} x \frac{\partial\Omega}{\partial x} - \nu_\mathrm{T} V^{(0)}\Omega \right) \right\} = -\frac{1}{\sin^3\theta} \frac{\partial}{\partial\theta} \left(\sin^3\theta \frac{\partial\Omega}{\partial\theta} \right) \qquad (3.23)$$

which must be solved with the stress-free boundary conditions $Q_{r\phi} = 0$ at $x = x_\mathrm{in}, 1$. Under the conditions that ν_T, ρ and $V^{(0)}$ do not depend on latitude one can show that the resulting rotation law also does *not* depend on latitude. The general solution of Eq. (3.23) reads

$$\Omega = \Omega_\odot \exp \int\limits_1^x V^{(0)}(x') \frac{\mathrm{d}x'}{x'}. \qquad (3.24)$$

This solution satisfies the stress-free condition not only at the outer boundary but everywhere in the solar convection zone. The angular momentum transport vanishes everywhere, but the sign of its nondiffusive part determines the sign of the radial gradient of Ω. Positive $V^{(0)}$ generate an angular velocity that increases outward ('superrotation') and negative $V^{(0)}$ generate one that increases inward ('subrotation'). Indeed, it seems that the latter situation is realized in the upper layers of the solar convection zone, since observational evidence has indicated that just beneath the surface the radial gradient is

$$\frac{\mathrm{d}\log\Omega}{\mathrm{d}\log x} \simeq -1.4. \qquad (3.25)$$

The observations can be explained by the existence of a *negative* $V^{(0)}$ of order unity.

Provided the functions V and H are given, one can use them to compute the rotation law with the eddy viscosity as a free parameter that can be fixed by comparison with the observations. The observed slope (3.25) only varies slightly with latitude (Fig. 3.2, left). In terms of the series development (3.8) we have $\omega_0' \simeq -1.4$. In fact, the boundary condition $Q_{r\phi} = 0$ for $x = 1$ is already enough to find the eddy viscosity, i.e.

$$\tilde{\nu} \frac{\mathrm{d}\omega_0}{\mathrm{d}x} \bigg|_1 = V_0, \qquad \tilde{\nu} \frac{\mathrm{d}\omega_2}{\mathrm{d}x} \bigg|_1 = V_2, \qquad (3.26)$$

where V_n are now the components of the function V expanded in terms of orthogonal polynomials, i.e. $V_0 \simeq V^{(0)} + 0.8V^{(1)} + 0.069V^{(2)}$ and $V_2 \simeq V^{(1)}$ (see Rüdiger 1989).

If the coefficients $V^{(l)}$ are known (from numerical simulations) we obtain $\tilde{\nu} = V_0/\omega_0'$ if the eddy viscosity is written as $\nu_{\mathrm{T}} = \tilde{\nu} u_{\mathrm{T}}^2/\Omega_\odot$, with the rms turbulence intensity

$$u_{\mathrm{T}} = \sqrt{\langle \boldsymbol{u'}^2 \rangle}. \tag{3.27}$$

In Rüdiger, Küker & Chan (2003) the simulations of Chan (2001) are used as the data source for the angular momentum transport. The expressions are normalized in the sense $Q_{r\phi} \propto V u_{\mathrm{T}}^2$ and $Q_{\theta\phi} \propto H u_{\mathrm{T}}^2$, so that V and H also denote the correlation coefficients. From the simulations the coefficients $(3.21)_1$ for V become $V^{(0)} = -0.3$, $V^{(1)} = 0.19$, $V^{(2)} = 0.02$, so that $V_0 = -0.15$, and the observed negative gradient of the angular velocity in the supergranulation layer can be explained with $\tilde{\nu} = 0.11$. With $u_{\mathrm{T}} \simeq 200$ m/s a relatively large eddy viscosity of about $\nu_{\mathrm{T}} \simeq 10^{13}$ cm^2/s is found.

Regarding the series expansion (3.8) we find a striking peculiarity of the function ω_0. The solar rotation law can be well described by the simple properties

$$\omega_0(x_{\mathrm{in}}) \simeq \omega_0(1), \qquad \omega_2(x_{\mathrm{in}}) = \omega_4(x_{\mathrm{in}}) = \cdots \simeq 0. \tag{3.28}$$

For the angular momentum expression we find

$$J = R^2 \int \rho x^2 \sin^2 \theta \Omega(x,\theta)\mathrm{d}\boldsymbol{x} = 4\pi R^2 \Omega_\odot \int_{x_{\mathrm{in}}}^1 \rho x^4 \omega_0(x)\mathrm{d}x \simeq 4\pi R^2 \Omega_\odot \int_{x_{\mathrm{in}}}^1 \rho x^4 \mathrm{d}x,$$

while the higher-order terms in Eq. (3.8) do not contribute. The property (3.28) can thus be reformulated so that the angular momentum of the convection zone should be the same as if it rotated rigidly with the interior.

The radial rotation law ω_0 is also exceptional insofar as conservation of angular momentum allows a radial integration, with the result

$$\frac{\mathrm{d}\omega_0}{\mathrm{d}x} = \left(V^{(0)} + \frac{4}{5}V^{(1)}\right)\frac{\omega_0}{x} - \frac{6A_2}{5\rho x^2 \nu_{\mathrm{T}} R}. \tag{3.29}$$

Obviously, the Λ-effect does not produce radial differential rotation if $V^{(0)} + 0.8V^{(1)} \simeq 0$, i.e. if $V^{(0)}$ and $V^{(1)}$ have opposite signs. Meridional flows toward the equator at the bottom of the convection zone and toward the pole at the surface have negative A_2. In this case the meridional circulation produces a positive $\mathrm{d}\omega_0/\mathrm{d}x$ at midlatitudes, which seems to be visible in Fig. 3.2. Here the meridional flow fulfills the restrictions of the Taylor–Proudman theorem, which for small viscosity requires the Ω-isolines to be parallel to the rotation axis.

Equation (3.16) without meridional flow and magnetic field can be solved analytically. It takes the form

$$\frac{\partial}{\partial x}\left(x^4 \frac{\partial \Omega}{\partial x}\right) + \frac{x^2}{\sin^3 \theta}\frac{\partial}{\partial \theta}\left(\sin^3 \theta \frac{\partial \Omega}{\partial \theta}\right) =$$
$$\frac{\partial}{\partial x}(x^3 V\Omega) + \frac{x^2}{\sin^3 \theta}\frac{\partial}{\partial \theta}(H \cos \theta \sin^2 \theta \Omega). \tag{3.30}$$

This equation is highly nonlinear in Ω because V and H depend on Ω. For slow rotation, however, the functions V and H lose their Ω-dependence. In this case the leading term on the

RHS of Eq. (3.30) contains combinations such as $V(\Omega_\odot)\Omega_\odot$, which provide the inhomogeneous part of a differential equation for Ω. Its solution yields the expression

$$\frac{\Omega(1,\theta)}{\Omega_\odot} \simeq C + \frac{1}{2}\sum_{l=1}\left(dV^{(l)} + \frac{H^{(l)}}{l}\right)\sin^{2l}\theta \tag{3.31}$$

for the surface rotation law, if the convection zone is sufficiently shallow ($d = 1 - x_{\mathrm{in}} \ll 1$). If $\Omega_\odot$ denotes the true value of the angular velocity on the polar axis, C is unity. The overall differential rotation is then given by

$$\delta\Omega^{\mathrm{lat}} = \Omega_{\mathrm{eq}}(1) - \Omega_{\mathrm{pole}}(1) \simeq \frac{1}{2}\sum_{l=1}\left(dV^{(l)} + \frac{H^{(l)}}{l}\right)\Omega_\odot \simeq \frac{1}{2}\left(dV^{(1)} + H^{(1)}\right)\Omega_\odot. \tag{3.32}$$

The absence of the density gradient here suggests that it does not have a strong influence on the rotation profile. One can formally produce the observed equatorial acceleration from just a single positive mode provided only that $l \geq 1$.

A similar procedure applied to the *radial* differences of the rotation coefficients leads to

$$\Omega(1) - \Omega(x_{\mathrm{in}}) \simeq d\,\Omega_\odot \sum_{l=0} V^{(l)}\sin^{2l}\theta \simeq d\,\Omega_\odot V(\theta). \tag{3.33}$$

This difference vanishes for $d \to 0$, which, on the other hand, according to (3.32) is not true for the difference in the latitudinal rotation rate. The main surprise of the helioseismological results, i.e. that the radial gradients are much smaller than the latitudinal gradients, can now be translated by these equations. Let us write out Eq. (3.33) for both the pole and the equator, i.e.

$$\omega_{\mathrm{pole}}(1) - \omega_{\mathrm{pole}}(x_{\mathrm{in}}) \simeq d \cdot V^{(0)} \simeq d \cdot V_{\mathrm{pole}},$$
$$\omega_{\mathrm{eq}}(1) - \omega_{\mathrm{eq}}(x_{\mathrm{in}}) \simeq d \cdot V_{\mathrm{eq}}. \tag{3.34}$$

Hence, the rather flat rotation profile beneath the solar equator can be explained with small V at the equator, while the strong subrotation along the rotation axis indicates large negative values of V for $\theta = 0°$ and $180°$. The following conclusions are therefore suggested

(i) equatorial acceleration requires the sum (3.32) to be positive. Positive $V^{(l)}$ and/or positive ('equatorward') $H^{(l)}$ are thus necessary for equatorial acceleration,

(ii) the radial differences of Ω at a given latitude θ reflect the value of $V(\theta)$. From the observations, V_{pole} must be negative and V_{eq} must be zero or slightly positive.

3.2.1   The Taylor Number Puzzle

Consider a rotating turbulent shell with stress-free boundaries. The correlation time of the turbulence may be so short that only the term $V^{(0)}$ exists of the Λ-effect so that Eq. (3.22) holds. Then in the hydrodynamic equation (3.12) the velocity vector is normalized with ν_{T}/R, so that apart from the normalized $V^{(0)}$ only the Taylor number Ta occurs. The solution of this problem is given by Brandenburg et al. (1991) for $V^{(0)} = 1$ and various Ta. As shown in Fig. 1

Ta	$\Omega_{\mathrm{eq}}/\Omega_0$	$\delta\Omega_{\mathrm{pole}}/\Omega_{\mathrm{eq}}$	Re
10^2	1.1	0.3	~ 0
10^4	1.2	0.2	1.3
10^6	1.3	0.03	5.2
10^8	1.3	0.01	16

Table 3.2: Internal rotation law characteristics for a model with fixed Λ-effect but increasing Taylor number. The isolines of angular velocity become more and more cylindrical. $\mathrm{Re} = u^{\mathrm{m}} R/\nu$ denotes the Reynolds number of the meridional flow. From Brandenburg et al. (1991).

of Brandenburg et al., for increasing Ta the Ω-isolines become more and more cylindrical, with $\partial\Omega/\partial r > 0$ at the equator and $\partial\Omega/\partial r \simeq 0$ at the pole. The numbers are given in Table 3.2. Independent of the Taylor number we always have the same superrotation of the equator. However, for increasing Ta the Ω-difference at the pole strongly decreases. One finds equatorial acceleration and polar deceleration. The Reynolds number of the (equatorward) meridional flow increases, but much slower than the Taylor number does. This means that for smaller and smaller eddy viscosity the meridional flow vanishes. When the isolines of Ω lie on cylinders the centrifugal force can always be expressed as the gradient of a potential, so that it can be balanced by the pressure gradient alone, without any circulation. As shown in Table 3.2 the Reynolds number of the flow apparently scales as $\mathrm{Ta}^{0.25}$, hence $u^{\mathrm{m}} \propto \sqrt{\nu}$. For small viscosity the meridional drift vanishes just as it does for very large viscosities[7].

Spherical 3D models by Gilman (1977), in which differential rotation is automatically generated by Reynolds stresses from the large-scale thermal convection, show cylindrical Ω contours in lower latitudes. Equatorial acceleration occurs only if the Rayleigh number is not too large, otherwise the profile at lower latitudes is reversed. The Taylor number in Gilman's models corresponds to our definition to $6 \cdot 10^7$. The meridional circulation in this model is always poleward; in lower latitudes are the results for the angular velocity in approximate agreement with ours.

The Taylor–Proudman structure of the flow pattern for high Taylor number is unavoidable for rapidly rotating, incompressible convection. This finding also holds for the more complicated structure of the Λ-effect ('Taylor number puzzle'). It is only the anisotropy of the convective heat transport that prevents the realization of the Taylor–Proudman state. The anisotropy produces a latitudinal temperature variation, with warm poles and cool equator. The resulting circulation opposes the meridional flow driven by the centrifugal force (Kitchatinov & Rüdiger 1995).

3.2.2   The Λ-Effect

Correctly defining the correlation tensor Q_{ij} through which the turbulent motions influence the mean flow is of primary importance in mean-field hydrodynamics. Only anisotropic and/or inhomogeneous turbulence under the influence of rotation can produce a Λ-effect. Both effects are included in the linearized equation of motion,

$$\frac{\partial \boldsymbol{m}}{\partial t} + \nabla P' - \nabla \pi' + 2\,\boldsymbol{\Omega} \times \boldsymbol{m} = \boldsymbol{f}'. \tag{3.35}$$

---

[7] Köhler (1970) finds a maximum of the meridional flow velocity for $\mathrm{Ta} \simeq 3 \cdot 10^7$

Here $\boldsymbol{m} = \rho \boldsymbol{u}'$ is the fluctuating momentum density, $\boldsymbol{f}'$ is the random body force driving the turbulence, P' is the fluctuating pressure and π' the viscous stress tensor

$$\pi'_{ij} = \rho \nu_t (u'_{i,j} + u'_{j,i}) + \rho \mu_t \nabla \cdot \boldsymbol{u}' \cdot \delta_{ij} . \tag{3.36}$$

The scalar coefficients ν_t and μ_t are due to the action of the small-scale background turbulence due to the action of smaller-scaled instabilities. They are assumed to be independent of the angular velocity. This implies that the background turbulence is short-lived enough to be insensitive to the Coriolis forces.

We assume the fluid to be anelastic, $\nabla \cdot \boldsymbol{m} = 0$. Rewriting π' in terms of $\boldsymbol{m}$ one obtains $\pi'_{ij} = \nu_t (m_{j,i} + m_{i,j} - G_i m_j - G_j m_i) - \mu_t (\boldsymbol{G} \cdot \boldsymbol{m}) \delta_{ij}$ with

$$\boldsymbol{G} = \nabla \log \rho . \tag{3.37}$$

All our derivations belong to the quasilinear approximation in which the linearized equations for fluctuating fields are applied to derive the second-order correlations of these fields. We assume the spatial scales of the fluctuating fields to be small compared to that of the mean fields.

It is convenient to use Fourier transforms, which are introduced according to

$$\boldsymbol{m}(\boldsymbol{x}, t) = \int \hat{\boldsymbol{m}}(\boldsymbol{k}, \omega) e^{i(\boldsymbol{k} \cdot \boldsymbol{x} - \omega t)} \, d\boldsymbol{k} \, d\omega . \tag{3.38}$$

The pressure will be eliminated with the condition $\nabla \cdot \boldsymbol{m} = 0$. The quantities $\boldsymbol{G}$ and $G_{ij} \equiv \partial^2 \log \rho / \partial x_i \partial x_j$ must be considered as spatially uniform.

As usual now the 'original turbulence' $\hat{\boldsymbol{m}}^{(0)}$ is introduced, which is defined as which that the force $\boldsymbol{f}'$ would produce in a nonrotating fluid. Then

$$\hat{m}_i(\boldsymbol{k}, \omega) = \left[D_{ij} - \frac{i \nu (\boldsymbol{G} \cdot \boldsymbol{k})}{-i\omega + \nu_t k^2} D_{ip} (D_{pj} - \delta_{pj}) \right] \hat{m}_j^{(0)}(\boldsymbol{k}, \omega), \tag{3.39}$$

where

$$D_{ij} = \left(\delta_{ij} + \frac{(2 \boldsymbol{k}^\circ \cdot \boldsymbol{\Omega})}{-i\omega + \nu_t k^2} \epsilon_{ijp} k_p^\circ \right) \bigg/ \left(1 + \frac{(2 \boldsymbol{k}^\circ \cdot \boldsymbol{\Omega})^2}{(-i\omega + \nu_t k^2)^2} \right) \tag{3.40}$$

and $\boldsymbol{k}^\circ = \boldsymbol{k}/k$ is a unit vector. The next step in the calculation produces the expression for $\hat{\boldsymbol{m}}$ valid to the second order in the scale-ratio. However, we shall not write out the rather complicated results (see Kitchatinov & Rüdiger 1993). In Sect. 4.2.2 the complete tensor is given up to the first order in the scale-ratio.

Equation (3.39) expresses the momentum density in terms of the nonrotating turbulence. It remains to define $\hat{\boldsymbol{m}}^{(0)}$. This is based on the application of the double-Fourier method of Roberts & Soward (1975) to handle the large-scale inhomogeneity in space, i.e.

$$\langle m_i^{(0)}(\boldsymbol{x}, t) m_j^{(0)}(\boldsymbol{x} + \boldsymbol{\xi}, t + \tau) \rangle = \int \hat{M}_{ij}^{(0)}(\boldsymbol{k}, \boldsymbol{\kappa}, \omega) e^{i\boldsymbol{\kappa} \cdot \boldsymbol{x}} e^{i((\boldsymbol{\kappa}/2 + \boldsymbol{k}) \boldsymbol{\xi} - \omega \tau)} \, d\boldsymbol{k} \, d\boldsymbol{\kappa} \, d\omega$$

with

$$\hat{M}_{ij}^{(0)} = \frac{\hat{E}(k, \omega, \boldsymbol{\kappa})}{16 \pi k^2} \left[\delta_{ij} - \left(1 + \frac{\kappa^2}{4k^2} \right) k_i^\circ k_j^\circ + \frac{1}{2k^2} (\kappa_i k_j - \kappa_j k_i) + \frac{\kappa_i \kappa_j}{4k^2} \right], \tag{3.41}$$

where $\boldsymbol{k}$ and $\boldsymbol{\kappa}$ are the wave vectors for the small and large scales (Kitchatinov 1987). The original turbulence will be assumed to be statistically steady but *not homogeneous*. If only the linear terms in $\boldsymbol{\kappa}$ are considered in Eq. (3.41), then the random field is locally isotropic, i.e. $\langle m_i^{(0)}(\boldsymbol{x},t)m_j^{(0)}(\boldsymbol{x},t)\rangle = \langle \boldsymbol{m}^{(0)2}\rangle\delta_{ij}/3$, but the gradient of $\langle \boldsymbol{m}^{(0)2}\rangle$ does not vanish.

Again the quantity $\hat{E}(k,\omega,\boldsymbol{\kappa})$ is the Fourier transform of the local spectrum $E(k,\omega,\boldsymbol{x})$, i.e.

$$E(k,\omega,\boldsymbol{x}) = \int \mathrm{e}^{i\boldsymbol{\kappa}\cdot\boldsymbol{x}}\hat{E}(k,\omega,\boldsymbol{\kappa})\,\mathrm{d}\boldsymbol{\kappa}\,, \qquad \langle \boldsymbol{m}^{(0)2}\rangle = \int\limits_0^\infty\int\limits_{-\infty}^\infty E(k,\omega,\boldsymbol{x})\mathrm{d}k\,\mathrm{d}\omega. \quad (3.42)$$

For homogeneous turbulence it is simply $\hat{E} = m^2\delta(\boldsymbol{\kappa})$. Due to the inhomogeneities of both the turbulence intensity and density the turbulence field

$$\langle u_\perp'^2\rangle - \langle u_\parallel'^2\rangle \simeq \frac{1}{8\rho^2}\frac{\mathrm{d}^2}{\mathrm{d}z^2}\left(\ell_{\mathrm{corr}}^2 u_{\mathrm{T}}^2\rho^2\right) \qquad\qquad (3.43)$$

is anisotropic on large scales, where $\perp$ and $\parallel$ are the horizontal and the vertical direction, and ℓ_{corr} is the correlation length of the turbulent motions. As long as $u_{\mathrm{T}}\ell_{\mathrm{corr}} \approx$ const., the density profile $\rho(z)$ determines the sign of (3.43). One can compute it with a standard convection zone model. It is positive in the upper part of the convection zone, hence the turbulence proves to be horizontal there but this effect is small. In the bottom layers the behavior of the turbulence is strictly of the vertical type (Fig. 3.14).

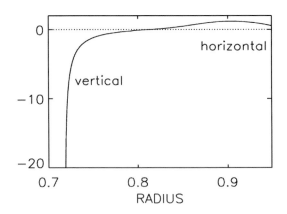

Figure 3.14: The normalized anisotropy excess (3.43) of the turbulence model (3.41) taken in a solar convection zone model by Stix (1989). Note that the horizontal velocity component dominates in the upper part while the vertical velocity component dominates in the lower part.

The influence of rotation is known to enable inhomogeneous and/or anisotropic turbulence to transport angular momentum even for rigid rotation. The angular momentum fluxes are proportional to that part of the velocity correlation tensor that is an odd function of the angular velocity but even in $\boldsymbol{\kappa}$. Application of the turbulence model (3.41) provides expressions of the form

$$
V^{(0)} = \int\limits_{0}^{\infty}\int\limits_{-\infty}^{\infty} \left(S(I_1 - I_2) + S_1 I_3\right)\frac{dk\,d\omega}{\omega^2 + \nu^2 k^4},
$$

$$
V^{(1)} = H^{(1)} = \int\limits_{0}^{\infty}\int\limits_{-\infty}^{\infty} \left(S\,I_2 + S_1 I_4\right)\frac{dk\,d\omega}{\omega^2 + \nu^2 k^4}, \tag{3.44}
$$

depending on stratification characteristics such as $S = r\rho^{-2}\,\partial/\partial r(r^{-1}\,\partial E/\partial r)$. The kernels I_n are rather complicated nonlinear functions of the angular velocity. In the slow-rotation limit they simplify to expressions like $I_1 = (1/30)(\nu^2 k^4 + 5\omega^2)/(\nu^2 k^4 + \omega^2)$ and $I_2 = I_3 = I_n = 0$.

In the slow-rotation limit only $V^{(0)}$ exists, so that the Λ-effect transports the angular momentum only along the radius. In slowly rotating stars one therefore expects the radial profile of the angular velocity to be much steeper than the latitudinal profile.

For rapid rotation only

$$
I_2 = \frac{\pi}{32\Omega}\,\frac{\left(\omega^2 + \nu^2 k^4\right)^2}{\nu^3 k^6} \tag{3.45}
$$

survives, while the other kernels are small. For these rapid rotators, if only I_2 is finite the expressions (3.44) lead to $V^{(1)} = H^{(1)} = -V^{(0)}$, so that

$$
Q^{\Lambda}_{r\phi} = -\nu_{\mathrm{t}}\,\Omega H^{(1)}\cos^2\theta\,\sin\theta, \qquad\qquad Q^{\Lambda}_{\theta\phi} = \nu_{\mathrm{t}}\,\Omega H^{(1)}\sin^2\theta\,\cos\theta \tag{3.46}
$$

results. The rotation laws given in Figs. 3.15 and 3.16 are obtained with Eq. (3.46) for $H^{(1)} = 1$. There is generally a flat rotation law beneath the equator and a clear subrotation beneath the poles[8]. Figures 3.15 and 3.16 are for stress-free upper boundary conditions, but they differ in the lower boundary condition. For the stress-free models the difference between thick and thin convection zones is very small, but it is not small if at the lower boundary the angular velocity is fixed. For the rigid-free models one has $\delta\Omega^{\mathrm{lat}}/\Omega \propto d\cdot V^{(1)}$, while for the free-free models $\delta\Omega^{\mathrm{lat}}/\Omega \propto H^{(1)}$, the latter being independent of the shell thickness d.

The expressions are still difficult to handle in models because they include a spectral function that is not yet known for stellar objects. We adopt the simplest representation for this function, which can be understood as a transition to the well-known 'τ-approximation', i.e.

$$
E = 2\rho^2 u_{\mathrm{T}}^2\,\delta\left(k - 1/\ell_{\mathrm{corr}}\right)\delta(\omega), \qquad\qquad \nu = \frac{\ell_{\mathrm{corr}}^2}{\tau_{\mathrm{corr}}}, \tag{3.47}
$$

where τ_{corr} is the convective turnover time $\tau_{\mathrm{corr}} \simeq \ell_{\mathrm{corr}}/u'$. By this procedure, the part $\partial u/\partial t - \nu\Delta u$ in the Navier-Stokes equation is approximately replaced by u/τ_{corr} (see Orszag 1970, Durney & Spruit 1979, Vainshtein & Kitchatinov 1983). This yields

$$
V^{(0)} = -\frac{r}{\rho^2}\frac{\partial}{\partial r}\left(\frac{1}{r}\frac{\partial}{\partial r}\tau_{\mathrm{corr}}^2 u_{\mathrm{T}}^2 \rho^2 I_2\right), \tag{3.48}
$$

---

[8] the results differ slightly from our above estimates in Eqs. (3.32) and (3.34) as the latter only hold for very weak differential rotation (first such solutions by Küker et al. (1993))

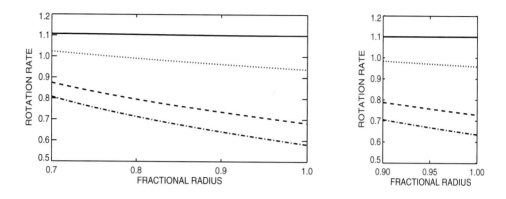

Figure 3.15: $\Omega(x)$ as solution of Eq. (3.30) for the Λ-effect Eq. (3.46) valid for rapid rotation. Both boundaries are stress-free. *Left*: Thick convection zone ($d = 0.3$). *Right*: Thin convection zone ($d = 0.1$, see also DeRosa & Toomre 2001). The equator is given by the solid line, the poles by the dot-dashed line. *Dotted line* 30°, dashed 60° solar latitude.

and

$$V^{(1)} = H^{(1)} = \frac{r}{\rho^2} \frac{\partial}{\partial r} \left(\frac{1}{r} \frac{\partial}{\partial r} \left(\tau_{\mathrm{corr}}^2 u_{\mathrm{T}}^2 \rho^2 I_2 \right) \right) , \qquad (3.49)$$

with $I_2 = \pi/16\Omega^*$ in the rapid-rotation case, i.e. for

$$\Omega^* = 2\tau_{\mathrm{corr}}\Omega > 1. \qquad (3.50)$$

The Coriolis number depends on the depth much less than the density does.

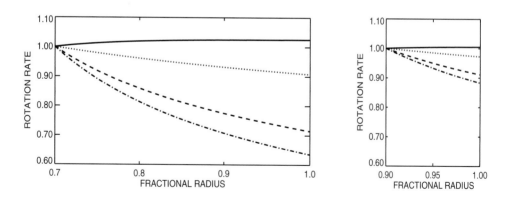

Figure 3.16: The same as Fig. 3.15 but with rigid lower boundary condition, and only the surface stress-free.

The general expressions for V and H in the τ-approximation can be simplified further by neglecting the contributions of all inhomogeneities except the density stratification. This yields for all Ω^*

$$V^{(0)} = \frac{\tau_{\mathrm{corr}}^2 \langle u'^2 \rangle}{H_\rho^2} \left[\mathcal{I}_0(\Omega^*) + \mathcal{I}_1(\Omega^*) \right], \quad V^{(1)} = H^{(1)} = -\frac{\tau_{\mathrm{corr}}^2 \langle u'^2 \rangle}{H_\rho^2} \mathcal{I}_1(\Omega^*), (3.51)$$

where $H_\rho = -\mathrm{d}r/\mathrm{d}\log\rho$ is the density scale height. Only two functions of the Coriolis number are still involved, i.e.

$$\mathcal{I}_0 = \frac{1}{2\Omega^{*4}} \left(9 - \frac{2\Omega^{*2}}{1 + \Omega^{*2}} - \frac{\Omega^{*2} + 9}{\Omega^*} \tan^{-1}\Omega^* \right),$$

$$\mathcal{I}_1 = -\frac{1}{2\Omega^{*4}} \left(45 + \Omega^{*2} - \frac{4\Omega^{*2}}{1 + \Omega^{*2}} + \frac{\Omega^{*4} - 12\Omega^{*2} - 45}{\Omega^*} \tan^{-1}\Omega^* \right). \quad (3.52)$$

The limits are $\mathcal{I}_0 = 4/15 - 16\Omega^{*2}/35$, $\mathcal{I}_1 = 16\Omega^{*2}/105$ for slow rotation, and $\mathcal{I}_0 = O(\Omega^{*-3})$, $\mathcal{I}_1 = -\pi/4\Omega^*$ for rapid rotation. The latter relation again leads to the result $V^{(1)} = H^{(1)} = -V^{(0)}$ with negative $V^{(0)}$. The function $H^{(1)}$ is thus positive and $V = V^{(0)} \cos^2\theta$. For fast rotation V is thus negative and *vanishes at the equator*. From Eq. (3.34) the radial rotation law at the equator must therefore be very flat compared with that along the polar axis.

In Fig. 3.17 the main results for the Λ-effect are shown. There are drastic differences for slow and fast rotation. For slow rotation we find *positive* $V^{(0)}$ and very small $V^{(1)}$ and $H^{(1)}$. According to Eqs. (3.51) and (3.52)

$$V^{(0)} \simeq \frac{4}{15} \frac{\tau_{\mathrm{corr}}^2 \langle u'^2 \rangle}{H_{\mathrm{p}}^2} \simeq \frac{4}{15} \frac{\ell_{\mathrm{corr}}^2}{H_{\mathrm{p}}^2} \simeq \frac{4}{15} \left(\frac{\alpha_{\mathrm{MLT}}}{\Gamma} \right)^2, \quad (3.53)$$

where the (mixing length) α_{MLT} comes from $\ell_{\mathrm{corr}} \simeq \alpha_{\mathrm{MLT}} H_{\mathrm{p}}$. Here ℓ_{corr} is the mixing length, H_{p} the pressure scale height and Γ the ratio of specific heats. We assume that $\alpha_{\mathrm{MLT}} = \Gamma = 5/3$.

The positivity of $V^{(0)}$ can be understood in the following way. One finds for slow rotation and for anisotropic one-mode turbulence models with the characteristic wave number K that

$$\Lambda_{\mathrm{V}} = \frac{2\tau_{\mathrm{corr}}}{K^2} \left((K_\theta^2 + K_\phi^2)\langle u_\phi'^2 \rangle - (K_r^2 + K_\theta^2)\langle u_r'^2 \rangle \right), \quad (3.54)$$

$$\Lambda_{\mathrm{H}} = \frac{2\tau_{\mathrm{corr}}}{K^2} \left((K_r^2 + K_\phi^2)(\langle u_\phi'^2 \rangle - \langle u_r'^2 \rangle) + (K_\phi^2 - K_r^2)\langle u_r'^2 \rangle \right), \quad (3.55)$$

which gives the rotationally generated Λ-effect for any mode of the original stochastic field. We must, in consequence, distinguish between two contributions, i.e. anisotropy in the intensities and/or anisotropy in the wave number vector components. If the latter are nearly equal the horizontal motions provide outward angular momentum transport and the radial motions provide inward transport, i.e.

$$\Lambda_{\mathrm{V}} \propto \tau_{\mathrm{corr}} \left(\langle u_\phi'^2 \rangle - \langle u_r'^2 \rangle \right). \quad (3.56)$$

Obviously, in the regime of slow rotation vertical motions lead to $d\Omega/dr < 0$, while horizontal motions lead to $d\Omega/dr > 0$. Rigid rotation results if both turbulent intensities are nearly equal.

Concerning the latitudinal transport Λ_H one finds

$$\Lambda_H = \frac{\tau_{corr}}{K^2}\left((K_r^2 + K_\phi^2)\langle u_\phi'^2\rangle - (K_\phi^2 + K_\theta^2)\langle u_\theta'^2\rangle)\right) \simeq 2\tau_{corr}\left(\langle u_\phi'^2\rangle - \langle u_\theta'^2\rangle\right) \quad (3.57)$$

the latter relation for turbulence fields with isotropic wave numbers. This quantity vanishes if the turbulence field is only anisotropic with respect to the vertical direction – which is the normal case (see Fig. 3.19, left). The horizontal Λ-effect, therefore, only exists for rapid rotation. It is positive in all the numerical simulations, as the rotation prefers u_ϕ' rather than u_θ' (see Fig. 3.19, right).

The positive H (Fig. 3.17) agrees well with the observations of the horizontal random motions of the large sunspot groups (see Sect. 3.1.3). From Eq. (3.46) we find that $Q_{r\phi}^\Lambda$ is then a negative function vanishing at the poles and the equator, and $Q_{\theta\phi}^\Lambda$ is a positive function of the same amplitude also vanishing at the poles and the equator.

Pulkkinen et al. (1993) find the function V negative and "increasing in magnitude between the equator and 55° latitude". Indeed, there is a distinct minimum at the equator (see their Fig. 10). According to their Fig. 7 $Q_{\theta\phi}$ is positive in the northern hemisphere, vanishing at the poles and the equator. The results of Rieutord et al. (1994) are similar with respect to the horizontal stress, but the function V ended up rather small and fluctuating in latitude. Very clear findings are reported by Chan (2001), with negative V, positive H, and V almost vanishing at the equator (Fig. 3.18).

Note that the pronounced changes in the functions occur at Coriolis numbers of order unity. For this reason Ω^* is a convenient parameter to distinguish between the slow and fast rotation regimes. The Sun is in effect a rapid rotator with $\Omega^* \simeq 6$ (Durney & Latour 1978, Durney & Spruit 1979).

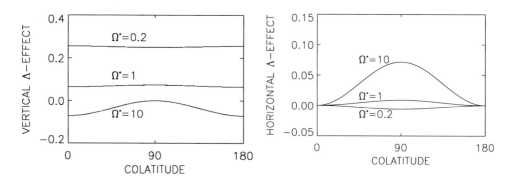

Figure 3.17: The functions $V(\theta)$ (*left*) and $H(\theta)$ (*right*) after Eqs. (3.51) and (3.52) for various basic rotation rates.

Rotating convection has also been studied with NIRVANA (see Sect. 3.5.1) for a fully compressible viscous gas obeying the ideal gas equation. The model considers a local Cartesian box placed tangentially on a rotating sphere, and involves a 3-layer planar polytrophic

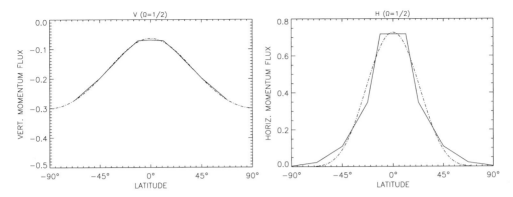

Figure 3.18: $V(\theta)$ (*left*) and $H(\theta)$ (*right*) from the simulations of Chan (2001) are very close to the results of the quasilinear theory for rapid rotation (see Fig. 3.17).

stratification. The middle layer of extent d is convectively unstable, whereas both the other layers are stable and serve to brake overshooting motions (see Hurlburt, Toomre & Massaguer 1986). The nonmagnetic hydrostatic 3-layer polytrophic state is realized by piecewise constant conductivity coefficients κ_{st}, κ_c and κ_{sb}, where the subscripts denote 'stable top', 'convective' and 'stable bottom'. With $d = 1$,

$$\kappa(z) = \begin{cases} \kappa_{st} & 0 < z < -1/4 \\ \kappa_c & -1/4 < z < -5/4 \\ \kappa_{sb} & -5/4 < z < -2. \end{cases} \tag{3.58}$$

Density and pressure, as well as the boundary conditions, are described by Ziegler (2002). We are here only interested in the results for the Λ-effect that arises from the anisotropies in the original velocity field. In Fig. 3.19 the turbulence intensities in the rotating box are given in comparison to the nonrotating one. The results clearly indicate that everywhere in the rotating box $\langle u_\phi'^2 \rangle$ dominates $\langle u_\theta'^2 \rangle$. On the other hand, the horizontal turbulence intensity $\langle u_\phi'^2 \rangle$ also dominates the vertical turbulence intensity $\langle u_r'^2 \rangle$, but only at the top and bottom of the box. In the bulk of the box the vertical intensity $\langle u_r'^2 \rangle$ dominates, but this effect is not too strong under the influence of the rotation[9]. Rüdiger, Tschäpe & Kitchatinov (2002) have shown with the quasilinear approximation that (at the equator) there is indeed a strong tendency of 'return-to-isotropy'. In the relation

$$\langle u_r'^2 \rangle - \langle u_\phi'^2 \rangle = A(\Omega^*) \left(\langle u_r^{(0)2} \rangle - \langle u_\phi^{(0)2} \rangle \right) \tag{3.59}$$

the 'transformation' factor $A(\Omega^*)$ sinks with increasing Ω^* and even changes its sign at a certain Ω^*. A rotating turbulence, therefore, proves to be much more isotropic than the same but without rotation (also Canuto, Minotti & Schilling 1994, Chan 2001). Figure 3.19 confirms these considerations.

In Fig. 3.20 (left) the behavior of the function $V(r, \theta)$ is given. It is always negative for rapid rotation, and it always vanishes at the equator. The maximal (negative) amplitude is in

---

[9] this behavior is also well described by our turbulence model (3.41), see Fig. 3.14

the bulk of the convection zone, where also $\langle u_r'^2 \rangle$ maximally dominates $\langle u_\phi'^2 \rangle$, see Eq. (3.56). The function $H(\theta)$ is never negative (Fig. 3.20, right). According to Eq. (3.56) this very clear result reflects the dominance of the azimuthal fluctuations (see Fig. 3.19, right). Note the weak dependence of H on the radius. The results confirm the analytical quasilinear computations for the Λ-effect surprisingly well. The question is whether the reported behavior persists for higher Ta. In particular, the vanishing of V at the equator is relevant here. It leads directly to the weak differential rotation observed beneath the solar equator. At first sight, the results of the global hydrodynamical simulation of turbulent convection under the influence of rotation by Miesch et al. (2000, their Fig. 9) and Brun & Toomre (2002) seem to indicate a positive V that peaks at the equator and vanishes at the poles (see Brun 2003, his Fig. 2).

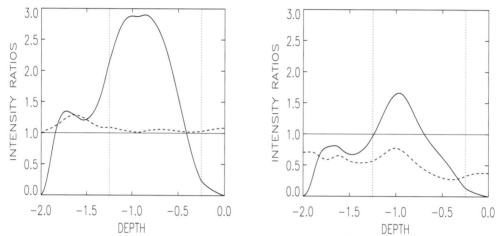

Figure 3.19: Ratios of the turbulence intensity within the box. *Solid:* $\langle u_r'^2 \rangle / \langle u_\phi'^2 \rangle$, dashed: $\langle u_\theta'^2 \rangle / \langle u_\phi'^2 \rangle$. *Left:* Without rotation; the vertical velocity dominates the horizontal velocity, there is no anisotropy in the horizontal plane. *Right:* With rotation (Ta $= 10^6$) at the equator; the vertical velocity dominates in the lower half but not in the upper. The azimuthal velocity always dominates the latitudinal one. The convection zone surface is at $z = -0.25$ and the bottom at -1.25.

3.2.3   The Eddy Viscosity Tensor

In general, the viscosity term in Eq. (3.5) must be replaced by the tensorial expression

$$Q_{ij} = \cdots - \mathcal{N}_{ijkl}\bar{u}_{k,l}, \tag{3.60}$$

with $\mathcal{N}$ as the viscosity tensor. It is also strongly influenced by the rotation. For isotropic turbulence (except the rotational influence) its structure is given by

$$\mathcal{N}_{ijkl} = \nu_1(\delta_{ik}\delta_{jl} + \delta_{jk}\delta_{il}) + \nu_2 \left(\delta_{il}\frac{\Omega_j\Omega_k}{\Omega^2} + \delta_{jl}\frac{\Omega_i\Omega_k}{\Omega^2} + \delta_{ik}\frac{\Omega_j\Omega_l}{\Omega^2} + \right.$$
$$\left. + \delta_{jk}\frac{\Omega_i\Omega_l}{\Omega^2} + \delta_{kl}\frac{\Omega_i\Omega_j}{\Omega_2} \right) + \nu_3\delta_{ij}\delta_{kl} + \nu_4\delta_{ij}\frac{\Omega_k\Omega_l}{\Omega^2} + \nu_5\frac{\Omega_i\Omega_j\Omega_k\Omega_l}{\Omega^4}, \tag{3.61}$$

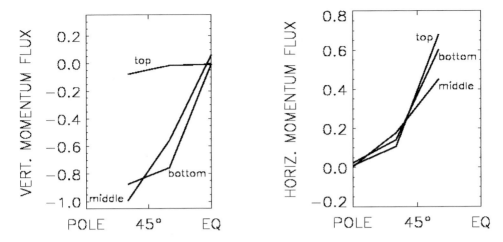

Figure 3.20: $V(\theta)$ (*left*) and $H(\theta)$ (*right*) for $\mathrm{Ta} = 10^6$, from the simulations of Egorov, Rüdiger & Ziegler (2004). Note that V is negative but vanishes at the equator, and H is positive but vanishes at the poles.

where $\nu_n = \nu_0 \Phi_n(\Omega^*)$, with $\nu_0 = 4/15 \, \tau_{\mathrm{corr}} u_{\mathrm{T}}^2$. Here ν_0 is the isotropic shear viscosity for the nonrotating fluid. The viscosity quenching functions $\Phi_n(\Omega^*)$ are known (Kitchatinov, Pipin & Rüdiger 1994). Only $\Phi_1(0) = \Phi_3(0) = 1$; the remaining functions parameterize the viscosity anisotropy due to the rotation, and must vanish for $\Omega^* = 0$.

In its full generality we only give the tensor for the nonrotating case, i.e. $\mathcal{N}_{ijkl} = \nu_{\mathrm{T}}(\delta_{ik}\delta_{jl} + \delta_{jk}\delta_{il}) + \mu_{\mathrm{T}}\delta_{ij}\delta_{kl}$, with

$$\nu_{\mathrm{T}} = \frac{4}{15} \int \frac{\nu^3 k^6 \hat{Q}_{ll}(k,\omega)}{(\omega^2 + \nu^2 k^4)^2} \, d\boldsymbol{k} \, d\omega,$$

$$\mu_{\mathrm{T}} = \frac{4}{15} \int \frac{\nu k^2 (\nu^2 k^4 + 5\omega^2)\hat{Q}_{ll}(k,\omega)}{(\omega^2 + \nu^2 k^4)^2} \, d\boldsymbol{k} \, d\omega. \tag{3.62}$$

The spectral function $\hat{Q}_{ll}$ is the positive-definite tensorial trace of the simplest spectral tensor

$$\hat{Q}_{ij} = q(k,\omega)(k^2 \delta_{ij} - k_i k_j) \tag{3.63}$$

for homogeneous and isotropic turbulence fields. Both expressions are equal in the τ-approximation but they are *not* equal for $\nu \to 0$:

$$\nu_{\mathrm{T}} = \frac{\mu_{\mathrm{T}}}{6} = \frac{2\pi}{15} \int \hat{Q}_{ll}(k,0) \, d\boldsymbol{k} \tag{3.64}$$

(see Stix et al. 1993). Here we have a good example of the consequences of various approximations. The τ-approximation is not identical with the inviscid limit or (later on) the high-conductivity limit, $\eta \to 0$.

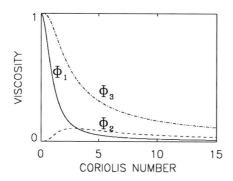
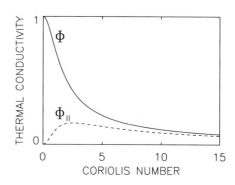

Figure 3.21: The rotational quenching of the most important viscosity coefficients (*left*) and the thermal conductivity coefficients defined in Eq. (3.79) (*right*). The τ-approximation is used, it is $\nu_1 = \nu_0 \Phi_1(\Omega^*), \nu_2 = \nu_0 \Phi_2(\Omega^*)$ and $\mu_T = \nu_0 \Phi_3(\Omega^*)$.

Only two viscosity coefficients contribute to the flux of the angular momentum, i.e.

$$Q_{r\phi}^{\nu} = -\nu_1 r \sin\theta \frac{\partial\Omega}{\partial r} - \nu_2 \sin\theta\cos\theta \left(r\cos\theta \frac{\partial\Omega}{\partial r} - \sin\theta \frac{\partial\Omega}{\partial\theta} \right),$$

$$Q_{\theta\phi}^{\nu} = -\nu_1 \sin\theta \frac{\partial\Omega}{\partial\theta} - \nu_2 \sin^2\theta \left(\sin\theta \frac{\partial\Omega}{\partial\theta} - r\cos\theta \frac{\partial\Omega}{\partial r} \right). \tag{3.65}$$

Note the presence in the radial flux of a latitudinal gradient in the angular velocity; in the latitudinal flux there is a radial gradient of Ω. This is the result of the anisotropy introduced by rotation (Kitchatinov 1986, Durney 1989). The rotational quenching effect for three coefficients is given in Fig. 3.21 (left). One finds that the rotational quenching of the eddy viscosity ν_1 is rather strong, while the Ω-quenching of μ_T is much smaller. This observation should be important for all effects where $\nabla \cdot \boldsymbol{u} \neq 0$, e.g. for the p-mode excitation in turbulent convection zones (see Stix et al. 1993). The ν_2-coefficient is zero for $\Omega \to 0$, and it remains small for larger Ω^*. The ν_2-effect does not seem to be an important effect. It is thus almost always neglected in theories of differential rotation.

3.2.4 Mean-Field Thermodynamics

Almost always differential rotation leads to meridional circulations. Its centrifugal force is not conservative (except if $\Omega = \Omega(s)$), so that a meridional flow must develop. According to Eq. (3.6) this flow then modifies the rotation law. From the early model of Köhler (1970) to the formulations of Glatzmaier (1985), Gilman & Miller (1986), Brandenburg et al. (1990), Miesch et al. (2000) and Brun & Toomre (2002) the inclusion of a meridional flow always led to the Taylor number puzzle: cylindrical isorotation contours are found, independently of the Λ-effect applied. The isolines of Ω in the solar convection zone, however, are *not* cylindrical (Fig. 3.4).

Only thermodynamics can solve the paradox. This involves the second term on the RHS of Eq. (3.13), which contains a large dimensionless factor. It is the working hypothesis that this term cancels the Taylor–Proudman state (Gilman 1986, Durney 1987). This is not the case when the turbulent viscosity and thermal conductivity are parameterized by scalars (Glatz-maier 1985, Gilman & Miller 1986). Only the anisotropy of the thermal conductivity produces a promising effect (Kitchatinov, Pipin & Rüdiger 1994).

Using the ideal gas law we find from Eq. (3.13) that $\nabla\rho \times \nabla P = -\rho T^{-1}\nabla\Delta T \times \nabla P$, where $\nabla\Delta T$ is the superadiabatic temperature gradient. This term is proportional to the deviation of the stratification from adiabaticity. The relative magnitude of this deviation is only 10^{-4}. The superadiabatic gradient can therefore be simplified by linearization, i.e. the superadiabatic part of the pressure gradient can be neglected. With $\nabla P = \rho g$ we then find that a relation of the form

$$\hat{e}_i\epsilon_{ijk}\frac{\partial}{\partial x_j}\frac{1}{\rho}\frac{\partial}{\partial x_l}\left(\rho\mathcal{N}_{klfn}\frac{\partial u_f^m}{\partial x_n}\right) = r\sin\theta\frac{\partial\Omega^2}{\partial z} - \frac{g}{T}\frac{\partial\Delta T}{\partial\theta} \tag{3.66}$$

represents the generation of the meridional flow due to the eddy viscosity ($\hat{e}$ is the eastward unit vector). Equatorial acceleration implies negative $\cos\theta\partial\Omega^2/\partial z$. The two terms in Eq. (3.66) can thus balance each other if $\cos\theta\partial\Delta T/\partial\theta$ negative. One needs, therefore, a 'hot pole' to solve the Taylor number puzzle (see Stix 1989).

The proper determination of the energy transport through a convective/turbulent medium is certainly the key problem of stellar hydrodynamics. Its complexity has been described in reviews by several authors (e.g. Gough 1976, Zahn 1979, Choudhuri 1998, Mestel 1999). Pioneering concepts such as nonlocality (Spiegel 1963, Maeder 1975, Ulrich 1976), the anelastic approximation (Latour et al., starting in 1976, see Lantz & Fan 1999), or that of Roxburgh (1978) or Canuto (1998) have led to interesting results.

The basic points can be taken from the fundamental entropy equation

$$\rho T\frac{dS}{dt} = \varepsilon^V - \nabla\cdot\mathbf{F}^D, \tag{3.67}$$

where the RHS represents the heat gained due to friction $\varepsilon^V = \frac{1}{2}\rho\nu(u_{i,k}+u_{k,i})^2$ and diffusion $\mathbf{F}^D = -\rho C_p\chi_D\nabla T$. Here χ_D is the sum of the thermal conductivities of molecular and radiative origins. From the thermodynamic relation $\rho TdS = \rho dH - dP$ (where H is the enthalpy) we readily obtain

$$\frac{\partial(\rho U)}{\partial t} + \nabla\cdot(\rho U\mathbf{u} + \mathbf{F}^D) = -P\nabla\cdot\mathbf{u} + \varepsilon^V, \tag{3.68}$$

with the internal energy $U = H - P/\rho$. From the Navier-Stokes equation one deduces an expression for $\mathbf{u}\cdot\nabla P$, which, together with Eq. (3.68), yields

$$\frac{\partial\left(\rho U + \frac{1}{2}\rho u^2\right)}{\partial t} + \nabla\cdot\left(\rho\mathbf{u}H + \mathbf{F}^D + \mathbf{F}^V + \frac{1}{2}\rho u^2\mathbf{u}\right) = \rho\mathbf{u}\cdot\mathbf{g}^{\text{eff}}, \tag{3.69}$$

with the friction term F^V being given by $F_j^V = -\rho\nu u_i(u_{i,j} + u_{j,i})$, and where $\mathbf{g}^{\text{eff}} = \mathbf{g} + \Omega^2\mathbf{s}$ is the effective gravity, which differs from $\mathbf{g}$ by the centrifugal acceleration. If this equation is applied to the steady flow of an ideal gas, using $H = C_pT$ with a constant C_p, and microscopic viscosity and conductivity are neglected, one obtains

$$\nabla\cdot\left(\rho C_p\mathbf{u}T + \mathbf{F}^D + \frac{1}{2}\rho u^2\mathbf{u}\right) = \rho\mathbf{u}\cdot\mathbf{g}^{\text{eff}}. \tag{3.70}$$

Frictional heating as a source of convection is excluded here. We shall also neglect radiative diffusion and mechanical energy flux, but the flow is allowed to be turbulent, the immediate consequence being

$$\nabla \cdot \left(\bar{\rho} \langle \boldsymbol{u}'T' \rangle + \langle \rho'T' \rangle \bar{\boldsymbol{u}} + \langle \rho'T'\boldsymbol{u}' \rangle \right) = \left(\bar{\rho}\bar{\boldsymbol{u}} + \langle \rho'\boldsymbol{u}' \rangle \right) \cdot \left(\frac{\boldsymbol{g}^{\mathrm{eff}}}{C_{\mathrm{p}}} - \nabla T \right). \tag{3.71}$$

The final bracket represents the superadiabatic rate, which will here be denoted by

$$\boldsymbol{\beta} = \frac{\boldsymbol{g}^{\mathrm{eff}}}{C_{\mathrm{p}}} - \nabla T. \tag{3.72}$$

Several different notations for this gradient excess may be found in the literature, e.g.

$$\boldsymbol{\beta} = \nabla T^{\mathrm{ad}} - \nabla T = \Delta \nabla T = \frac{T}{H_{\mathrm{p}}} (\nabla - \nabla^{\mathrm{ad}}) = -\frac{T}{C_{\mathrm{p}}} \nabla S, \tag{3.73}$$

where H_{p} is the pressure scale height, with S, for an ideal gas, given by $S = C_{\mathrm{v}} \log(P\rho^{-\Gamma})$. Since the entropy decreases outward in a convectively unstable layer, the vertical component of $\boldsymbol{\beta}$ is positive, but this vertical component is rather small. When evaluated using the values appropriate to the Sun, the adiabatic temperature gradient is of the order of 10^{-4} K/cm. The magnitude of the temperature excess is much smaller, namely about $2 \cdot 10^{-10}$ K/cm. If in unstable layers the vector $\boldsymbol{\beta}$ points outward, then the actual radial temperature profile must be steeper than the adiabatic. That is also the essence of the Schwarzschild criterion.

Neglecting the viscous heating terms in the energy transport equation, we finally obtain

$$\nabla \cdot (\boldsymbol{F}^{\mathrm{conv}} + \boldsymbol{F}^{\mathrm{rad}}) = C_{\mathrm{p}}\rho \boldsymbol{u}^{\mathrm{m}} \cdot \boldsymbol{\beta}, \tag{3.74}$$

where $\boldsymbol{F}^{\mathrm{conv}} = \rho C_{\mathrm{p}} \langle \boldsymbol{u}'T' \rangle$ and $\boldsymbol{F}^{\mathrm{rad}} = -C_{\mathrm{p}}\rho\chi_{\mathrm{rad}}\nabla T$ are the convective and radiative heat fluxes, and $\chi_{\mathrm{rad}} = 16\sigma T^3/3\kappa\rho^2 C_{\mathrm{p}}$.

If gravity is the only preferred direction, the eddy-heat transport will be radial. As the eddy-heat flux must vanish in an adiabatically stratified medium we write

$$\langle \boldsymbol{u}'T' \rangle = -\chi_{\mathrm{T}} \left(\nabla T - \frac{\boldsymbol{g}}{C_{\mathrm{p}}} \right) = -\frac{\chi_{\mathrm{T}}}{C_{\mathrm{p}}} T \nabla S = \chi_{\mathrm{T}} \boldsymbol{\beta}. \tag{3.75}$$

The eddy-heat diffusivity χ_{T} should be of order $\ell_{\mathrm{corr}} u_{\mathrm{T}}$. Wasiutynski (1946) introduced a tensorial conductivity, i.e.

$$F_i^{\mathrm{conv}} = \rho C_{\mathrm{p}} \langle u_i'T' \rangle = \rho C_{\mathrm{p}} \chi_{ij} \beta_j, \tag{3.76}$$

or in component form

$$F_r^{\mathrm{conv}} = \rho C_{\mathrm{p}} (\chi_{rr}\beta_r + \chi_{r\theta}\beta_\theta), \qquad F_\theta^{\mathrm{conv}} = \rho C_{\mathrm{p}} (\chi_{\theta r}\beta_r + \chi_{\theta\theta}\beta_\theta). \tag{3.77}$$

If rotation is involved, the eddy conductivity tensor may contain the following elementary tensors

$$\delta_{ij}, \quad \mathring{g}_i \mathring{g}_j, \quad (\mathring{\boldsymbol{g}} \cdot \boldsymbol{\Omega})^2 \delta_{ij}, \quad (\mathring{\boldsymbol{g}} \cdot \boldsymbol{\Omega})^2 \mathring{g}_i \mathring{g}_j, \quad (\mathring{\boldsymbol{g}} \cdot \boldsymbol{\Omega})(\mathring{g}_i \Omega_j + \mathring{g}_j \Omega_i), \quad \Omega_i \Omega_j, \tag{3.78}$$

all of which are polar combinations, symmetric in i and j. The tensor χ_{ij} is not necessarily symmetric, however, so that other terms may also appear, even linear in Ω. Therefore, like the α-effect and the $V^{(0)}$-effect, there are first-order effects even for F_ϕ^{conv}. One can show that F_ϕ^{conv} is closely related to the Λ-effect: $F_\phi^{\mathrm{conv}} \propto \Omega^* V \sin\theta$. It is thus positive for slow rotation and negative for rapid rotation.

Let us here concentrate on the expression

$$\chi_{ij} = \chi_{\mathrm{T}}\delta_{ij} + \chi_\| \frac{\Omega_i \Omega_j}{\Omega^2}, \tag{3.79}$$

with $\chi_{\mathrm{T}} = \chi_0 \Phi(\Omega^*)$ and $\chi_\| = \chi_0 \Phi_\|(\Omega^*)$, where $\chi_0 = \tau_{\mathrm{corr}} u_{\mathrm{T}}^2/3$ is the isotropic thermal conductivity for a nonrotating fluid. The functions Φ and $\Phi_\|$ are positive, and represent the effects of rotation; $\Phi(0) = 1$ and $\Phi_\|(0) = 0$. Explicit expressions in the τ-approximation are given by Kitchatinov, Pipin & Rüdiger (1994), see Fig. 3.21 (right).

As a result of the $\chi_\|$-term in Eq. (3.79), the radial thermal conductivity is larger at the poles than at the equator; a poleward latitudinal component in the convective heat flux Eq. (3.76) results. A hot pole can therefore indeed be expected to resolve the Taylor number puzzle (see Miesch et al. 2000, their Fig. 13). Belvedere, Paternò & Stix (1980) with their latitude-dependent heat transport, i.e. χ_{T} as a function of $(g\Omega)^2$, developed a similar concept. Figure 3.22 presents for the northern hemisphere the correlation $\langle u'_\theta T' \rangle$ which vanishes at the pole and the equator and which is negative so that indeed the heat flows towards the poles. Only this effect in the very last analysis is responsible for the existence of the solar surface rotation law.

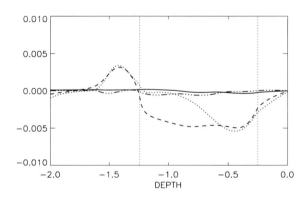

Figure 3.22: The latitudinal heat transport $\langle u'_\theta T' \rangle$ vanishes at the pole and the equator and is negative at the midlatitudes $30°$ (dashed) and $60°$ (dotted) of the northern hemisphere. Courtesy P. Egorov.

3.3 Differential Rotation and Meridional Circulation for Solar-Type Stars

In order to compute the differential rotation of a solar-type star the stratifications of pressure, density and temperature are needed as input parameters. The entropy gradient not only determines the baroclinic term in the equation for the meridional flow, but also the convection velocity u_{T} that according to the standard mixing-length theory of stellar convection, is

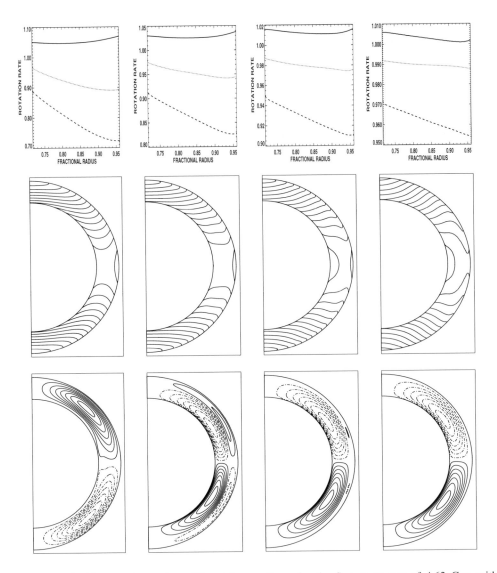

Figure 3.23: The rotation and meridional flow patterns for the Sun at an age of 4.62 Gyr, with $\alpha_{\mathrm{MLT}} = 5/3$, for rotation periods 56 days, 28 days, 14 days and 7 days (from left to right). *Top*: The normalized rotation rate at the equator (solid lines), $45°$ latitude (dotted), and the poles (dashed). *Middle*: Contour plots of the rotation rate. *Bottom*: Contours of the stream function. Dash-dotted lines denote counterclockwise circulation. From Küker & Stix (2001).

determined by the stratifications of pressure and entropy through the relation

$$u_{\mathrm{T}}^2 = -\frac{\ell_{\mathrm{corr}}^2 g}{4C_{\mathrm{p}}} \frac{\mathrm{d}S}{\mathrm{d}r}, \qquad (3.80)$$

where the mixing length $\ell_{\mathrm{corr}} = \alpha_{\mathrm{MLT}} H_{\mathrm{p}}$ is determined by the pressure scale height. The eddy viscosity and heat conductivity coefficients then read

$$\nu_{\mathrm{T}} = -\frac{\tau_{\mathrm{corr}} g \alpha_{\mathrm{MLT}}^2 H_{\mathrm{p}}^2}{15 C_{\mathrm{p}}} \frac{\mathrm{d}S}{\mathrm{d}r}, \qquad\qquad \chi_{\mathrm{T}} = -\frac{\tau_{\mathrm{corr}} g \alpha_{\mathrm{MLT}}^2 H_{\mathrm{p}}^2}{12 C_{\mathrm{p}}} \frac{\mathrm{d}S}{\mathrm{d}r}. \qquad (3.81)$$

Note that for a given pressure stratification the convective turnover time τ_{corr} is a function of the entropy gradient as well. The convective heat transport is thus a nonlinear function of $\mathrm{d}S/\mathrm{d}r$. Since both the viscosity coefficient and the convective turnover time (and hence the Coriolis number) depend on the entropy gradient, rotationally inhibited heat transport will affect the transport of momentum as well.

Figure 3.23 shows the rotation law and meridional flow pattern resulting for the Sun for rotation periods ranging from 56 days to 7 days (Küker & Stix 2001). The stratifications of density, temperature, and luminosity were taken from the solar model by Ahrens, Stix & Thorn (1992). The model does not include an atmosphere. It ends 35,000 km below the photosphere, and the lower boundary is located at the bottom of the convection zone. Convective overshooting at the boundaries is also not considered. In the uppermost layers of the convection zone and atmosphere the gas density is a rapidly decreasing function of radius. No significant stress can therefore be maintained on the convection zone by the surrounding medium. Magnetic torques exerted by the magnetic field, though essential for the solar angular momentum evolution, can be neglected as well since the timescale for the spin-down of the solar rotation is Gyr while that of internal angular momentum transport by Reynolds stress is years. The microscopic viscosity is about ten orders of magnitude smaller than the eddy viscosity in the convection zone. The stress on the lower boundary of the convection zone is therefore negligible in the problem of solar differential rotation. Both the upper and lower boundaries are thus required to be stress-free. For a solar-type star the total heat flux through the boundaries is required to maintain the total luminosity, i.e.

$$F_r^{\mathrm{tot}} = L(r)/4\pi r^2. \qquad (3.82)$$

With the above boundary conditions, F_r and hence the gradient of the entropy does not vary with latitude on the boundaries. The entropy itself does vary though.

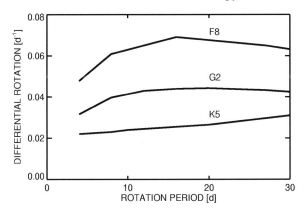

Figure 3.24: The horizontal shear as a function of the (average) rotation rate for an F star, a solar-type star, and a K5 star, as predicted by theory. The F star by M. Küker (2004), G & K stars by Kitchatinov & Rüdiger (1999).

The first column in Fig. 3.23 shows the rotation and flow patterns for a slowly rotating Sun, with a rotation period of 56 days. A value of 33 per cent is found for the normalized horizontal shear $\delta\Omega^{\mathrm{lat}}/\Omega_\odot$, and the meridional flow shows one cell per hemisphere, toward the equator at the surface and toward the pole at the bottom of the convection zone. The maximum flow velocity is 9.2 m/s at the surface and 0.3 m/s at the bottom of the convection zone. In the $\tau_{\mathrm{rot}} = 28$ d case (second column), the equator rotates about 20 per cent faster than the poles at the top of the convection zone. The meridional flow pattern consists of two cells per hemisphere, with the flow directed toward the equator at both the bottom and the top of the convection zone and poleward at intermediate depths. The maximum speed of the horizontal motion is 5.4 m/s at the surface and 2 m/s at the bottom of the convection zone. These results are in good agreement with those of the fully-compressible nonlinear simulations of Miesch et al. (2000). For a convection zone rotating with a period of 14 days the normalized pole-equator difference decreases to 10 per cent. There are still two flow cells per hemisphere, but the outer cell is restricted to a very shallow surface layer. The surface flow is still poleward with an amplitude of 3.8 m/s. At the bottom of the convection zone the flow reaches a speed of 2.8 m/s. Decreasing the rotation period to 7 days (last column) leads to a further reduction of $\delta\Omega^{\mathrm{lat}}/\Omega_\odot$ to less than five per cent. The meridional flow is now directed poleward at the top and equatorward at the bottom of the convection zone, with flow speeds of 2.7 m/s at the top and 3.6 m/s at the bottom of the convection zone. For the model of the present Sun, the total latitudinal shear $\delta\Omega^{\mathrm{lat}}$ is roughly the same for the rotation periods of 28 days, 14 days and 7 days, but decreases for slow rotation as the result for $\tau_{\mathrm{rot}} = 56$ d shows. At the bottom of the convection zone the meridional flow drifts to the equator, except for the very old Sun with a 56-day rotation period.

In Fig. 3.24 the dependence of the latitudinal differential rotation on the average rotation period is shown for the same model of a solar-type star as in Fig. 3.23, a MS star with 1.2 solar masses (spectral type F8 results by Küker 2004), and a model of a K5 dwarf with 0.7 solar masses. In Fig. 3.12 the rhombs indicate the observed rotational shear for the K0 stars AB Dor, PZ Tel, and for the Sun. Theory and observations both suggest that the surface shear does not depend very strongly on the rotation rate. For the stellar types considered the theory instead predicts a strong dependence on the spectral type, with the most luminous star showing the strongest shear. The dependence of the differential rotation on the properties of the star has, however, not been systematically studied yet. As a function of the rotation period the differential rotation of the F-type star shows a maximum at a period of about one week. The differential rotation of the solar-type star is roughly constant for periods between 10 and 30 days, with a maximum at about two weeks. The K dwarf shows a monotonic increase of the shear with the rotation period in the interval shown, reaching a maximum at a period longer than 30 days.

Figure 3.25 shows the speed of the meridional flow at the top and bottom of the convection zone, as a function of the rotation period for the stars of Fig. 3.24. For the K dwarf the surface flow is always directed toward the poles. In the case of the G2 star the surface flow is toward the pole for fast rotation and toward the equator for slow rotation, while for the F stars it is always toward the equator. All stellar models show an increase of the values with increasing rotation rate.

Figure 3.25 (right) shows the flow at the bottom of the convection zone. For a single-cell flow it would always have the opposite direction of the surface flow. The plot shows, however,

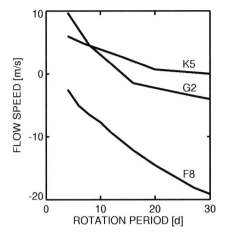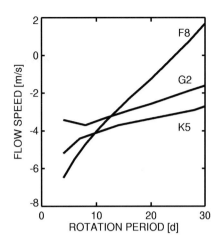

Figure 3.25: The meridional flow speed at the top (*left*) and bottom (*right*) of the convection zone for the same models as in Fig. 3.24. Negative values mean that the flow is directed toward the equator.

that in the case of rapid rotation the flow is directed toward the equator for all three stars. In all cases the values are larger for slow rotation, i.e. decrease with increasing rotation rate (or increase with increasing rotation period), but only the flow speed of the F8 star reaches positive values. Due to the density stratification the bottom flow is usually slower than the surface flow.

Figures 3.23–3.25 cover a limited range of rotation rates. We know, however, the rotation pattern in the limiting cases of very slow and very fast rotation. In the case of very fast rotation the baroclinic term in the equation of motion becomes negligible and the Taylor–Proudman theorem applies. As the boundary conditions require that the vertical stress vanishes on both boundaries, the rotation must be rigid. In the opposite limit of very slow rotation the horizontal Λ-effect vanishes and the turbulent heat transport becomes isotropic. Since the Taylor number is small the radial Λ-effect is the only effect that generates any shear, and the resulting rotation law shows (positive) radial shear only. As a function of the rotation rate the latitudinal shear must therefore have a maximum value at intermediate rotation rates, as the curves representing the G2 and F8 stars in Fig. 3.24 indeed show.

3.4 Kinetic Helicity and the DIV-CURL-Correlation

Let us ask whether the rotational influence on the turbulence can already be directly observed at the solar surface. Granulation and mesogranulation are the main candidates due to their nonmagnetic character. We shall show that the *horizontal* motions alone suffice to estimate the rotational influence.

Finite kinetic helicity $\langle \boldsymbol{u}' \cdot \nabla \times \boldsymbol{u}' \rangle$ indicates basic rotation. Here that part of the helicity is considered that results from the vertical components of velocity and vorticity, i.e.

$$\mathcal{H}_{\mathrm{kin}} = \left\langle w \left(\frac{\partial v}{\partial x} - \frac{\partial u}{\partial y} \right) \right\rangle, \tag{3.83}$$

where the turbulence pattern $\boldsymbol{u}' = (u, v, w)$ is taken in a Cartesian coordinate system, where x points east, y north, and z points radially outward. In a stratified convection zone rising material expands and rotates because of the action of the Coriolis force. On the northern hemisphere the results are *left-handed* helical motions, i.e. $\mathcal{H}_{\mathrm{kin}} < 0$ (see Fig. 3.26, left, also Miesch et al. 2000, their Fig. 22). Expansion results in clockwise rotation and vice versa. Instead of Eq. (3.83) we consider now the DIV-CURL-correlation

$$\mathcal{C} = \left\langle \left(\frac{\partial u}{\partial x} + \frac{\partial v}{\partial y} \right) \left(\frac{\partial v}{\partial x} - \frac{\partial u}{\partial y} \right) \right\rangle \tag{3.84}$$

(Wang et al. 1995).

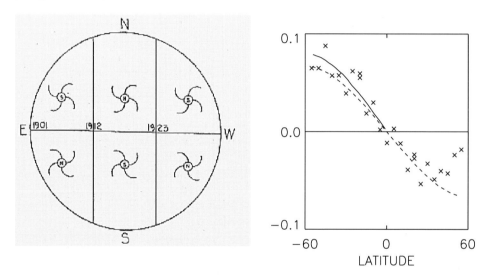

Figure 3.26: *Left:* During the solar cycles the magnetic polarity of the spots changes but not the helicity. Hale (1927) found that the dominant orientation of the cyclones around sunspots only changes for both hemispheres. *Right:* The correlation of the radial component of curl and the horizontal divergence for supergranulation. The solid line gives the $\cos \theta$-profile. From Duvall & Gizon (2000).

The anelastic approximation, $\nabla \cdot \rho \boldsymbol{u} = 0$, is adopted hence

$$\mathcal{C} \approx \left\langle \frac{w}{H_{\mathrm{m}}} \left(\frac{\partial v}{\partial x} - \frac{\partial u}{\partial y} \right) \right\rangle = \frac{\mathcal{H}_{\mathrm{kin}}}{H_{\mathrm{m}}}, \tag{3.85}$$

where $H_{\mathrm{m}} = -\partial z / \partial \log |\rho w|$ is the scale height for the vertical momentum fluctuations. We assume $H_{\mathrm{m}} > 0$, i.e. the vertical momentum fluctuations decrease with height (see Simon & Weiss 1997). Close to the bottom of the cells, however, H_{m} becomes negative. We see that in the top of the cells a positive horizontal divergence corresponds to an updraft ($w > 0$) and a negative horizontal divergence corresponds to a downdraft ($w < 0$).

Like the helicity $\mathcal{H}_{\mathrm{kin}}$, so also is $\mathcal{C}$ a pseudoscalar, and its sign depends on the coordinate system. The only pseudoscalar that can be constructed in anisotropic turbulence with the radial anisotropy direction $\overset{\circ}{g}$ is the scalar product $\overset{\circ}{g} \cdot \Omega$. A nonvanishing helicity proxy $\mathcal{C}$, of course, can only exist for rotating turbulence.

As the typical mesogranulation pattern only lives for a few hours, Ω^* is estimated to be of order of 0.1. Hence, the correlation effect $\mathcal{C}$ should be very small. Brandt et al. (1988) and Simon et al. (1994) present the first results of an overall inspection of horizontal flow patterns on mesoscales. The maximum velocities are ~ 750 m/s and the maximum vertical vorticity is about $2 \cdot 10^{-4}$ s^{-1} (see also Simon et al. 1988). There are indications for a negative (positive) correlation $\mathcal{C}$ on the northern (southern) hemisphere (Wang et al. 1995). The correlation presented by Duvall & Gizon (2000) seems to be well-established (Fig. 3.26, right).

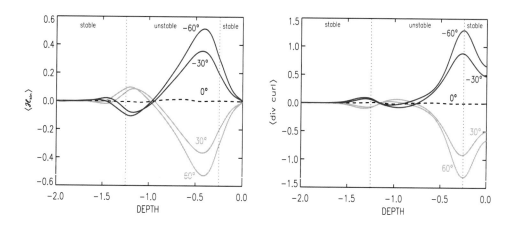

Figure 3.27: The helicity $\mathcal{H}_{\text{kin}}$ (*left*) and the DIV-CURL correlation $\mathcal{C}$ (*right*) after time-averaging vs. latitude and depth. The convection zone surface is at $z = -0.25$ and the bottom at -1.25. From Egorov, Rüdiger & Ziegler (2004).

The desired expression can generally be written as the z-component of the axial vector $\mathcal{C} = \langle \nabla \cdot \tilde{u} \, \nabla \times u' \rangle$, with $\tilde{u} = u' - (\overset{\circ}{g} \cdot u') \overset{\circ}{g}$ as the horizontal component of the random flow field. With the random momentum field $m = \rho u'$ and with $G = \nabla \log \rho$ it follows that

$$\mathcal{C} = -\frac{1}{\rho^2} \langle (\overset{\circ}{g} \cdot \nabla)(\overset{\circ}{g} \cdot m)(\nabla \times m + m \times G) \rangle. \tag{3.86}$$

With the Fourier transform (3.38) one obtains

$$\rho^2 \, \mathcal{C}_i = \epsilon_{ijn} \, \overset{\circ}{g}_f \, \overset{\circ}{g}_m$$
$$\int k'_j k_m \langle \hat{m}_f(k, \omega) \hat{m}_n(k', \omega') \rangle e^{i(k+k') \cdot x - (\omega + \omega')t} \, dk \, dk' \, d\omega \, d\omega' \tag{3.87}$$

for the components of $\mathcal{C}$. To first order in Ω the vector $\mathcal{C}$ has only the component (3.84).

It remains to compute the tensor M_{ij}. In a stratified turbulent medium the spectral tensor in Eq. (3.87) is (3.41). Only the influence of the basic rotation on the turbulence produces a finite effect, see Eq. (4.29). The results must be introduced into Eq. (3.87), which after massive manipulations with the τ-approximation Eq. (3.47) leads to

$$\mathcal{C} = -\frac{8}{35}\frac{\alpha_{\mathrm{MLT}}^2}{\gamma^2}\frac{\Omega}{\tau_{\mathrm{corr}}}\cos\theta \tag{3.88}$$

for the radial component of the vector $\mathcal{C}$, if only the density stratification is taken into account. The resulting expression proves to be *negative* in the northern hemisphere, is proportional to $\Omega\cos\theta$, and is of order $10^{-10}\,\mathrm{s}^{-2}$ (Rüdiger, Brandenburg & Pipin 1999).

Using NIRVANA (see Sect. 3.5.1) numerical simulations for the relationship between the correlation $\mathcal{C}$ and the kinetic helicity $\mathcal{H}_{\mathrm{kin}}$ have been carried out (Fig. 3.27). In accordance with the analytical results both the correlation $\mathcal{C}$ and the helicity $\mathcal{H}_{\mathrm{kin}}$ are negative in the northern hemisphere and positive in the southern. We find a very clear distribution for the helicity. In the northern hemisphere it is negative at the top of the convection zone and positive but smaller at the bottom. The helicity distribution given in Fig. 3.27 can be understood with the estimate $\mathcal{H}_{\mathrm{kin}} \propto \mathrm{d}(u_{\mathrm{T}}\rho)/\mathrm{d}z$, very similar to the α-effect formulations in Sect. 4.2.2. Obviously, the resulting helicity is dominated by the stratification of the turbulence intensity rather than the density stratification.

As also shown in Fig. 3.27, the correlation $\mathcal{C}$ forms an interesting proxy for the helicity. It is also negative at the top of the convection zone (in the northern hemisphere) but it does *not* *vanish* at $z = 0$ where the helicity vanishes. All in all, the quasilinear SOCA-theory[10] and numerical simulations always lead to the same result concerning the correlation (3.84). The DIV-CURL-correlation at the solar surface should be observed carefully in the future. With his numerical modeling of the turbulence in the solar tachocline Miesch (2003) with a randomly driven turbulence in the rotating and (stably) stratified medium also obtains negative (positive) values for $\mathcal{C}$ at the northern (southern) hemisphere. Egorov, Rüdiger & Ziegler (2004) correctly reproduced the DIV-CURL correlation for supergranulation observed by Duvall & Gizon (2000) for a $\mathrm{Ta} = 10^3$ which corresponds to the Coriolis number of $\lesssim 0.20$.

3.5   Overshoot Region and the Tachocline

Overshooting regions are the natural envelopes of stellar convection zones. In convectively unstable shells of late-type stars penetrative convection is believed to enable the downward transport of poloidal magnetic field into the overshoot region. Whereas the existence of shear flows at the base of the convection zone in the Sun has indeed been deduced from helioseismology (Gough & Toomre 1991, Thompson et al. 1996), our knowledge about the magnetic-field pumping must come from simulations. Any estimates of the pumping effect therefore require one to study penetrative convection and its dependence on important parameters like the rotation rate and magnetic field strength.

The mixing-length theory has often been used to describe transport phenomena. If rotational effects and magnetic fields come into play such parameterizations meet their limits. Direct numerical simulation of convective penetration is the only alternative to estimate penetration depths. Such calculations have been performed by several authors in 2D (Cattaneo, Hurlburt & Toomre 1990, Xie & Toomre 1993, Hurlburt et al. 1994) and, more recently, in 3D (Stein & Nordlund 1989, Singh, Roxburgh & Chan 1998, Saikia et al. 2000, Ziegler & Rüdiger 2003).

---

[10] SOCA=Second Order Correlation Approximation

3.5.1   The NIRVANA Code

NIRVANA is a public code for astrophysical gas dynamics developed by Ziegler (1998). It numerically integrates the nonrelativistic MHD equations in 2D or 3D. NIRVANA assumes local thermodynamic equilibrium with constant heat capacities. Radiation transport is not treated but diffusive processes can be included. The equations

$$\rho\left(\frac{\partial \boldsymbol{u}}{\partial t} + (\boldsymbol{u}\nabla)\boldsymbol{u}\right) = -\nabla P + \nabla \cdot \pi + \boldsymbol{F}_{\text{rot}} + \frac{1}{\mu_0}(\nabla\times\boldsymbol{B})\times\boldsymbol{B} + \rho\boldsymbol{g},$$

$$\frac{\partial \rho U}{\partial t} = -\nabla \cdot (\rho U\boldsymbol{u}) - P\nabla\cdot\boldsymbol{u} + \varepsilon^V + \nabla \cdot (\rho C_{\text{p}}\chi_{\text{D}}\nabla T) + \frac{\eta}{\mu_0}|\nabla\times\boldsymbol{B}|^2,$$

$$\frac{\partial \rho}{\partial t} = -\nabla \cdot (\rho\boldsymbol{u}), \qquad\qquad \frac{\partial \boldsymbol{B}}{\partial t} = \nabla\times(\boldsymbol{u}\times\boldsymbol{B} - \eta\nabla\times\boldsymbol{B}) \qquad (3.89)$$

are solved with $\pi_{ij} = \rho\nu(u_{i,j} + u_{j,i})$ as the viscous stress tensor. NIRVANA allows simulations in a rotating frame of reference in which case the centrifugal and Coriolis force terms,

$$\boldsymbol{F}_{\text{rot}} = -2\rho\boldsymbol{\Omega}\times\boldsymbol{u} - \rho\boldsymbol{\Omega}\times(\boldsymbol{\Omega}\times\boldsymbol{x}), \qquad\qquad\qquad (3.90)$$

appears in the momentum equation. χ denotes the thermal conductivity. The thermodynamic variables are related by an ideal gas law $P = (\mathcal{R}/\mu) \cdot \rho T$, with μ the mean molecular weight of the gas.

NIRVANA is based on the ZEUS code by Stone & Norman (1992a,b), Stone, Mihalas & Norman (1992) and Clarke, Norman & Fiedler (1994). Momentum and total energy are not exactly conserved, which, without modification of the basic scheme, would lead to an incorrect treatment of shocks. This problem is handled by the technique of artificial viscosity, which modifies the momentum and energy equations by adding a stress term and/or a viscous heating term. The purpose of artificial viscosity is to provide the correct jump relations and shock velocities when the flow becomes supersonic. Artificial viscosity must, therefore, be sensitive only to compression. The viscosity tensor formulation of Tscharnuter & Winkler (1979) is applied.

The code makes use of the *staggered-grid* formalism. The MHD variables are defined at different locations in a cell, which permits a compact space discretization of second-order accuracy. Scalar variables and diagonal elements of tensors are defined in the cell center, vector components on the cell faces and off-diagonal elements of tensors at the cell edges.

NIRVANA is a *time-explicit* Eulerian grid code. The maximum numerical time step that can be used by an explicit code to advance the solution in time is controlled by the Courant-Friedrich-Lewy condition. This states that during a timestep information cannot be transported along distances larger than the dimension of a cell to ensure numerical stability.

The hydrodynamic *advection* part of the equations is solved with finite-volume techniques using the reconstruction scheme of van Leer (1977) with flux limiter. Magnetic field transport described by the induction equation is solved in a similar fashion. The transport ansatz of Evans & Hawley (1988) is used to maintain $\nabla\cdot\boldsymbol{B} = 0$ to within machine precision. To ensure a (numerically) stable propagation of magnetic shear waves the method of characteristics is applied as an intermediate step to estimate the electric field after a half-time step to be used then in the finite-volume updating scheme (Stone & Norman 1992a,b).

3.5.2 Penetration into the Stable Layer

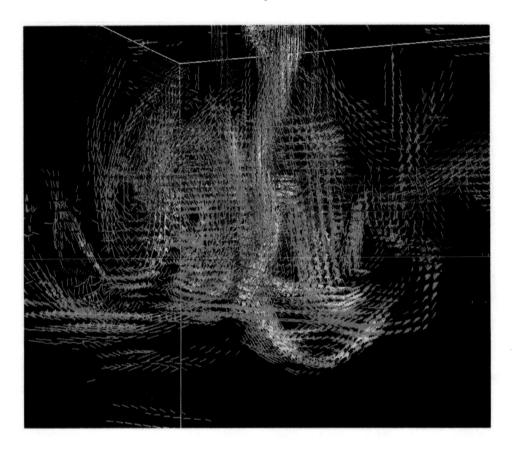

Figure 3.28: The magnetohydrodynamics of the overshoot phenomenon for rotating convection (Nordlund et al. 1992). Both the maxima of magnetic fields (yellow) and vorticity (gray) are given. The lower part of the box is convectively stable. The magnetic tubes are oriented horizontally there. Note the existence of rotation-induced twisted (left-handed) structures. $\Omega^* \simeq 3$, penetration depth $\sim 0.1~H_{\mathrm{p}}$. Courtesy A. Brandenburg.

Rotating and stratified convection has been numerically tested for dynamo action by Gilman & Miller (1981), Glatzmaier (1985) and Meneguzzi & Pouquet (1989). With a grid of 63^3 points the fully compressible code by Stein & Nordlund (1989) simulated rotating convection with a Taylor number of 10^5 and a magnetic Prandtl number between 0.2 and 20 (Brandenburg & Tuominen 1991, Nordlund et al. 1992). In the simulations the 'downdraft' phenomenon has been observed as characteristic for stratified MHD turbulence (Fig. 3.28, see Nordlund & Stein 1989, also Grossmann-Doerth, Schüssler & Steiner 1998). Energy saturation only resulted for magnetic Prandtl numbers exceeding unity. Although it is widely believed that the magnetic buoyancy quickly transports the magnetic flux upward, the opposite was observed in the simulations. The field is transported downward (see Dorch & Nordlund 2001).

Overshoot and penetration are also important for the tachocline problem. NIRVANA is used to study penetrative convection for a fully compressible, viscous gas threaded by a toroidal magnetic field with a profile with maximum magnetic field at the stable/unstable interface at $z = z_{\mathrm{in}}$. The problem is characterized by the Rayleigh number Ra, the Prandtl numbers Pr and Pm, the Taylor number Ta and the plasma $\beta^* = 2\mu_0 P/B^2$ taken at z_{in}. The magnetic boundary conditions in z are $\partial B_x/\partial z = \partial B_y/\partial z = 0$, and B_z is then obtained from $\nabla \cdot \boldsymbol{B} = 0$.

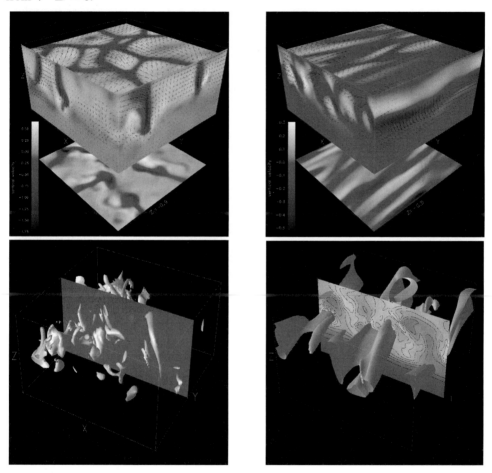

Figure 3.29: Magnetoconvection for the extreme models M1 (weak field, no rotation, *left*) and the model M4 (strong field, rapid rotation, *right*). *Top:* Velocity u_z as grayscale along the walls and beneath the box. *Bottom:* Magnetic energy isocontours. Note the trend to two-dimensionality along the azimuth due to the strong horizontal magnetic field and the basic differences between flow and field.

A series of runs was made for fixed $\mathrm{Ra} = 3 \cdot 10^5$, $\mathrm{Pr} = 0.1$ and $\mathrm{Pm} = 1$. A summary of the simulations is given in Table 3.3. The differences for weak and strong-field magnetocon-

Table 3.3: The influence of rotation and magnetic field in the simulations. The last column gives the downward directed pumping velocity of the mean magnetic field discussed in Sect. 4.2.1.

model	Ta	Ω^*	β	$u_\mathrm{T}(z_\mathrm{in})$	$\triangle$	γ
M1	0	0	5000	26.2	0.44	-7.1
M2	0	0	5	18.9	0.17	-6.8
M3	$6 \cdot 10^5$	4.2	5000	19.0	0.28	-2.3
M4	$6 \cdot 10^5$	9.1	5	6.5	0.05	-1.0

vection are given in Fig. 3.29, presenting snapshots of the velocity field and magnetic energy distribution for the (extreme) models M1 and M4. In the weak-field case the convection essentially shows the magnetic field treated as a passive ingredient advected with the flow. Such convection is dominated by narrow regions of strong, downward-directed plumes embedded in broader regions of upflowing material. This spatial asymmetry between narrow downflow regions and broader upflow regions is a typical feature of compressible convection in stratified media observed in all numerical simulations (also Cattaneo, Hughes & Weiss 1991, Brandenburg et al. 1996, Weiss et al. 1996, Brummell, Hurlburt & Toomre 1996, Steiner et al. 1998). The flow is highly time-dependent, occasionally generating fast plumes that are able to penetrate deep into the stable zone. The turbulent nature of convection also manifests itself in the topology of the magnetic field. Magnetic flux accumulates in small-scale structures distributed over the whole computational domain.

When the magnetic field strength is increased, the structure of the convection evolves toward two-dimensionality along the azimuthal direction. For strong magnetic fields the convective motions occur in the form of cylindrical rolls aligned in the ϕ-direction. The turbulence intensity u_T is substantially reduced to only 30% of that of weak-field convection.

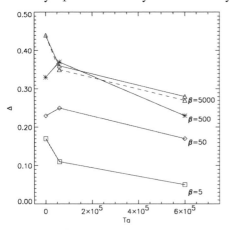

Figure 3.30: Penetration depth $\triangle$ as a function of Ta and β. The dashed line gives the nonmagnetic case. $\triangle = 0.1$ corresponds to an overshoot layer of 3300 km. From Ziegler & Rüdiger (2003).

When the gas enters the stable layer it suffers 'negative buoyancy' and is finally stopped, transferring energy to the stable layer. The extent of penetration is estimated with the help of the averaged vertical kinetic energy flux $\langle F_\mathrm{kin} \rangle$, where $F_\mathrm{kin} = \rho \boldsymbol{u}^2 u_z$ and the brackets denote time and horizontal averaging. $\langle F_\mathrm{kin} \rangle$ is negative throughout the convection zone. At some depth $\langle F_{\mathrm{kin}} \rangle$ becomes zero, providing a measure for the degree of convective penetration. The resulting $\triangle$ is depicted in Fig. 3.30 as a function of the parameters. Without rotation and without magnetic field we find $\triangle = 0.44$ (see Table 3.3). In the absence of magnetic fields $\triangle$ is reduced with increasing Ta. The penetration depth for Ta $= 6 \cdot 10^5$ is $\triangle = 0.27$, indicating rotational quenching. For strong-field convection, however, the extent of penetration is significantly reduced. The penetration depth is smallest for model M4, where both rotational and magnetic quenching yield the very small value $\triangle = 0.05$.

We are particularly interested in the depth of the penetration zone that, according to the results of helioseismology, only covers 10% of the pressure scale height at the bottom of the convection zone (Montero et al. 1994, Basu & Antia 1994, Christensen-Dalsgaard, Monteiro & Thompson 1995). Our result is that the depth of the overshoot region depends *strongly* on the rotation and the magnetic field. In our model the (dimensionless) pressure scale height at the stable/unstable interface is $H_{\mathrm{p}} = |\mathrm{d}z/\mathrm{d}\log P| = 0.55$. From the simulations we find the extreme values $\triangle = 0.8 H_{\mathrm{p}}$ for model M1 and $\triangle = 0.09 H_{\mathrm{p}}$ for M4. The latter is in agreement with observations, suggesting in this way that a strong magnetic field is present in the rotating tachocline. Using a nonlocal formulation Stix (1989) finds for the nonrotating and nonmagnetic layer a pressure scale height of 60,000 km and an overshooting of about 10,000 km, i.e. 16% of the pressure scale height. The latter number is also in accordance with van Ballegooi-jen (1982). From 2D hydrodynamic simulations Freytag, Ludwig & Steffen (1996) derived a nonuniform diffusion coefficient describing the mixing properties of the overshooting. The diffusion coefficient decays exponentially at a depth of only 3500 km, which is enough to explain the solar Li decay. A more detailed discussion of the overshooting approach of the solar Li problem is given by Schlattl & Weiss (1999). The diffusion processes for lithium and beryllium during the MS evolution have been discussed by Piau, Randich & Palla (2003) basing on new data for open clusters older than 0.6 Gyr. Quite another approach for the Li problem is given by Rüdiger & Pipin (2001) who constructed the eddy diffusivity tensor for rotating anisotropic turbulence and solved the mean-field diffusion equation for such turbulence below the convection zone (see Fig. 3.31).

3.5.3 A Magnetic Theory of the Solar Tachocline

The transition from differential rotation in the solar convection zone to rigid rotation in the solar interior is rather sharp. The thickness of the transition layer is certainly smaller than (say) ten per cent of the solar radius, i.e. smaller than 70,000 km. From helioseismological data Charbonneau, Dikpati & Gilman (1999) derived a thickness of only 28,000 km for the tachocline (4% of solar radius), without any indication of a variation of this value between the equator and 60°. Following Spiegel & Zahn (1992) we use the term *tachocline* for this transition layer. The shape of the tachocline has been reported as prolate, i.e. its radius is least (by 14,000 km) at the equator (see also Gough & Kosovichev 1995). Basu & Antia (2001) report an extremely thin transition layer of about 10,000 km at the equator, and growing to the poles. The dissipation time of such a layer with $\nu \simeq 10 \ \mathrm{cm}^2/\mathrm{s}$ is about 2–3 Gyr. After the solar lifetime of 4.6 Gyr a delta-like tachocline would be spread to the value of 50,000 km. A theory is thus necessary to explain the present-day extreme sharpness of the tachocline.

The screening of given differences in the angular velocity in the convective zone is very ineffective (Stix 1981). Velocity differences produced at the base of the convection zone are still visible at the surface. They should also be 'visible' deep in the solar interior. The latter, however, is not found by helioseismology.

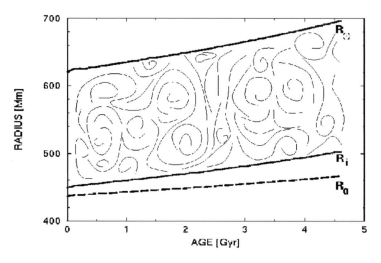

Figure 3.31: The temporal evolution of the location of the bottom of the convection zone (R_i) and the location R_0 where the Li burns. The solar model used is by Stix & Skaley (1990).

Moreover, the convection zone changed its thickness over the course of the Sun's evolution, and the Sun also decreased its rotation rate by one order of magnitude over its MS life due to continuous loss of the angular momentum to the solar wind (Stauffer & Soderblom 1991, Stępień 1991, see Mestel 1999 for full details). The spin-down of the solar-type stars is accompanied by Li depletion and Ca-activity decay (Skumanich 1972). Today the core rotates roughly with the same angular velocity as the surface. This implies a highly effective angular momentum exchange between core and convection zone that cannot be provided by the microscopic viscosity. That would require a (turbulent) viscosity of $\gtrsim 10^5$ cm^2/s (Rüdiger & Kitchatinov 1996) that is far beyond the maximum value of 10^3 cm^2/s for the turbulent diffusion coefficient imposed by the observed Li abundance (Baglin, Morel & Schatzman 1985, Spruit 1987). Spiegel & Zahn suggested an explanation of the problem by introducing a strong *anisotropy* of the turbulent viscosity produced by horizontal mixing of the radiative interior (see Michaud & Zahn 1998, Brun, Turck-Chièze & Zahn 1999). In Canuto's (1998) theory the tachocline is so thick that its upper half is within the convection zone itself. If this is true, a much more complicated discussion of Reynolds stress and turbulent heat conductivity must be applied.

However, a weak internal magnetic field may also solve the problem (Spruit 1987, Mestel & Weiss 1987). A weak poloidal field may well survive in the radiative interior against the Ohmic decay. The differential rotation then winds up a toroidal field. The resulting angular momentum transport by the Maxwell stress suppresses the differential rotation. An internal magnetic field was found to produce a very efficient coupling of the solar convection zone and the radiative interior (Charbonneau & MacGregor 1993). One can also demonstrate that it is equally robust in explaining the geometry of the solar tachocline (Rüdiger & Kitchatinov 1997) .

The surface rotation law is taken as the rotation law $\Omega(\theta)$ at the bottom of the convection zone, r_{in}. The axisymmetric magnetic field is represented by

$$\boldsymbol{B} = \left(\frac{1}{r^2 \sin\theta} \frac{\partial A}{\partial \theta} \ , \ -\frac{1}{r \sin\theta} \frac{\partial A}{\partial r} \ , \ B \right), \tag{3.91}$$

where $A = \text{const.}$ are the poloidal field lines, assumed to be given. The toroidal field B and the internal rotation law are provided by the equations

$$\eta \frac{\partial}{\partial\theta} \left(\frac{1}{\sin\theta} \frac{\partial \sin\theta B}{\partial\theta} \right) + r \frac{\partial}{\partial r} \left(\eta \frac{\partial Br}{\partial r} \right) = r \left(\frac{\partial\Omega}{\partial\theta} \frac{\partial A}{\partial r} - \frac{\partial\Omega}{\partial r} \frac{\partial A}{\partial\theta} \right),$$

$$\frac{\rho\nu}{\sin^3\theta} \frac{\partial}{\partial\theta} \left(\sin^3\theta \frac{\partial\Omega}{\partial\theta} \right) + \frac{1}{r^2} \frac{\partial}{\partial r} \left(r^4 \rho\nu \frac{\partial\Omega}{\partial r} \right) =$$

$$= \frac{1}{\mu_0 r^2 \sin^3\theta} \left(r \frac{\partial A}{\partial r} \frac{\partial \sin\theta B}{\partial\theta} - \sin\theta \frac{\partial A}{\partial\theta} \frac{\partial Br}{\partial r} \right). \tag{3.92}$$

The meridional flow is neglected. The typical circulation time for the baroclinic or Eddington–Sweet flow is about 10^{12} yr (Spruit 1987), much longer than the solar age, but we do not know the amplitude of the Lorentz force-induced meridional flow.

The boundary conditions are $\partial\Omega/\partial\theta = B = 0$ at $r = 0$ and $B = 0$ at $r = r_{\text{in}}$. We assume the internal poloidal field is located completely within the radiative interior. We do not assume any turbulence in the radiative zone, and adopt the microscopic diffusivities,

$$\eta = 10^{13}\, T^{-3/2}, \quad \nu_{\text{micro}} = 1.2 \cdot 10^{-16} \frac{T^{5/2}}{\rho}, \quad \nu_{\text{rad}} = 2.5 \cdot 10^{-25} \frac{T^4}{\kappa\rho} \tag{3.93}$$

(all in cm^2/s), with $\nu = \nu_{\text{micro}} + \nu_{\text{rad}}$, where ν_{micro} and ν_{rad} are the molecular and radiative viscosities (Spitzer 1978, Parker 1979). The diffusivities are specified according to the solar model of Stix & Skaley (1990), see also Fig. 3.32.

The magnetic field is considered as a dipole concentrated in the radiative interior, i.e.

$$A = B_0 \frac{r^2}{2} \left(1 - \frac{r}{r_{\text{b}}} \right)^2 \sin^2\theta, \tag{3.94}$$

where B_0 is the field amplitude.

Without the magnetic field the 'diffusive tachocline' is much too thick. However, as B_0 grows the tachocline becomes increasingly concentrated toward the top boundary, and becomes very thin for $B_0 \simeq 0.1$ mG (see Fig. 3.33). The toroidal field amplitude remains nearly constant as B_0 changes. It does not exceed a value of about 200 G. Gough & McIntyre (1998) estimate a field strength of only 1 G. There is no chance, however, to observe such small fields. Goode & Dziembowski (1993) argue for 1000 kG as an upper limit for toroidal magnetic fields while Basu (1997) and Antia (2002) provide 300 kG for this value. It by far even exceeds the expected amplitudes for a dynamo-induced field at the base of the convection zone of (say) $\lesssim 10$ kG.

The polar cap is an essential feature of the present model, i.e. the tachocline thickness at the poles is systematically larger than at lower latitudes ('Ferraro's law, see MacGregor & Charbonneau 1999). This effect seems indeed to exist in the helioseismological data (see Antia, Basu & Chitre 1998).

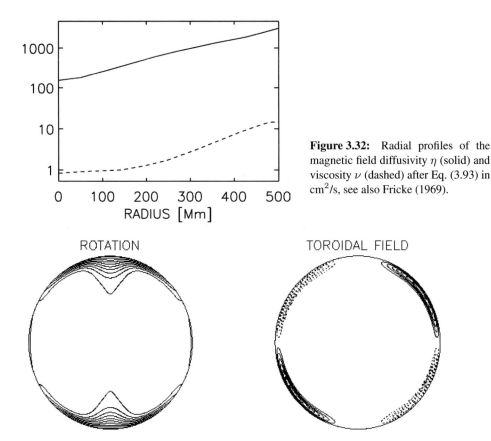

Figure 3.32: Radial profiles of the magnetic field diffusivity η (solid) and viscosity ν (dashed) after Eq. (3.93) in cm^2/s, see also Fricke (1969).

Figure 3.33: The angular velocity (*left*) and toroidal field isolines (*right*). Different line styles mean different signs of the toroidal field. The poloidal field strength is 10^{-4} G. The corresponding value of the maximum toroidal field is 209 G. From Rüdiger & Kitchatinov (1997).

The presented model ignores any meridional flow that should exist in the shear layer. For small viscosity the flow is only small if the geometry of the lines $\Omega = $ const. is cylindrical. This is obviously not the case. A well-developed meridional flow, however, will always modify the rotation law to fulfill the Taylor–Proudman theorem, i.e. $\Omega = \Omega(z)$. With other words, the solution of the tachocline problem consists in the search for the possibility to avoid both the constraints of Ferraro's law and the Taylor–Proudman theorem (Gough & McIntyre 1998, Garaud 2002). On the other hand, the resulting meridional flow should be slow enough and/or highly concentrated to the overshoot region to avoid the conflict with the Li observations.

The possibility of turbulence is another open question. The thin overshoot layer at the top of the tachocline is certainly turbulent, with a corresponding magnetic Prandtl number slightly smaller than unity (see Yousef, Brandenburg & Rüdiger 2003). The presented model also ignores the existence of internal instabilities within the tachocline. The tachocline of Spiegel & Zahn is turbulent with dominating horizontal intensities due to the negative buoyancy in the

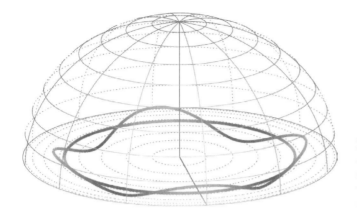

Figure 3.34: Illustration of an azimuthal Parker-type instability of the toroidal field with $m = 4$. From Schüssler (1996).

stable shear. Such turbulence, however, would produce a Λ-effect, so that new Ω-gradients will be maintained against the viscosity (see below).

The situation is probably even more complicated. Watson (1981) showed that the latitudinal shear between equator and pole is hydrodynamically stable (according to the Rayleigh criterion) only if it does not exceed 29%, which is remarkably close to the true value. Indeed, the *stability* of the shear could be the key to understanding the solar tachocline (see Dziembowski & Kosovichev 1987, Charbonneau, Dikpati & Gilman 1999 or Garaud 2001 for inviscid hydrodynamic instability studies).

Gilman & Fox (1997), working within the ideal MHD regime, include a toroidal magnetic field of free amplitude but without any radial dependence. The result is that a globally unstable magnetic mode with $m = 1$ appears. More recently also higher modes ($m < 7$) are reported as unstable (Dikpati & Gilman 2000). Similar results for nonaxisymmetric disturbances have been found earlier, within the thin flux tube approximation, by Schüssler et al. (1994), Caligari, Moreno-Insertis & Schüssler (1995), Caligari, Schüssler & Moreno-Insertis (1998) and for pre-MS and giant stars by Granzer (2002), see Fig. 3.34. If the strong toroidal fields are to be be kept in the tachocline, then the gas pressure and the centrifugal force must compensate the Lorentz force, so that consequences for the stellar structure and even for the shape of the star are unavoidable (Rempel, Schüssler & Tóth 2000). The main assumption is that the systematical empirical data for sunspot groups, i.e. orientation, tilt angle and proper motions, require toroidal field amplitudes of 10^5 G at the bottom of the convection zone. The equipartition value there or in the overshoot region under the convection zone is only 10^4 G (Schüssler 1996). The calculations show that the 10^5 G are strong enough to trigger an undulatory instability of Parker-type of the global toroidal flux-tube system in the tachocline. The magnetic tachocline theory presented above is too simple in order to include these instabilities. It must be extended to include axisymmetry and temperature stratifications.

With 3D box simulations of a thermally stably stratified but rotating flow Brandenburg & Schmitt (1998) have also attacked the interaction of toroidal magnetic fields, Coriolis force and (negative) buoyancy. The resulting magnetic-buoyancy instability under the influence of rotation yields an α-tensor that will be discussed in Chapt. 4.

With his hydrodynamical simulations Miesch (2003) probed the concept of Spiegel & Zahn (1992) that in the tachocline with its *anisotropic* viscosity the latitudinal differential rotation is reduced to zero over (say) 2–9% of the solar radius. Additionally, in his code the action of driven turbulence and/or the resulting meridional flow is included. The shear between equator and pole, $\partial\Omega/\partial\theta$, proved to be stable and did not lead to an instability. The turbulence, therefore, could only arise from artificially introduced random forcing.

If a rotation law is imposed on the layer with a finite equator-pole difference then it will be influenced by the meridional circulation, the forced turbulence in the density stratified layer *and* the sub-grid viscosities. It is known, however, that anisotropic viscosity in spherical coordinates with a dominating horizontal component does not suppress nonuniform radial rotation but generates a local superrotation (Wasiutynski 1946, Biermann 1951, Kippenhahn 1963). A latitudinal dependence of Ω, on the other hand, is smoothed by such anisotropic turbulence. This can be found exactly in the multitude of Miesch's results. That it is not possible to dissipate radial differential rotation with 2D horizontal turbulence is an old result (see Rüdiger 1977, McIntyre 1994). An important finding, however, is the nonexistence of hydrodynamic shear instabilities. Whether this is also true in the MHD-regime is still an open question.

4 The Stellar Dynamo

4.1 The Solar-Stellar Connection

Schwabe's 11-year sunspot cycle, Carrington's differential rotation (of the solar surface) and Spörer's law (of the equatorward migration of sunspots during the cycle, see Fig. 4.1) are the basic properties of the solar activity that must be explained by the dynamo theory. The parameters of the turbulence in the solar convection zone do not provide us with simple explanations of the 11-year timescale. It might be understood, however, as a diffusion time scale across the convection zone ($L \simeq 200,000$ km) if the magnetic diffusivity was $\eta_T \simeq 10^{12}$ cm^2/s. The diffusion time is reduced to 1 yr for the frequently used value of 10^{13} cm^2/s, and grows to 100 yr for the value 10^{11} cm^2/s (known from the observed rate of sunspot decay).

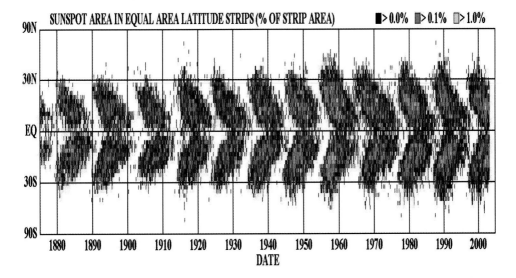

Figure 4.1: The butterfly diagram of sunspot statistics. Note in particular that the spots only exist very close to the equator; only a few spots appear at latitudes higher than (say) $33°$. There is a high symmetry of the phenomenon with respect to the equator but both cycle amplitude and length are obvious functions of time. Courtesy D. Hathaway.

Solar dynamo theory is reviewed here in the special context of the cycletime problem. A notable number of interesting phenomena have been investigated in the search for the solution

The Magnetic Universe: Geophysical and Astrophysical Dynamo Theory.
Günther Rüdiger, Rainer Hollerbach
Copyright © 2004 Wiley-VCH Verlag GmbH & Co. KGaA
ISBN: 3-527-40409-0

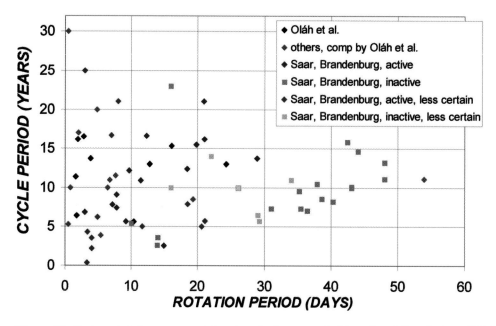

Figure 4.2: Cycle periods in years vs. rotation times in days, compiled from Saar & Brandenburg (1999) and Oláh, Kolláth & Strassmeier (2000). From Rüdiger & Arlt (2003).

of this problem: flux-tube dynamics, magnetic quenching, parity breaking and even chaos. Nevertheless, even the simplest observation – the solar cycle period of 11 yr – is hard to explain (see DeLuca & Gilman 1991, Stix 1991, Gilman 1992, Levy 1992, Schmitt 1993, Brandenburg 1994a, Weiss 1994, Rüdiger & Arlt 1996, Ossendrijver 2003). How can we understand the existence of the large ratio of the mean cycle period and the correlation time of the turbulence? The basic 'observations' are

- there is a factor of about 300 between the solar cycletime and the Sun's rotation period,
- this finding is confirmed by stellar observations (see Fig. 4.2),
- the convective turnover time near the base of the convection zone is very similar to the solar rotation period.

The problem of the large observed ratio of cycle and correlation times, $\tau_{\mathrm{cyc}}/\tau_{\mathrm{corr}} \gtrsim 10^2$, constitutes the primary concern of dynamo models. In a thick convection shell this number reflects the square of the ratio of the stellar radius to the correlation length, so that numbers of the order 100 are easily possible. For the thin boundary layer dynamo, however, the problem becomes more dramatic and is in need of an extra hypothesis.

4.1.1   The Phase Relation

Only the antisymmetric part of the poloidal component of the solar magnetic field oscillates and reverses its sign in the 11-year cycle (Stenflo & Vogel 1986). The magnetic reversal

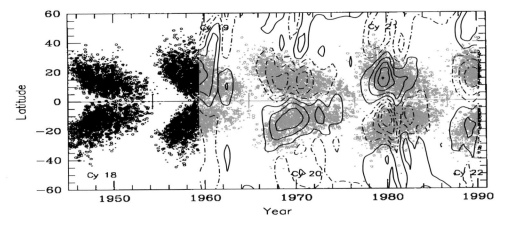

Figure 4.3: The anticorrelation of the radial magnetic field (solid: positive, dashed: negative) and the toroidal magnetic field (the negative polarity is marked with the cycle number). From Schlichenmaier & Stix (1995).

happens at the poles during the maximum of the sunspot activity. As the latter is located rather close to the equator one can be sure that the solar activity phenomenon is a global one.

Stix (1976) pointed out that at the solar surface a fundamental phase relation between the components of the global magnetic field can be observed. If the observed signs of the toroidal magnetic fields and the poloidal magnetic fields are correlated, a distinct antiphase relation

$$\bar{B}_r \cdot \bar{B}_\phi < 0 \tag{4.1}$$

appears in each cycle (Fig. 4.3). Here the sign of the toroidal magnetic field has been derived from the magnetic signature of the sunspots, which is conserved during the whole cycle. This relation forms a basic 'phase dilemma'. If a poloidal field component is sheared by a radial differential rotation then a correlation $\bar{B}_r \cdot \bar{B}_\phi < 0$ results only for $d\Omega/dr < 0$ (which is true in all disk dynamos). However, helioseismology leads to $d\Omega/dr > 0$ below the domain of the butterfly. The observed antiphase relation $\bar{B}_r \cdot \bar{B}_\phi < 0$ is thus a very strong argument against the idea that the Sun is a simple $\alpha\Omega$-dynamo. All $\alpha\Omega$-dynamos with the observed $d\Omega/dr > 0$ (in the butterfly region) lead to positive $\bar{B}_r \cdot \bar{B}_\phi$ – in contrast to the observations.

4.1.2   The Nonlinear Cycle

A linear theory is only concerned with the mean value of the oscillation frequency. The activity period of the Sun, however, varies strikingly about its average from one cycle to another (Fig. 4.4). Only a nonlinear theory will be able to explain the nonsinusoidal (chaotic or not) character of the activity cycle (Fig. 4.5). The variability of the cycle period can be expressed by the 'quality', $\omega_{cyc}/\delta\omega_{cyc}$, which is as low as 5 for the Sun (Wittmann 1978, Hoyng 1993).

The period of the solar cycle and its amplitude are far from constant. The most prominent activity drop was the Maunder minimum between 1670 and 1715 (see Spörer 1887). Measurements of ^{14}C abundances in sediments and long-lived trees provide much longer time series

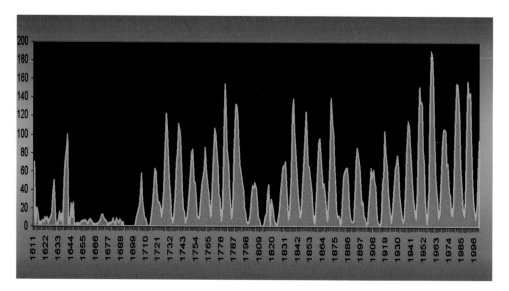

Figure 4.4: The time series of the sunspot numbers indicate a highly nonlinear behavior. From 1670–1715 a grand minimum ('Maunder minimum', see Eddy 1976) is seen.

than sunspot datasets. Schwarz (1994), Voss, Kurths & Schwarz (1996) and Voss et al. (1997) found a secular periodicity of 80–90 yr as well as a long-duration period of about 210 yr. The measurements of atmospheric ^{14}C abundances by Hood & Jirikowic (1990) suggested a periodicity of 2400 yr, which is also associated with a long-term variation of solar activity. The variety of frequencies found in solar activity may even indicate chaotic behavior, as discussed by Kurths et al. (1993, 1997), Rozelot (1995) and Knobloch & Landsberg (1996).

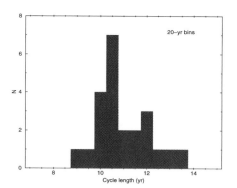
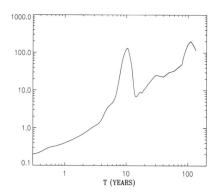

Figure 4.5: *Left*: The distribution of the solar cycle length does not approach a Dirac function, the 'quality' of the cycle only gives values of about 5. *Right*: The wavelet spectrum of the sunspot number time series shows two peaks for both 10 yr and 100 yr. From Frick et al. (1997).

The short-term cycle period appears to decrease at the end of a grand minimum according to a wavelet analysis of sunspot data by Frick et al. (1997). There is even empirical evidence that the magnetic cycles persisted through the solar Maunder minimum, as first found by Wittmann (1978, old sunspot data) and more recently by Beer, Tobias & Weiss (1998, Be data in ice cores, Fig. 4.6).

Charbonneau (2001) reports the 'odd-even effect' of the cycle amplitudes.   Since Schwabe's times (but not before) the odd-numbered cycles are stronger than their neighboring even-numbered cycles (his Fig. 1 or our Fig. 4.8) or – which is the same – the sunspot number of the complete 22-year cycle does not fluctuate too much from cycle to cycle.

4.1.3   Parity

The latitudinal distribution of the few sunspots observed during the end of the Maunder minimum was highly asymmetric (Spörer 1887, Ribes & Nesme-Ribes 1993, Sokoloff & Nesme-Ribes 1994). Short-term deviations from the north-south symmetry in regular solar activity are readily observable (Verma 1993), yet a 30-year period of asymmetry in sunspot positions as seen during the Maunder minimum remains a unique property of grand minima and should be associated with a parity change of the internal magnetic fields. All spots except two or three in this period appeared on the southern hemisphere (Spörer 1887). The first cycle after the minimum, with maximum in 1706, existed almost exclusively in one hemisphere (Knobloch, Tobias & Weiss 1998). Pulkkinen et al. (1999) studied the north-south asymmetry on the basis of a sunspot data set ranging from 1853 to 1996. They introduced a quantity for the mean latitude of sunspots taken as spatial averages over the hemispheres. The sum of both values defines a magnetic equator by mean latitudes. As shown in Fig. 4.7 (left) these calculations lead to a systematic long-term activity variation with a distinct north-south asymmetry. The period of this variation is about 90 yr.

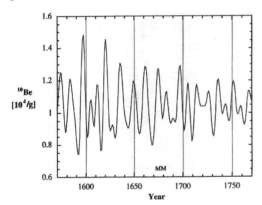

Figure 4.6: The variation of ^{10}Be in ice cores does not show any phase discontinuity of the cycle during the Maunder minimum (Beer, Tobias & Weiss 1998).

An interesting question concerns a possible systematic north-south asymmetry of the cycle length. Indeed, there are theories explaining the existence of grand minima with the simultaneous oscillation of modes with equatorial symmetry and antisymmetry. If this is true then systematic differences of the cycle length in the two hemispheres must appear. One has to check the sunspot data for the cycletimes in both hemispheres. Pelt et al. (2000) have considered the solar cycle as a traveling wave (see their Fig. 5) and define a mean cycletime in

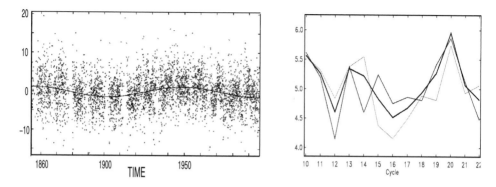

Figure 4.7: *Left*: The variation of the magnetic equator according to Pulkkinen et al. (1999). *Right*: The length of the sunspot maxima, with data separated for the northern and the southern (thin line) hemisphere. Also given is the average length (*bold line*). Courtesy J. Pelt.

this manner with an optimum of 10.84 yr. The length of the individual cycles around their maximum is given in Fig. 4.7 (right). Note the crossovers between 14 and 15 and at cycle number 21. Between the two crossovers the northern cycles were always longer than the southern ones.

Both cycle lengths and amplitudes in the dataset of Hathaway et al. (2003) are given in Fig. 4.8. There is no clear (anti-)correlation visible between cycle length and cycle amplitude as it is described by Hoyng (1993) and Ossendrijver, Hoyng & Schmitt (1996). As noted by Otmianowska-Mazur et al. (1997), the correlation coefficient, 0.39 in this case, is indeed rather small. Solanki et al. (2002) with 0.35 find a very similar value.

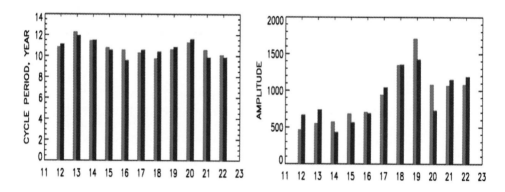

Figure 4.8: Cycle length (*left*) and cycle amplitudes (*right*) for 11 cycles separated for northern (green) and southern (blue) hemispheres. The data are the same as used in Fig. 3.10. Courtesy D. Hathaway.

4.1.4   Dynamo-related Stellar Observations

The activity cycle of the Sun is not exceptional: The observation of chromospheric Ca-emission of solar-type stars yields activity periods between 3 and 30 yr (Noyes, Weiss & Vaughan 1984, Baliunas & Vaughan 1985, Saar & Baliunas 1992, Baliunas et al. 1995). Up to 15% of the solar-type stars, however, do not show any significant activity. This suggests that even the existence of the grand minima is a typical property of cool MS stars like the Sun (Saar 1998). From ROSAT X-ray data Hempelmann, Schmitt & Stępień (1996) find that up to 70% of the stars with a constant level of activity exhibit a rather low level of coronal X-ray emission. HD 142373 with its X-ray luminosity of only $\log F_X = 3.8$ is a typical candidate. Saar (1998) studied UV and X emission for a sample of the 'flat activity' stars. Their Ca emission seems to reach a temperature-dependent basal value. Their coronal fluxes, however, are lower than those of the cyclic stars. The same is true for the emission of the transition region. We conclude that during a grand minimum the magnetic field is weaker than usual not only in the activity belts, but overall.

On the other hand, Hempelmann et al. (2003) have shown by comparison of CaII data from Baliunas et al. (1995) with ROSAT X-ray data for 61 Cyg A and B that their chromospheric periodic activity (A: 7.3 yr, B: 11.7 yr) is consistently followed by a coronal activity evolution with much higher amplitude. Unlike chromospheric activity, the X-ray data do not show any basal value of nonmagnetic origin, so that X-ray observations seem to be a proper tool to observe magnetic-induced stellar activity as suggested by Linsky (1994).

The cycle statistics are not yet very clear. Saar & Brandenburg (1999) report that for the relation

$$\omega_{\mathrm{cyc}} \propto \Omega^n \tag{4.2}$$

the full data set leads to the very small value $n \simeq -0.09$. The value discussed by Noyes, Weiss & Vaughan (1984) was $n = 1.28$. Baliunas et al. (1996) find $n \simeq 0.47$ for young stars and $n \simeq 1.97$ for old stars, and Lanza & Rodonò (1999) derive a value of $n \simeq 0.36$ for RS CVn systems. Saar & Brandenburg (2001) report values of 1.15, 0.8 and 0.4 for their groupings 'inactive', 'active' and 'superactive'. The steepest relation (4.2) with $n \approx 2$–2.5 is by Ossendrijver (1997) for slowly rotating stars with well-defined periods. All the reported exponents except the smallest are positive.

The situation can be seen in Fig. 4.2, where the data of Saar & Brandenburg (1999) and Oláh, Kolláth & Strassmeier (2000) have been used. Only the slow-rotating 'inactive' stars of Saar & Brandenburg (1999) form an own branch, while there is no rule for fast-rotating stars. We take from this plot a value of (say) $n \simeq 0.7$. The overall finding, however, seems to be that the rotation itself does not seem the only determinant of the cycletime. The dependence of the cycle period on the rotation period is also rather weak. Below we shall demonstrate that such a weak dependence is characteristic of the influence of the meridional flow on the cycletime.

EK Dra is – as a member of the Pleiades group – a very young version of the Sun with its rapid rotation (2.7 d). Flares, and X-ray and microwave emission are already observed. Doppler tomography revealed the existence of a nearly polar spot (Fig. 4.9, Strassmeier & Rice 1998). The main question is whether such a young sun already exhibits cyclic activity or not, and if yes whether the cycletime is long or short. The EK Dra light curve indeed suggests a secular dimming since 1975 (Fröhlich et al. 2002). This is not exceptional: in Fig. 4.10 a

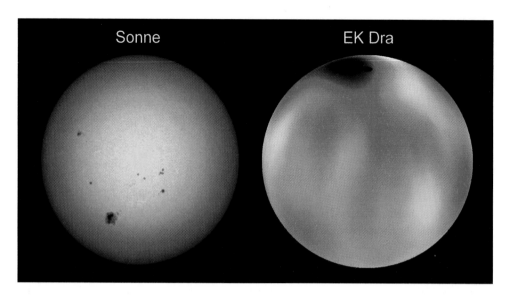

Figure 4.9: Comparison of a young sun (Doppler image of EK Dra, *right*) and our present-day Sun (*left*). Note the different positions and sizes of the spots. The rotation period of EK Dra is 2.7 d. Courtesy K. G. Strassmeier.

similar long-term analysis is given for the K0 giant HK Lac. There is also a secular dimming, from 1990 on followed by an impressive rise of the brightness.

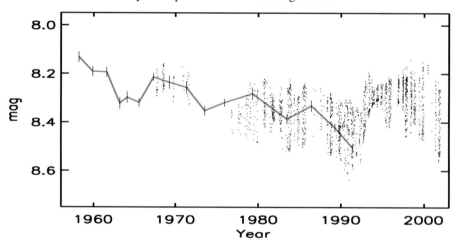

Figure 4.10: Light curve of HK Lac ($\tau_{rot} = 24$ d) derived from 540 blue plates of the Sonneberg Sky Patrol archive. The dots are photoelectric data. The 10-year variation is statistically significant. Note the grand minimum of 1990. Courtesy H.-E. Fröhlich.

Generally, the rotational brightness modulation of many active stars indicates the existence of large cool starspots with sizes differing strongly from the solar case. An impressive example is given in Fig. 4.11, detected by Strassmeier (1999). As we shall show, for many solar-type dynamo models the toroidal field belts are formed in the polar regions, so that the existence of polar spots does not constitute a surprise. The question is rather why the spots observed at the solar surface are located so close to the equator. On the other hand, Schüssler (1996) argues that the different latitude distribution of the spots is due to the difference of the Sun and the active stars. The latter are fast-rotating giants or subgiants in RS CVn systems or T Tauri systems with deep convection zones. Schüssler suggests the existence of different types of dynamos in the different types of stars (see Fig. 4.12). His main point, however, is to look for different stability regimes of magnetic flux tubes in these stars (see Caligari, Moreno-Insertis & Schüssler 1995).

The basic question whether the starspots are magnetic or not has been discussed in detail by Solanki (2002). There are many arguments in favor of a magnetic character of the starspots but the direct measurement of the magnetic field of individual spots proves to be difficult.

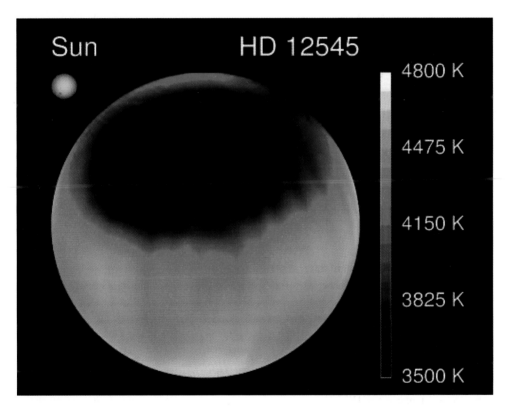

Figure 4.11: The largest starspot ever observed belongs to the K0III RS CVn star HD 12545. The Sun (left corner, with spot!) is given for a direct comparison. Note the temperature scale on the right. Courtesy K. G. Strassmeier.

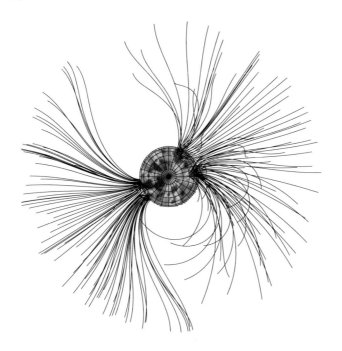

Figure 4.12: Pole-on representation of the magnetic field of AB Dor. Note the dominance of the nonaxisymmetric magnetic field component. Rotation period of AB Dor is 0.5 d. From Jardine et al. (2002).

4.1.5   The Flip-Flop Phenomenon

Photometry and surface imaging can be used to obtain the spot distribution at stellar surfaces. The light curves, however, only contain the longitudinal brightness informations. If this method is applied to active RS CVn variables then the existence of permanent active longitudes is found. Sometimes there is a sudden change of the active longitude to the opposite side of the star. This 'flip-flop' phenomenon appears to be the most innovative observational finding in the present-day stellar-magnetism research (see Tuominen, Berdyugina & Korpi 2002). It is a change of the spot position by 180° in longitude in a rather short time interval. Mathematically speaking, it is the simultaneous existence of an oscillating axisymmetric mode and a stationary nonaxisymmetric mode. Originally it was discovered in the light curve of FK Com (Jetsu et al. 1991, cycletime 3.2 yr, see Fig. 4.13) but now there is already a (small) sample of stars known, including also the single dwarf star LQ Hya (cycletime 2.6 yr, Table 4.1). Korhonen et al. (2001) have shown, with surface images before and after the flip-flop, that the spot groups at opposite longitudes do indeed change their strength rather than drift across the surface (Fig. 4.14).

Photometrically monitored active longitude structures are known to exist over many years (e.g. Henry et al. 1995). There is a long-standing discussion about the existence of active longitudes in the solar activity. Recently, Mordvinov & Willson (2003) found patterns of radiative excesses and deficits in the total solar irradiance reflecting the existence of long-lived magnetic-active longitudes. The pattern with radiative excess proved to be concentrated around two active longitudes separated by 180°. Moreover, Berdyugina & Usoskin (2003),

Table 4.1: List of stars with active longitudes and flip-flop phenomenon (Berdyugina, Korhonen & Tuominen 2001).

	Sp	$M/M_\odot$	$R/R_\odot$	$\Omega/\Omega_\odot$	remarks
II Peg	K2 IV	0.8	3.4	3.8	RS CVn
IM Peg	K3 III	1.5	13.3	0.95	RS CVn
El Eri	G5 IV	> 1.4	> 3.4	13.3	RS CVn
σ Gem	K2 III	~ 2	~ 13		RS CVn
HR 7275	K2 IV	~ 1	> 8	0.90	RS CVn
FK Com	G5 III	~ 1	~ 10	10.8	single?
LQ Hya	K2 V	~ 0.8	~ 0.8	16.3	single, young

with sunspot group data for 120 yr, find the active longitude drifting with the rotation at the place of the mean latitude of the sunspots. The most dramatic finding, however, is the alternating dominance of the two active longitudes, oscillating with periods of 3.8 yr (north) and 3.65 yr (south). The flip-flop phenomenon seems, therefore, also to exist in the Sun. Flip-flop frequency and main cycle frequency have here rather different values.

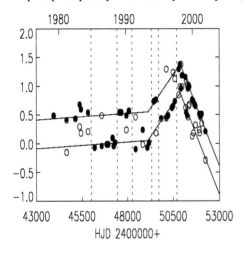

Figure 4.13: The flip-flop phenomen of FK Com demonstrated with the phases of large spots (filled circles) and small spots (open circles). There are two active longitudes with about 180° separation (solid lines). Courtesy I. Tuominen.

4.1.6   More Cyclicities

Coronal Solar Cycle

The transition layer between chromosphere and corona itself as well as the solar corona are subject to the magnetic-activity cycle. The cyclic variation of the 10-cm radio flux is shown in Fig. 4.15, and also the X-ray images of the coronal gas reflect the basic solar cycle. The X-ray corona basically exists in the activity maximum apart from a few coronal bright points that are also present in the minimum. The high degree of inhomogeneity of the coronal temperature distribution seems to exclude heating sources such as compression-heat produced by acoustic

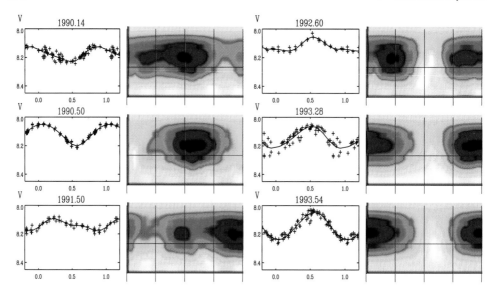

Figure 4.14: FK Com light-curves and spot maps for 1990–1993 from Korhonen, Berdyugina & Tuominen (2002). Within two years the minimum at phase 0.5 developes to a maximum. The activity suddenly changes by 180°.

waves. It is the magnetic field[1] that obviously heats the corona and determines its geometrical structure (see Solanki et al. 2003, Rosner 2003).

Solar p-Mode Oscillations

The enormous progress in measuring the solar p-mode frequencies, up to an accuracy of 10^{-5}, forms new restrictions on the theoretical models. It has been established that the computed frequencies for the solar p-mode oscillations systematically exceed the observed values (Brown 1984, Zhugzhda & Stix 1994, Gabriel & Carlier 1997, Rosenthal 1998, see Fig. 4.15). The observed frequencies are therefore considered as *redshifted*. The theoretical frequency values are too high and an extra effect is necessary to lower them (Christensen-Dalsgaard & Thompson 1997). As shown in Fig. 4.15, the details of the phenomenon are more complicated. The redshift only exists for high frequencies. For frequencies smaller than 2.5–2.6 mHz there is a transition to *blueshift* with a maximum of a few μHz (see also Guzik & Swenson 1997). Two opposite effects seem to influence the eigenmodes and dominate in different domains of the frequency diagram.

There are indications that the corrections must be searched for in the outer layers of the Sun. First, the difference increases with increasing radial order n. The higher the radial order n, the closer to the surface the upper turning point of the waves lies. Because the frequencies depend very sensitively on the location of this reflection point, this is a first hint that the solar model has to be improved in this outer region. Moreover it can be found that

---

[1] E. Priest (1999): "YOHKOH has revealed a whole new MHD world"

the difference between the observed and the calculated frequencies increases with increasing harmonic degree l, i.e. the more the waves are confined in the outer 'convective' part of the Sun. Specifically, the difference is as low as a few μHz for low degree modes ($l < 10$) and increases up to the order of a few tens of μHz for high degrees ($l > 100$) (Guzik & Swenson 1997).

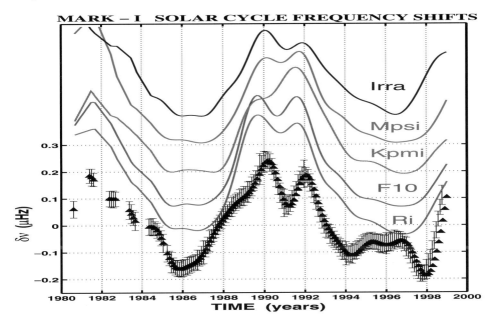

Figure 4.15: The frequency shift of low degree modes over the cycle 22 (black). For comparison also other solar cycle indicators are plotted (Ri: sunspot number; F10: radio flux; Kpmi: Kit Peak magnetic index; Mpsi: magnetic plage strength; Irra: irradiance from SOHO). Courtesy P.L. Pallé.

Therefore, turbulence has become one of the explanations for the p-mode frequency shifts (Stein, Nordlund & Kuhn 1988, Murawski & Roberts 1993, Rosenthal 1998, Böhmer & Rüdiger 1998, Stix & Zhugzhda 2004). The turbulent pressure modifies the pressure in the convective region, and this modification is largest in the outermost part of the Sun where the convection cells have the highest velocities.

The lineshift phenomenon varies over the cycle, as observed by Régulo et al. (1994) and Jiménez-Reyes et al. (1998). The redshift disappears in the activity minimum and is largest in the maximum epoch (see Libbrecht & Woddard 1990). In the maximum a strong magnetic suppression of the turbulent pressure may produce the frequency redshift of the low-l p-modes. Its time evolution can be followed in Fig. 4.15.

Climate Research

Satellite observations have shown that the total irradiance from solar minimum to solar maximum increases by about 0.1% (Fig. 4.16), resulting in a temperature change of $\sim 0.2°$C (Lean 1994).

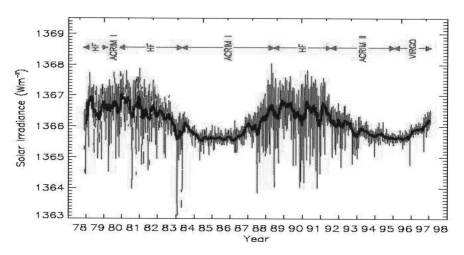

Figure 4.16: The variation of solar irradiance over the activity cycle. At the solar activity minimum the solar radiation production is also minimum – with consequences for the irradiance during the Maunder minimum. Courtesy C. Fröhlich.

Besides this 11-year timescale there is also a timescale of centuries in the solar radiation output, which might be even more interesting. Lean (1994) and Hoyt, Schatten & Nesme-Ribes (1994) focused the general interest of climate research on the long-term reconstructions of solar variability. Isotopes in ice cores and tree rings provide much longer time series of data related to the solar activity. During the grand minima the solar luminosity decreases by (say) 0.2%, so that by this effect the temperature reduction is about 0.3–0.5°C (see Weiss 1997). A time series of temperature in central England was investigated by Baliunas et al. (1997), who indeed found a cooling by 0.3°C during the Maunder minimum (Fig. 4.17).

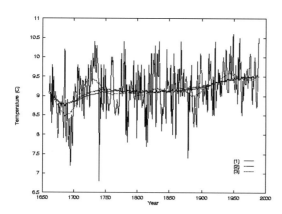

Figure 4.17: A wavelet reconstruction of a temperature record in central England by Baliunas et al. (1997). The peak-to-peak variation of the mean temperature profile is 0.8°C. Courtesy P. Frick

Solanki & Fligge (1998, 1999) reconstructed the solar irradiance over decades and centuries in great detail for a comparison with terrestrial temperature records. Indeed, a fluctuation of 0.2% in the irradiance led to a temperature *increase* of 0.5°C (with ∼ 10 yr delay, Fig. 4.18).

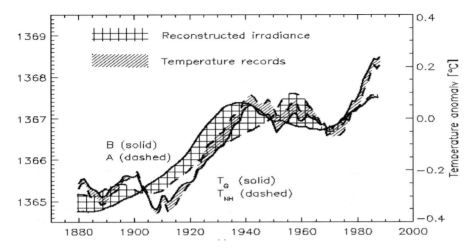

Figure 4.18: The temporal variation of total solar irradiance and terrestrial temperature for the past 100 years. The solid line represents irradiance reconstructions from the activity. Until 1980 the Earth's temperature follows the solar activity-dominated forcing. After 1980 the global warming exceeds that by the solar forcing (Solanki & Fligge 1998).

RS CVn Activity Cycles

The RS CVn systems as defined by Hall (1976) are characterized by solar-like activity that has also been observed in X-rays, UV, optical and radio spectral domains (Linsky 1988, Rodonò 1992, Guinan & Giménez 1992, Strassmeier et al. 1993).

On the other hand, the eclipse times of the (detached) close binary systems[2] of (say) 4–5 days orbital period varies by about 10 s with the timescale of an activity cycle (Fig. 4.19). The modulation timescale for RS CVn binaries goes from a few to several decades, with a median value of around 50 yr. The normalized amplitude of the modulation is of order 10^{-5} (Hall 1989, 1990) and its period P_{mod} is twice the spot-cycle length (Lanza & Rodonò 2004).

The possibility that a time-variable angular momentum loss, due to magnetic activity, may be responsible for alternate orbital period changes was proposed by DeCampli & Baliunas (1979). They concluded that this explanation was implausible because of the large mass loss required and also because of the long timescale needed to couple the variation of the stellar rotation to the orbital motion through tidal effects. The characteristic timescale for spin-orbit coupling in RS CVn binaries turns out to be of the order of 10^3 yr, i.e., about two orders of magnitude longer than the periods of the observed short-term modulations (Zahn 1989).

Matese & Whitmire (1983) proposed a mechanism on the basis of a cyclic change of the gravitational quadrupole moment of active stars. Linsky (1999) reports the direct and indirect indications for observable magnetic fields at the surface of the stars. The orbital period modulation has then been viewed as due to the magnetic stress in an oscillating dynamo (Lanza, Rodonò & Rosner 1998, Rüdiger et al. 2002).

---

[2] with a mass-donating late-type component

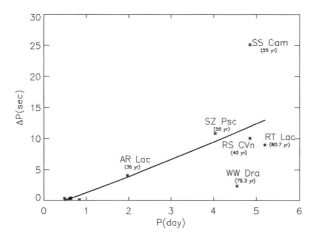

Figure 4.19: Data for close binary systems: period modulation (a few seconds per rotation over dozens of years), orbital period and modulation cycletimes P_{mod}. Data from Lanza & Rodonò (1999).

Lanza & Rodonò (1999) for the orbital modulation of close binaries due to a modulation of the quadrupole moment[3] J_2 find

$$\frac{\delta P}{P} = -3 \left(\frac{R}{a}\right)^2 \delta J_2,\tag{4.3}$$

where a is the semi-major axis of the orbit. Note that the orbital period decreases when the quadrupole moment increases and the star becomes more oblate (J_2 increases). The typical value $a \simeq 4R$ holds for RS CVn stars. If the potential series of Legendre polynomials, $\psi = \psi_0 + \psi_2 P_2 + \ldots$, is introduced one finds

$$\frac{\delta P}{P} = \frac{3}{16} \frac{R}{GM} \delta\psi_2(R) \simeq 4 \times 10^{-16} \delta\psi_2(R),\tag{4.4}$$

with ψ_2 in c.g.s. units. Positive (negative) $\psi_2(R)$ describe prolate (oblate) spheroids (see Ulrich & Hawkins 1981).

In order to obtain an effect of $\delta P/P \simeq 10^{-5}$ we thus have to find the magnetic field amplitude that produces a potential modulation of $\delta\psi_2 \simeq 2.5 \times 10^{10}$ c.g.s. It is shown by Rüdiger et al. (2002) that an oscillating α^2-dynamo (an exotic case) needs an amplitude of more than 100 kG to produce such an $O(10^{-5})$ effect. As stressed by Lanza, Rodonò & Rosner (1998) the detailed theory of the RS CVn period modulation may easily lead to a better understanding of even the geometry of the stellar dynamo.

Following Applegate (1992), one can also argue that small magnetic-induced changes of the rotation law lead to small changes of the centrifugal force leading to small quadrupole moments. As is known from the theory of the solar torsional oscillations, one then additionally has to solve the time-dependent equation for the angular momentum (Schüssler 1981, Yoshimura 1981).

---

[3] of the gravitation potential of the active component

Not surprisingly, also for another class of almost fully convective stars solar-type activity cycles are reported (Ak, Ozkan & Mattei 2001). Cyclical variations may exist in dwarf novae due to possible activity cycles (~ 10 yr) of the late-type companions. For SS Cyg with data known over 100 yr a cycletime was found of about 7.2 yr (see also Hempelmann & Kurths 1990).

4.2   The α-Tensor

Kinematic dynamo theory in turbulent media utilizes only one equation to advance the large-scale magnetic field in time, i.e. the mean-field induction equation

$$\frac{\partial \bar{B}}{\partial t} = \nabla \times \left(\bar{u} \times \bar{B} + \mathcal{E} \right).$$ (4.5)

Often only a differential rotation represents the mean flow $\bar{u}$; any meridional flow shall be introduced later. The turbulent electromotive force (EMF),

$$\mathcal{E} = \langle u' \times B' \rangle,$$ (4.6)

simultaneously contains induction, α_{ij}, and dissipation, η_{ijk}, which are the coefficient tensors in the series expansion

$$\mathcal{E}_i = \alpha_{ij} \bar{B}_j + \eta_{ijk} \bar{B}_{j,k} + \dots .$$ (4.7)

Both tensors are pseudotensors. While for η_{ijk} an elementary isotropic pseudotensor exists ('ϵ_{ijk}'), the same is not true for α_{ij}. The simplest possibility appears for inhomogeneous turbulence characterized by only one preferred direction g, say. The product $\epsilon_{ijk} g_k$ provides an antisymmetric contribution to α_{ij} that, however, only plays the role of a magnetic transport[4] along g. More interesting is the symmetric part of α_{ij}. As it must also be a pseudotensor, it can only exist in connection with global rotation. An odd number of Ω's is, therefore, required for the α-tensor that is only possible with an odd number of g. The α-effect can thus only exist in stratified and rotating turbulences (Parker 1955, Steenbeck, Krause & Rädler 1966). The first formula reflecting this situation,

$$\alpha = c_\alpha \frac{\ell_{\text{corr}}^2}{H_\rho} \Omega \cos \theta,$$ (4.8)

was given by Steenbeck & Krause (1966) and Krause (1967) with H_ρ the density scale height. Evidently, α is a complicated effect where the effective value might really be very small; the unknown factor c_α in (4.8) may be smaller than unity. The strength of this effect was estimated in many numerical and analytical studies, for both convectively unstable and stable stratifications. The magnitudes of the α-effects do not reach the given estimate in many cases, i.e. the dimensionless factor c_α seems really to be much smaller than unity.

In the framework of the Second Order Correlation Approximation (SOCA) the sign of the mean kinetic helicity

$$\mathcal{H}_{\text{kin}} = \langle u' \cdot \nabla \times u' \rangle$$ (4.9)

---

[4] compare this expression with $\bar{u} \times \bar{B}$ in Eq. (4.5)

is related to the sign of the α-effect. For isotropic α-effect negative kinetic helicity leads to positive α and vice versa, i.e.

$$\alpha \simeq -\tau_{\mathrm{corr}} \mathcal{H}_{\mathrm{kin}} \tag{4.10}$$

(Krause & Rädler 1980). The situation is more complicated if the tensor character of the α-effect is taken into account, but in this case also the important azimuthal component of the α-tensor, $\alpha_{\phi\phi}$, is positive in the northern hemisphere.

The α-tensor is not isotropic, and not even symmetric by definition. By comparison with the mean-field EMF $\bar{\boldsymbol{u}} \times \bar{\boldsymbol{B}}$ in Eq. (4.5) one finds that the antisymmetric parts of the α-tensor form advection terms. Let us thus split the α-tensor in the two parts $\alpha_{ij} = \alpha_{ij}^{\mathrm{S}} + \alpha_{ij}^{\mathrm{A}}$, with $\alpha_{ij}^{\mathrm{S}} = \alpha_{ji}^{\mathrm{S}}$ and $\alpha_{ij}^{\mathrm{A}} = -\alpha_{ij}^{\mathrm{A}}$. Then we can write $\alpha_{ij}^{\mathrm{A}} = \epsilon_{ijk}\gamma_k$, where γ denotes the advection velocity. In spherical coordinates (r, θ, ϕ) the radial advection can be written as $\gamma_r = \alpha_{\phi\theta} = -\alpha_{\theta\phi}$, i.e.

$$\alpha^{\mathrm{A}} = \begin{pmatrix} 0 & 0 & 0 \\ 0 & 0 & -\gamma_r \\ 0 & \gamma_r & 0 \end{pmatrix}. \tag{4.11}$$

Cylindrical coordinates are considered later in Eq. (4.41). The determination of both the symmetric and the antisymmetric part of the α-tensor is the essence of the present section.

4.2.1   The Magnetic-Field Advection

The mean magnetic field can here be considered as uniform. For simplicity we restrict the calculations to the first-order terms in the scale ratio ℓ_{corr}/L, with L being a typical spatial scale of mean fields. In this section the rotation rate is assumed to be zero.

Working with Fourier transforms we derive from the linearized induction equation

$$\frac{\partial \boldsymbol{B}'}{\partial t} - \eta\Delta\boldsymbol{B}' = \nabla \times \left(\boldsymbol{m}' \times \frac{\bar{\boldsymbol{B}}}{\rho(x)} \right) \tag{4.12}$$

the relation

$$(-\mathrm{i}\omega + \eta k^2)\hat{\boldsymbol{B}}(\boldsymbol{k}, \omega) = \mathrm{i}k_j \int \hat{\rho}^{-1}(\boldsymbol{q}) \left\{ \bar{B}_j \hat{\boldsymbol{m}}(\boldsymbol{k} - \boldsymbol{q}, \omega) - \hat{m}_j(\boldsymbol{k} - \boldsymbol{q}, \omega)\bar{\boldsymbol{B}} \right\} \mathrm{d}\boldsymbol{q}, \tag{4.13}$$

where the divergence-free momentum density $\boldsymbol{m} = \rho\boldsymbol{u}'$ has been used instead of the random velocity $\boldsymbol{u}'$. Equation (4.12) is written without Hall effect; its inclusion, of course, would lead to much more complicated expressions, which are not yet discussed in detail (Helmis 1968, Mininni, Gómez & Mahajan 2002). The Fourier transforms again are defined in accordance with Eq. (3.38). After multiplication and averaging Eq. (4.13) directly leads to

$$\rho^2 \alpha_{ij} = -\epsilon_{ipe} \int \frac{\mathrm{i}k_j}{-\mathrm{i}\omega + \eta k^2} \langle \hat{m}_p(\boldsymbol{k}, \omega)\hat{m}_e(\boldsymbol{k}', \omega') \rangle \mathrm{d}\boldsymbol{k}\mathrm{d}\boldsymbol{k}'\mathrm{d}\omega\mathrm{d}\omega' +$$

$$+ \epsilon_{ipe}G_f \int \frac{\partial}{\partial k_f} \left(\frac{k_j}{-\mathrm{i}\omega + \eta k^2} \right) M_{pe}^0(\boldsymbol{k}, \omega) \mathrm{d}\boldsymbol{k}\mathrm{d}\omega + \epsilon_{ipj}G_f \int \frac{M_{fp}^0(\boldsymbol{k}, \omega)}{-\mathrm{i}\omega + \eta k^2} \mathrm{d}\boldsymbol{k} \, \mathrm{d}\omega.$$

$\hat{M}^0$ is the homogeneous part of the spectral tensor for the momentum fluctuations.

We must now turn to the momentum equation

$$\rho\left(\frac{\partial \boldsymbol{u}}{\partial t} + (\boldsymbol{u}\nabla)\boldsymbol{u}\right) = -\nabla P + \nabla\cdot\pi + \frac{1}{\mu_0}(\nabla\times\boldsymbol{B})\times\boldsymbol{B} + \boldsymbol{f}. \tag{4.14}$$

For $\Omega = 0$ the Fourier transform of its linearized form reads

$$\left(-\mathrm{i}\omega + \nu k^2 + \mathrm{i}\nu(\boldsymbol{G}\cdot\boldsymbol{k})\right)\hat{\boldsymbol{m}} - \frac{\mathrm{i}}{\mu_0}(\boldsymbol{k}\cdot\bar{\boldsymbol{B}})\hat{\boldsymbol{B}} = \hat{\boldsymbol{f}}^{\mathrm{s}}, \tag{4.15}$$

where $\boldsymbol{f}^{\mathrm{s}}$ is the solenoidal part of the *fluctuating force* $\boldsymbol{f}'$ driving the turbulence: $\hat{f}_i^{\mathrm{s}} = (\delta_{ij} - k_i k_j/k^2)\hat{f}_j$. Density fluctuations as the source of buoyancy have been neglected here.

Note the derivation of $\mathcal{E}$ as linear in the scale ratio ℓ_{corr}/L. This is why $\boldsymbol{G}$ from Eq. (3.37) is considered as constant; allowance for spatial variations of $\boldsymbol{G}$ would involve higher orders of ℓ_{corr}/L. The magnetic field fluctuations can be replaced in Eq. (4.15) by using Eq. (4.13). This yields

$$\left(-\mathrm{i}\omega + \nu k + \frac{(\boldsymbol{k}\cdot\boldsymbol{V}_{\mathrm{A}})^2}{(-\mathrm{i}\omega + \eta k^2)} + \mathrm{i}\nu\boldsymbol{G}\cdot\boldsymbol{k}\right)\hat{\boldsymbol{m}}-$$

$$-G_j\frac{(\boldsymbol{k}\cdot\boldsymbol{V}_{\mathrm{A}})^2}{-\mathrm{i}\omega + \eta k^2}\frac{\partial\hat{\boldsymbol{m}}}{\partial k_j} - \frac{\mathrm{i}(\boldsymbol{k}\cdot\boldsymbol{V}_{\mathrm{A}})\boldsymbol{V}_{\mathrm{A}}}{-\mathrm{i}\omega + \eta k^2}(\boldsymbol{G}\cdot\hat{\boldsymbol{m}}) = \hat{\boldsymbol{f}}^{\mathrm{s}}, \tag{4.16}$$

with $\boldsymbol{V}_{\mathrm{A}} = \bar{\boldsymbol{B}}/\sqrt{\mu_0\rho}$ as the Alfvén velocity. Equation (4.16) must be solved with the scale ratio ℓ_{corr}/L as a small parameter. At lowest order one finds that $\hat{\boldsymbol{m}} = \hat{\boldsymbol{m}}^{(0)}/N$, with

$$N = 1 + \frac{(\boldsymbol{k}\cdot\boldsymbol{V}_{\mathrm{A}})^2}{(-\mathrm{i}\omega + \nu k^2)(-\mathrm{i}\omega + \eta k^2)}, \tag{4.17}$$

in which $\hat{\boldsymbol{m}}^{(0)} = \hat{\boldsymbol{f}}^{\mathrm{s}}/(-\mathrm{i}\omega + \nu k^2)$ is thought of as the momentum density for an 'original' turbulence existing without any magnetic field.

In the next step all the first-order terms in the scale ratio are collected, i.e.

$$\hat{\boldsymbol{m}} = \left\{1 - \frac{\mathrm{i}\nu(\boldsymbol{G}\cdot\boldsymbol{k})}{N(-\mathrm{i}\omega + \nu k^2)}\right\}\frac{\hat{\boldsymbol{m}}^{(0)}}{N} + \frac{\mathrm{i}(\boldsymbol{k}\cdot\boldsymbol{V}_{\mathrm{A}})(\boldsymbol{G}\cdot\hat{\boldsymbol{m}}^{(0)})}{(-\mathrm{i}\omega + \nu k^2)(-\mathrm{i}\omega + \eta k^2)}\frac{\boldsymbol{V}_{\mathrm{A}}}{N^2}+$$

$$+\frac{\mathrm{i}(\boldsymbol{k}\cdot\boldsymbol{V}_{\mathrm{A}})^2 G_j}{(-\mathrm{i}\omega + \nu k^2)(-\mathrm{i}\omega + \eta k^2)}\frac{1}{N}\frac{\partial}{\partial k_j}\frac{\hat{\boldsymbol{m}}^{(0)}}{N}. \tag{4.18}$$

With this relation the spectral tensor for the momentum density can be expressed in terms of the 'original' spectral tensor (3.41) through a linear relation. After such reductions we find $\mathcal{E} = \boldsymbol{\gamma}\times\bar{\boldsymbol{B}}$ with

$$\boldsymbol{\gamma} = \boldsymbol{U}^{\mathrm{dia}} + \boldsymbol{U}^{\mathrm{buo}}, \tag{4.19}$$

where[5]

$$\boldsymbol{U}^{\mathrm{dia}} = -\nabla\int\mathcal{K}^{\mathrm{dia}}(k,\omega,\bar{\boldsymbol{B}})\frac{\eta k^2\hat{Q}_{ll}}{\omega^2 + \eta^2 k^4}\,\mathrm{d}k\,\mathrm{d}\omega \tag{4.20}$$

---

[5] the notation of the spectral tensor (3.63) is used with the transformation $E = 8\pi k^2\rho^2\hat{Q}_{ll}$

and

$$U^{\mathrm{buo}} = G \int \mathcal{K}^{\mathrm{buo}}(k, \omega, \bar{B}) \frac{\eta k^2 \hat{Q}_{ll}}{\omega^2 + \eta^2 k^4} \mathrm{d}\boldsymbol{k} \, \mathrm{d}\omega. \tag{4.21}$$

The effective velocities U^{dia} and U^{buo} are consequences of the nonuniformity of the turbulence intensity and density where the former is attributed to the so-called diamagnetic pumping. The velocities (4.20) and (4.21) depend on the magnetic field through the kernels $\mathcal{K}^{\mathrm{dia}}$ and $\mathcal{K}^{\mathrm{buo}}$, which are given by Kitchatinov & Rüdiger (1992). We are here particularly interested in the discussion of the limiting cases of weak and strong magnetic fields.

For $\bar{B} = 0$ the known linear expression for the velocity of the diamagnetic pumping is reproduced

$$U^{\mathrm{dia}} = -\frac{1}{2} \nabla \eta_{\mathrm{T}}, \tag{4.22}$$

with

$$\eta_{\mathrm{T}} = \frac{1}{3} \int \frac{\eta k^2 \hat{Q}_{ll}}{\omega^2 + \eta^2 k^4} \mathrm{d}\boldsymbol{k} \, \mathrm{d}\omega \tag{4.23}$$

as the linear turbulent magnetic diffusivity.

Series expansion of $\mathcal{K}^{\mathrm{buo}}$ in terms of the magnetic field yields

$$U^{\mathrm{buo}} = -G \frac{2\bar{B}^2}{15\mu_0\rho} \int \frac{\eta k^4 (\nu\eta k^4 - \omega^2)\hat{Q}_{ll}}{(\omega^2 + \nu^2 k^4)(\omega^2 + \eta^2 k^4)^2} \mathrm{d}\boldsymbol{k} \, \mathrm{d}\omega \tag{4.24}$$

as the weak-field representation of the transport velocity. As the integral in this expression is almost always positive, the turbulent buoyancy is directed toward the lower density, i.e. *upward*, amplifying the usual buoyancy. This is an unexpected result. For sufficiently weak magnetic fields mean-field buoyancy and flux-tube buoyancy are acting in parallel, both transporting the magnetic flux upward.

In the strong-field limit, the turbulence approaches two-dimensionality. The diamagnetic pumping is weak in this case and the density effect dominates. It is suppressed by the magnetic field as $\mathcal{K}^{\mathrm{buo}} \simeq \bar{B}^{-1}$. The positivity of this expression indicates that the transport is now *downward*, in contrast to the above result for weak fields. The variation of U^{buo} with $1/\bar{B}$ is quite similar to that of the turbulence intensity $\langle u'^2 \rangle$ in the strong-field approach (see Eq. (6.43)).

Let us generally write

$$U^{\mathrm{dia}} = -\Phi^{\mathrm{dia}}(\Omega, \bar{B}) \, \nabla \eta_{\mathrm{T}}, \qquad\qquad U^{\mathrm{buo}} = \Phi^{\mathrm{buo}}(\Omega, \bar{B}) \eta_{\mathrm{T}} \nabla \log \rho. \tag{4.25}$$

The function Φ^{dia} depends on the rotation and the magnetic field; it is always positive and does not vanish for vanishing rotation and vanishing magnetic field. It is thus clear that in the overshoot region the transport is always downward. Φ^{buo} also depends on the angular velocity of the basic rotation and on the magnetic field, but for nonrotating fluids it vanishes for vanishing magnetic field. The effect only arises due to the interaction of the density stratification and the magnetic field, so its sign is not quite obvious.

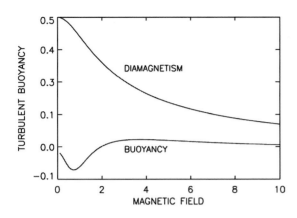

Figure 4.20: Magnetic quenching of the diamagnetic effect according to Kitchatinov (1991) and Kitchatinov & Rüdiger (1992). Φ^{dia} is always positive, but Φ^{buo} changes sign around $\beta = \bar{B}/B_{\mathrm{eq}} \simeq 2$.

For strong magnetic fields the functions Φ^{dia} and Φ^{buo} both approach zero. Even in the frame of quasilinear theories only approximations are known for the Φ's. Within the τ-approximation (introduced in MHD by Pouquet, Frisch & Leorat 1976) both of the quenching functions are known for rotation *or* for magnetism, i.e. $\Phi^{\mathrm{dia}}(0, \bar{B})$ and $\Phi^{\mathrm{dia}}(\Omega, 0)$. Only the magnetic quenching functions are given in Fig. 4.20. The function Φ^{buo} vanishes for vanishing $\bar{B}$, which means that the density stratification only produces a turbulent advection in magnetized turbulence. Note that the function Φ^{buo} has no definite sign: it is positive for very strong magnetic fields but negative for weak fields.

We have to ask whether situations exist where the upward-directed turbulence buoyancy dominates the downward-directed turbulence diamagnetism. From Fig. 4.20 it is obvious that at the bottom of a convection zone (where $\mathrm{d}\eta_{\mathrm{T}}/\mathrm{d}r > 0$) the diamagnetic advection is always downward. This is also true for the overshoot region below the convection zone. The magnetic quenching of this effect, however, is very strong. In the bulk of the convection zone the turbulence intensity is nearly uniform so that the downward advection should be smaller than in the overshoot regime. On the other hand, if the magnetic field is strong, the downward advection in the convection zone should be changed to a magnetic-dominated upward advection. The reason for this is that the density stratification $\mathrm{d}\log\rho/\mathrm{d}r$ is much stronger than in the overshoot region, so that the turbulent buoyancy can dominate the turbulent diamagnetism if the magnetic field is strong enough. One finds this result in the behavior of the solid line $\Phi^{\mathrm{buo}}(\beta)$ in Fig. 4.20. For magnetic fields smaller than $\beta \simeq 2$ the function $\Phi^{\mathrm{buo}}(\beta)$ is negative, representing an upward-directed (buoyancy) effect. It does not appear in the overshoot region as the quantity $|\rho^{-1}\mathrm{d}\log\rho/\mathrm{d}r|$ is much smaller there than in the convection zone.

In summary, the mean field transport in the stable layer should be downward and dominated by turbulent diamagnetism (U^{dia}), whereas in the convection zone for (not too) strong magnetic fields the transport is upward, and dominated by the turbulent buoyancy (U^{buo}). The magnetoconvection simulations by Ziegler & Rüdiger (2003) confirm these results. The last column in Table 3.3 gives the radial component of the advection vector γ in the overshoot region. The simulations always lead to a downward pumping of the magnetic field (see Dorch & Nordlund 2001, Thomas et al. 2002, see Fig. 4.21). That the transport of magnetic fields at the bottom of the convection zone is always downward was first found by Brandenburg et al. (1996).

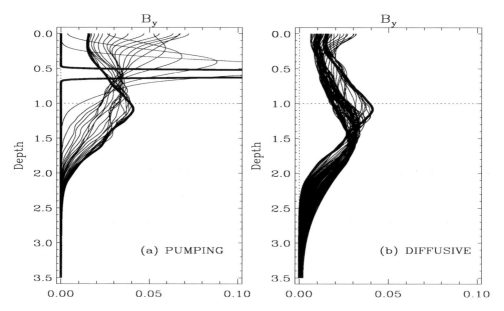

Figure 4.21: Downward magnetic flux pumping in the solar granulation layer. The original field is of the form of a toroidal thin slab (*left*) above the stable layer that ends at $z = 1$ (dotted line). It is transported downwards and then it diffuses into the stable layer with $z > 1$ (*right*). Thomas et al. (2002).

4.2.2   The Highly Anisotropic α-Effect

Now we have to include the influence of the basic rotation. Equation (4.15) is replaced by

$$\left(-\mathrm{i}\omega + \nu k^2 + \mathrm{i}\nu(\boldsymbol{G} \cdot \boldsymbol{k}) \right)\hat{\boldsymbol{m}} + 2(\boldsymbol{k}^\circ \cdot \boldsymbol{\Omega})(\boldsymbol{k}^\circ \times \hat{\boldsymbol{m}}) - \frac{\mathrm{i}}{\mu_0}(\boldsymbol{k} \cdot \bar{\boldsymbol{B}})\hat{\boldsymbol{B}} = \hat{\boldsymbol{f}}^{\mathrm{s}} \qquad (4.26)$$

$(\boldsymbol{k}^\circ = \boldsymbol{k}/k)$. If the Lorentz force is neglected in Eq. (4.26), we obtain

$$D_{ij} = \left(\delta_{ij} + \frac{2(\boldsymbol{k}^\circ \cdot \boldsymbol{\Omega})}{-\mathrm{i}\omega + \nu k^2 + \mathrm{i}\nu(\boldsymbol{G} \cdot \boldsymbol{k})}\epsilon_{ijp}k_p^\circ \right) \Big/ \left(1 + \frac{4(\boldsymbol{k}^\circ \cdot \boldsymbol{\Omega})^2}{(-\mathrm{i}\omega + \nu k^2 + \mathrm{i}\nu(\boldsymbol{G} \cdot \boldsymbol{k}))^2} \right)$$

for the tensor D, from the definition, Eq. (3.40). If, on the other hand, the influence of rotation and the mean magnetic field on the turbulence is considered, D changes to

$$D_{ij}^{\mathrm{mag}} = \frac{1}{N}\left(\delta_{ij} + 2\frac{(\boldsymbol{k}^\circ \cdot \boldsymbol{\Omega})}{(-\mathrm{i}\omega + \nu k^2)N}\epsilon_{ijp}k_p^\circ \right), \qquad (4.27)$$

with N given by Eq. (4.17). Equation (4.27) only holds for slow rotation and density stratification neglected. With Eq. (4.27) the spectral tensor for the momentum density can be expressed in terms of the spectral tensor (3.41) for the background turbulence after processing the transformation

$$\langle \hat{m}_i(\boldsymbol{k}, \omega)\hat{m}_j(\boldsymbol{k}', \omega') \rangle = D_{in}(\boldsymbol{k}, \omega)D_{jp}(\boldsymbol{k}', \omega')\langle \hat{m}_n^{(0)}(\boldsymbol{k}, \omega)\hat{m}_p^{(0)}(\boldsymbol{k}', \omega') \rangle, \qquad (4.28)$$

where all the effects that magnetic fields or rotation produce are involved in the tensor D.

We proceed with the consideration of the kinematic α-effect where the feedback of the magnetic field on the turbulence is neglected. The α-effect is both odd in $\boldsymbol{\Omega}$ and in $\boldsymbol{G}$. In the language of the spectral formulation this mean that the corresponding spectral tensor is odd in $\boldsymbol{\Omega}$ and in $\boldsymbol{\kappa}$. As an example the spectral tensor

$$\hat{M}_{ij} = \hat{M}_{ij}^{(0)} + \frac{\rho^2 \hat{Q}_{ll}}{2}\left[\frac{2\nu k \epsilon_{ijp}}{\omega^2 + \nu^2 k^4}\left((\boldsymbol{\kappa\Omega})k_p^\circ + (\boldsymbol{k}^\circ\boldsymbol{\Omega})\kappa_p\right)+\right.$$

$$+\frac{2(\boldsymbol{k}^\circ\boldsymbol{\Omega})\kappa_p k_m^\circ}{k}\left(\frac{\epsilon_{jmp}k_i^\circ}{i\omega + \nu k^2} - \frac{\epsilon_{imp}k_j^\circ}{-i\omega + \nu k^2}\right) + \frac{4(\boldsymbol{k}^\circ\boldsymbol{\Omega})\epsilon_{ijp}k_p^\circ}{(\omega^2 + \nu^2 k^4)^2}\times$$

$$\left.\times\left(-2\nu^3 k^5(\boldsymbol{k}^\circ\boldsymbol{\kappa}) + i\omega(\omega^2 + \nu^2 k^4) + i\nu(\omega^2 - \nu^2 k^4)(\boldsymbol{Gk})\right)\right], \qquad (4.29)$$

is given following from Eqs. (3.39) and (3.40). It directly leads to the expressions for the kinetic helicity and the α-effect for slow rotation. Already its structure demonstrates that the quasilinear approach can be handled – but not with ease.

The 'nonmagnetic' representation (4.27) for the tensor D is used valid for arbitrary rotation rate Ω. The resulting α-tensor splits into two parts that separately involve the effects of the inhomogeneities of turbulence intensity and density, i.e. $\alpha = \alpha^\rho + \alpha^u$. Both tensors have the same structure, which is given here only for the density stratification, i.e.

$$\alpha_{ij}^\rho = -(\boldsymbol{G}\cdot\boldsymbol{\Omega})\left(\alpha_1^\rho\delta_{ij} + \alpha_4^\rho\frac{\Omega_i\Omega_j}{\Omega^2}\right) - \alpha_2^\rho(G_i\Omega_j + G_i\Omega_i) + \alpha_3^\rho(G_i\Omega_j - G_i\Omega_i), \quad (4.30)$$

with

$$\alpha_n^\rho = \int \frac{\nu\eta k^4 \hat{Q}_{ll}}{(\omega^2 + \nu^2 k^4)(\omega^2 + \eta^2 k^4)} A_n^\rho(\Omega, k, \omega)\mathrm{d}\boldsymbol{k}\,\mathrm{d}\omega, \qquad (4.31)$$

for $n = 1, 2, 4$ and

$$\alpha_3^\rho = \int \frac{\omega^2 \hat{Q}_{ll}}{(\omega^2 + \nu^2 k^4)(\omega^2 + \eta^2 k^4)} B_3^u(\Omega, k, \omega)\mathrm{d}\boldsymbol{k}\,\mathrm{d}\omega. \qquad (4.32)$$

Detailed calculations of the α-tensor (4.30) on the basis of different turbulence models are also presented by Moffatt (1978), Roberts & Soward (1975) and Wälder, Deinzer & Stix (1980). The dependence on the angular velocity enters the relations through the kernels A and B, which are rather complicated expressions (Rüdiger & Kitchatinov 1993). Only special realizations shall be discussed here.

In cylindrical coordinates (s, ϕ, z) the symmetric part of the α-tensor is

$$\alpha^{\mathrm{S}} \sim \begin{pmatrix} -\alpha_1\cos\theta & 0 & -\alpha_2\sin\theta \\ 0 & -\alpha_1\cos\theta & 0 \\ -\alpha_2\sin\theta & 0 & -(\alpha_1 + 2\alpha_2 + \alpha_4)\cos\theta \end{pmatrix}, \qquad (4.33)$$

while in spherical coordinates (r, θ, ϕ) we have the structure

$$\alpha^{\mathrm{S}} \sim \begin{pmatrix} \alpha_{rr} & \alpha_{r\theta} & 0 \\ \alpha_{\theta r} & \alpha_{\theta\theta} & 0 \\ 0 & 0 & \alpha_{\phi\phi} \end{pmatrix}, \qquad (4.34)$$

with $\alpha_{rr} = -(\alpha_1 + 2\alpha_2 + \alpha_4 \cos^2 \theta) \cos \theta$, $\alpha_{r\theta} = \alpha_{\theta r} = (\alpha_2 + \alpha_4 \cos^2 \theta) \sin \theta$, $\alpha_{\theta\theta} = -(\alpha_1 + \alpha_4 \sin^2 \theta) \cos \theta$ and $\alpha_{\phi\phi} = -\alpha_1 \cos \theta$.

The combination of the inhomogeneities of the turbulence intensity and the density into a common gradient, $\nabla \log(\rho u_T)$, was believed to appear for the entire α-effect (Krause 1967). In general, however, this is not exactly true and the relative contribution of the two basic inhomogeneities depends on the angular velocity. We introduce a weight factor, S, which characterizes the relative contribution of the density stratification such as in

$$\alpha_{\phi\phi} = -\alpha_1^u \Omega \nabla \log(\rho^S u_T), \tag{4.35}$$

where S also depends (slightly) on the angular velocity.

Slow Rotation

Explicit representations of the α^u for the slow-rotation case are given in Rüdiger (1978) and we do not reproduce them here. We concentrate upon the contribution of the density stratification, i.e.

$$\alpha_1^\rho = \frac{4}{15} \int \frac{\nu \eta k^4 (3\nu^2 k^4 + 5\omega^2) \hat{Q}_{ll}}{(\omega^2 + \nu^2 k^4)^2 (\omega^2 + \eta^2 k^4)} \mathrm{d}k \, \mathrm{d}\omega,$$

$$\alpha_2^\rho = -\frac{8}{15} \int \frac{\nu^3 \eta k^8 \hat{Q}_{ll}}{(\omega^2 + \nu^2 k^4)^2 (\omega^2 + \eta^2 k^4)} \mathrm{d}k \, \mathrm{d}\omega \tag{4.36}$$

and $\alpha_4^\rho = 0$. Note the opposite signs of α_1^ρ and α_2^ρ, to which we shall return below.

The α-effect also exists in the high-conductivity limit. The relations provide for $\eta \to 0$

$$\alpha_1^\rho = \frac{32\pi^2}{5\nu} \int_0^\infty k^2 \, q(k, 0, \boldsymbol{x}) \, \mathrm{d}k, \qquad \alpha_2^\rho = -\frac{64\pi^2}{15\nu} \int_0^\infty k^2 \, q(k, 0, \boldsymbol{x}) \, \mathrm{d}k, \tag{4.37}$$

which remains finite for finite viscosity. The same is not true, however, for the expression describing the magnetic quenching of the α-effect (see below).

From here the value $S = 1.5$ can be found for the weight factor. The difference from unity is not large. Therefore, the importance of the two inhomogeneities for the α-effect depends mainly upon the gradients themselves. The τ-approximation leads to

$$\alpha_{rr} = \hat{\alpha} \left(\boldsymbol{U} + \frac{\boldsymbol{G}}{4} \right) \boldsymbol{\Omega}, \qquad \alpha_{\phi\phi} = \alpha_{\theta\theta} = -\hat{\alpha} \left(\boldsymbol{U} + \frac{3\boldsymbol{G}}{2} \right) \boldsymbol{\Omega}. \tag{4.38}$$

While in the northern hemisphere the most important component $\alpha_{\phi\phi}$ becomes positive (if density stratification dominates), the component α_{rr} becomes negative. In contrast to the standard formulations, the α_2-components in Eq. (4.30) are dominant, as confirmed by numerical simulations. Our quasilinear theory of the α-effect provides $\hat{\alpha} = (8/15)\tau_{\mathrm{corr}}^2 u_T^2$. The small value $S = 0.25$ for the radial α-component in Eq. (4.38) is also a surprise.

Rapid Rotation

For rapid rotation $\alpha_1 = -\alpha_4$ and $\alpha_2 = \alpha_3 = 0$. One then finds the rather simple relation

$$\alpha_{ij} = -\hat{\alpha}\frac{(\boldsymbol{G}+\boldsymbol{U})\cdot\boldsymbol{\Omega}}{\Omega}\left(\delta_{ij} - \frac{\Omega_i\Omega_j}{\Omega^2}\right),\tag{4.39}$$

with

$$\hat{\alpha} = \frac{\pi\eta}{4\nu^2}\int\frac{(\omega^2+\nu^2k^4)\hat{Q}_{ll}}{k^2(\omega^2+\eta^2k^4)}\,\mathrm{d}\boldsymbol{k}\,\mathrm{d}\omega.\tag{4.40}$$

Both basic inhomogeneities combine into a common gradient in Eq. (4.39), leading to $S = 1$ as the weight factor. Since the $\boldsymbol{U}$ dominates the $\boldsymbol{G}$ in the bottom layers of the convection zone – and in the overshoot region below the convection zone – one expects that the α-effect becomes negative in these layers (Fig. 4.22, right).

The α-effect for inhomogeneous intensity is known to become two-dimensional for rapid rotation (Moffatt 1970, Rüdiger 1978, see Busse & Miin 1979). We notice from Eq. (4.39) that this remains valid with density stratification included. The α-effect vanishes in the z-direction in cylindrical coordinates, i.e.

$$\alpha = c_\alpha\begin{pmatrix} -\frac{3\pi}{8}\cos\theta & \frac{3\pi}{8\Omega^*}\cos\theta & 0 \\ -\frac{3\pi}{8\Omega^*}\cos\theta & -\frac{3\pi}{8}\cos\theta & 0 \\ 0 & -\frac{3\pi}{8\Omega^*}\sin\theta & 0 \end{pmatrix}u_{\mathrm{T}}^2\tau_{\mathrm{corr}}\frac{\mathrm{d}\log(u_{\mathrm{T}}\rho)}{\mathrm{d}r}.\tag{4.41}$$

Note that

- all components with index z disappear,
- the remaining diagonal terms do not vanish for rapid rotation,
- the α-terms are negative in the northern hemisphere if the turbulence intensity increases outward (as in the overshoot region),
- the advection terms vanish for rapid rotation[6].

In Fig. 4.22 (left) the rotational behavior of the α-tensor components $\alpha_{\phi\phi}$ and α_{zz} is given taken from the expressions by Rüdiger & Kitchatinov (1993, in τ-approximation). Note that indeed there is no rotational Ω-quenching of $\alpha_{\phi\phi}$. Contrary to this the α_{zz} linearly grows for small Ω^* but after its maximum around $\Omega^* \simeq 1$ it is strongly reduced by the rotation.

The Sign of the α-Effect

In Fig. 4.23 the results of box calculations are given for increasing influence of the density stratification. A toroidal magnetic field has been applied to a convectively unstable temperature profile. The mean vertical density profile can be prescribed as very smooth (left) or rather steep (right). The resulting numbers are averaged over the horizontal plane (and in time). The turbulent EMF is normalized with $\bar{B}_y u_{\mathrm{T}}$, the latitude is 45° and the Taylor number is about 10^7.

---

[6] $\alpha_{\phi s} = -\alpha_{s\phi}$ is also called the escape velocity V_{esc}

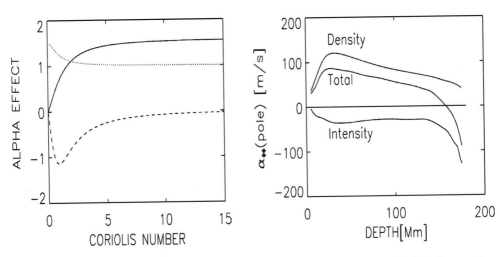

Figure 4.22: *Left*: The influence of the rotation on $\alpha_{\phi\phi}$ (solid) and α_{zz} (dashed). No Ω-quenching of the $\alpha_{\phi\phi}$ exists. The weight-factor S (dotted) approaches unity for fast rotation. *Right*: The depth-dependence of the α-effect in the solar convection zone with $c_\alpha = 1$. Note the different contributions due to the density gradient and turbulence intensity distribution. The α-effect becomes negative only at the bottom of the convection zone.

Without density stratification the α-effect proves to be a boundary phenomenon (Soward 1974). It is zero in the middle of the box and it reflects the vertical profile of the turbulence intensity, $\alpha \propto -\mathrm{d}u_{\mathrm{T}}/\mathrm{d}z$, see Eq. (4.35). At the end of Sect. 4.4.1 we note that oscillatory α^2-dynamos are possible for profiles as shown in Fig. 4.23 (left).

Figure 4.23 also demonstrates the influence of the density stratification upon the α-effect. As the density decreases in the vertical direction, $\mathrm{d}\rho/\mathrm{d}z$ is negative, and the resulting α-effect becomes more and more positive (Fig. 4.23, right). Negative values only survive at the bottom of the box. Also, the temporal fluctuations of the α-effect are reduced by the density stratification. Note that the density ratio in the solar convection zone between $x = 0.7$ and $x = 0.9$ is about 15, only in comparison with the surface values does one obtain numbers of order 10^3 (see Stix 2002).

The simulations reveal the resulting α-effect to be rather small. Only a small fraction of the turbulence intensity actually results in a toroidal α-effect. The unknown parameter c_α in Eq. (4.41) appears to be of order 10^{-2}.

Krivodubskij & Schultz (1993) (with the inclusion of the depth-dependence of the Coriolis number Ω^* and using a mixing-length model of the solar convection zone) derived a profile of the α-effect with a magnitude of 100 m/s, positive in the convection zone and negative in the overshoot layer (Fig. 4.22, right).

Ossendrijver, Stix & Brandenburg (2001) and Ossendrijver et al. (2002) present simulations for all the components of the α-tensor in spherical coordinates. Also, these simulations were done in a Cartesian box including a convectively stable overshoot layer. All simulations concern the southern (!) hemisphere, and the angle between the vertical (radial) direction and the axis of rotation was varied between the south pole and the equator. The radial dependence

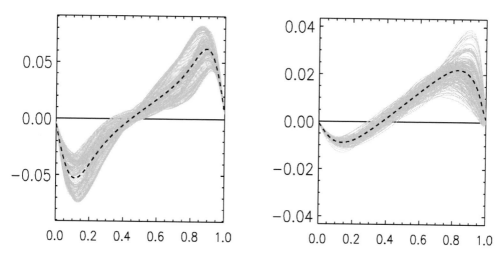

Figure 4.23: The $\alpha_{\phi\phi}$ normalized with u_T in boxes at 45° north from $z = 0$ (bottom) to $z = 1$ (top) for low Mach number of order 10^{-2}. *Left*: No density stratification (like in the Earth core). *Right*: $\rho_{\text{bottom}}/\rho_{\text{top}} \simeq 33$ as in the supergranulation layer. Without density stratification the crossover in the middle of the box is very characteristic (see Soward 1974). The stronger the density stratification the more the positive sign dominates. Ta $= 10^7$, $T \simeq 10^4$ K. Compare with the helicity profiles in Fig. 3.27 (left). Courtesy A. Giesecke.

of $\alpha_{\theta\theta}$ and $\alpha_{\phi\phi}$ has a typical shape, namely a negative sign in the bulk of the convection zone, and a positive sign in the thin overshooting layer. The amplitudes of $\alpha_{\theta\theta}$ and $\alpha_{\phi\phi}$ for various latitudes follow the commonly assumed $\cos\theta$-function. The α-effect, therefore, does *not* vanish at the poles. For weak rotation the component α_{rr} has a larger amplitude than the other two diagonal components, and it has *the opposite sign*. If the rotation increases α_{rr} becomes small and strongly fluctuating, unlike $\alpha_{\theta\theta}$ and $\alpha_{\phi\phi}$. This is the rotational quenching of the vertical α-effect reported by Ossendrijver, Stix & Brandenburg (2001). In the overshoot region $\alpha_{\phi\phi}$ becomes positive (at the south pole, see Fig. 4.24).

It has been argued that the short rise times of flux tubes in the convection zone prevent the formation of the α-effect (Spiegel & Weiss 1980, Schüssler 1987, Stix 1991). Indeed, the rise times may be much longer in the overshoot region (Ferriz-Mas & Schüssler 1993, 1995, van Ballegooijen 1998). Note, however, the agreement of the results of the surprising numerical simulations with those of the SOCA calculations. Sign and latitudinal dependence are very similar in both approaches. For the amplitudes of the $\alpha_{\phi\phi}$ we have to compare the 100 m/s obtained by Krivodubskij & Schultz (Fig. 4.22, right) with the result $\alpha/u_T \simeq 0.05$–0.1 of the box simulations (see Figs. 4.23, 4.24 and 4.26, also Brandenburg 1994b) leading to 10 m/s, (say). It must be stressed, however, that the basic parameters used in the simulations (Rayleigh number, Taylor number, Prandtl numbers) are still far from the values of reality. It seems that the presented nonlinear simulations more reflect the quasilinear SOCA rather than the turbulent convection but it is certainly too early for further speculations.

It is this uncertainty that led Blackman & Field (2000), Brandenburg (2001) and Blackman & Brandenburg (2002) to more dynamical formulations of the α-effect problem. It is not the

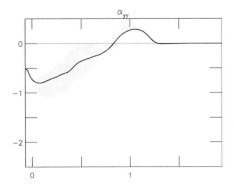
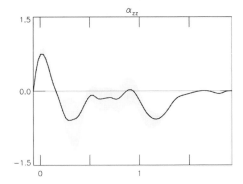

Figure 4.24: The α-tensor components $\alpha_{\phi\phi}$ (*left*) and α_{rr} (*right*) as a function of depth, for a box at the south pole measured in units of 0.01 $\sqrt{dg}$. The domain consists of a cooling layer ($z < 0$), a convectively unstable layer ($0 < z < 1$) and a stably stratified layer with overshooting convection ($z > 1$). $\Omega^* = 2.4$. The component $\alpha_{\phi\phi}$ does *not* vanish at the pole. From Ossendrijver, Stix & Brandenburg (2001).

place here, however, for a discussion of this development which recently has been reported in detail by Brandenburg & Subramanian (2004).

4.2.3   The Magnetic Quenching of the α-Effect

The influence of the magnetic fields on the α-effect is now considered. For simplicity the rotation has to be assumed to be slow. We shall also restrict consideration to the τ-approximation, while some more general expressions can be found in the original paper by Rüdiger & Kitchatinov (1993). This approximation yields for the nondiffusive part of the mean EMF the expression

$$\mathcal{E} = -\tau_{\text{corr}}^2 u_{\text{T}}^2 \left(\frac{8}{15}\Psi(\beta)\,(\boldsymbol{U\Omega})\,\bar{\boldsymbol{B}} + \Psi_1(\beta)\frac{(\boldsymbol{\Omega}\bar{\boldsymbol{B}})\,(\boldsymbol{U}\bar{\boldsymbol{B}})}{\bar{B}^2}\bar{\boldsymbol{B}} - \right.$$
$$\left. -\frac{4}{5}\Psi_2(\beta)\,(\boldsymbol{\Omega}\bar{\boldsymbol{B}})\,\boldsymbol{U} - \frac{4}{5}\Psi_3(\beta)\,(\boldsymbol{U}\bar{\boldsymbol{B}})\,\boldsymbol{\Omega} \right). \tag{4.42}$$

Here $\beta = \bar{B}/B_{\text{eq}}$ is the field strength[7] normalized with the equipartition value

$$B_{\text{eq}} = \sqrt{\mu_0\rho}\,u_{\text{T}}; \tag{4.43}$$

the numerical factors were introduced to normalize the quenching functions Ψ, Ψ_2, Ψ_3 to unity at $\beta = 0$ (the function Ψ_1 vanishes at the origin). It is

$$\Psi = \frac{15}{32\beta^4}\left(1 - \frac{4\beta^2}{3(1+\beta^2)^2} - \frac{1-\beta^2}{\beta}\tan^{-1}\beta \right). \tag{4.44}$$

For weak fields it scales as $\Psi \sim 1 - (12/7)\beta^2$, and for strong fields it goes to zero as $15\pi/64\beta^3$. This cubic quenching was found by Moffatt (1972), Rüdiger (1974) and Gilbert

---

[7] do not confuse with the plasma beta that is named below β^*

& Sulem (1990) but it is missing in the relations of Rogachevskii & Kleeorin (2001). The remaining functions are

$$\Psi_1 = \frac{1}{4\beta^4}\left(\frac{3\beta^8 + 8\beta^6 - 36\beta^4 - 40\beta^2 - 15}{3(1+\beta^2)^3} + \frac{5+\beta^4}{\beta}\tan^{-1}\beta\right),$$

$$\Psi_2 = \frac{5}{16\beta^4}\left(\frac{3\beta^4 - 8\beta^2 - 3}{3(1+\beta^2)^2} + \frac{1+\beta^2}{\beta}\tan^{-1}\beta\right),$$

$$\Psi_3 = \frac{5}{16\beta^4}\left(\frac{(\beta^2-1)(3\beta^4 + 8\beta^2 + 3)}{3(1+\beta^2)^2} + \frac{1+\beta^4}{\beta}\tan^{-1}\beta\right). \tag{4.45}$$

For weak fields the series expansions start with $\Psi_1 = (40/21)\beta^2$, $\Psi_2 = 1 - (13/7)\beta^2$ and $\Psi_3 = 1 - (25/21)\beta^2$. For strong fields it is $\Psi_1 \simeq \pi/8\beta$, $\Psi_2 \simeq 5\pi/32\beta$ and Ψ_2 varies as β^{-3}.

It is possible to rearrange the given expressions. The four terms can be written as a scalar α-effect and an advection vector $\boldsymbol{V}^{\mathrm{mag}}$ with three components, i.e.

$$\boldsymbol{\mathcal{E}} = \alpha(\bar{B})\bar{\boldsymbol{B}} + \boldsymbol{V}^{\mathrm{mag}} \times \bar{\boldsymbol{B}}. \tag{4.46}$$

The α-term follows as

$$\alpha = -\tau_{\mathrm{corr}}^2 u_{\mathrm{T}}^2\left(\frac{8}{15}\Psi(\boldsymbol{U}\boldsymbol{\Omega}) + (\Psi_1 - \frac{4}{5}\Psi_2 - \frac{4}{5}\Psi_3)\frac{(\boldsymbol{\Omega}\bar{\boldsymbol{B}})(\boldsymbol{U}\bar{\boldsymbol{B}})}{\bar{B}^2}\right), \tag{4.47}$$

which obviously depends on the direction of $\bar{\boldsymbol{B}}$. This dependence is in fact rather strong. The α-effect for fields parallel to $\boldsymbol{U}$ is called $\alpha_\parallel$, and $\alpha_\perp$ is used for the perpendicular case. One obtains

$$\alpha_\parallel = -\Omega^*\eta_{\mathrm{T}}U\cos\theta\Psi_\parallel(\beta), \qquad \alpha_\perp = -\Omega^*\eta_{\mathrm{T}}U\cos\theta\Psi_\perp(\beta), \tag{4.48}$$

where $\eta_{\mathrm{T}} = \tau_{\mathrm{corr}}u_{\mathrm{T}}^2$ is the magnetic eddy diffusivity for the nonmagnetic case, and θ is the angle between the vectors $\boldsymbol{U}$ and $\boldsymbol{\Omega}$. The functions $\Psi_\parallel$ and $\Psi_\perp$ are linear combinations of the quenching functions in Eq. (4.42). They are plotted in Fig. 4.25 together with the standard quenching function $1/(1+\beta^2)$ that is often used by dynamo modelers.

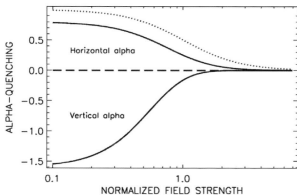

Figure 4.25: Magnetic quenching of the vertical (by a poloidal field) and horizontal (by a toroidal field) α-effect. The dotted line represents the heuristic function $1/(1+\beta^2)$, which underestimates the magnetic quenching in the quasilinear approximation.

We note that strong magnetic fields suppress the α-effect more strongly than β^{-2}. For the strong-field case one finds at leading order in β

$$\alpha_\parallel = O(\beta^{-5}), \qquad \alpha_\perp = -\Omega^*\eta_{\mathrm{T}}U\cos\theta\frac{3\pi}{16\beta^3}. \tag{4.49}$$

The α-effect is not only anisotropic but is even anisotropically quenched. The vertical α-effect hardly exists for $\beta > 1$. The remaining α-effect (for toroidal fields) is suppressed as β^{-3}.

The magnetic quenching of $\alpha_{\phi\phi}$ can be observed in the simulations plotted in Fig. 4.26 for density-stratified ($\rho_{\mathrm{bottom}}/\rho_{\mathrm{top}} \simeq 11$, the solar value) and rotating (Ta $= 10^7$) turbulent convection. The magnetic field in both simulations differs by a factor of 5. This is also true for the maximum of the $\alpha_{\phi\phi}$. Here, obviously, the quenching law is linear.

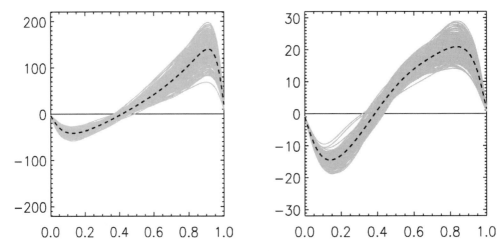

Figure 4.26: The same as in Fig. 4.23 but for $\rho_{\mathrm{bottom}}/\rho_{\mathrm{top}} \simeq 11$ and for weak (*left*) and strong (*right*) magnetic field and in units of m/s. Courtesy A. Giesecke.

It may be noted that the effective velocity $\boldsymbol{V}^{\mathrm{mag}}$ is much less quenched by the strong magnetic fields than the α-effect,

$$\boldsymbol{V}^{\mathrm{mag}} = \frac{3}{8} \frac{\pi \tau_{\mathrm{corr}} \eta_{\mathrm{T}}}{\beta} \frac{(\boldsymbol{U}\bar{\boldsymbol{B}})}{\bar{B}^2} \left(\boldsymbol{\Omega} \times \bar{\boldsymbol{B}}\right). \qquad (4.50)$$

This velocity may wind up a toroidal field from a poloidal one, just like a differential rotation would, even if the rotation is uniform.

Ossendrijver, Stix & Brandenburg (2001) also studied the magnetic quenching of the α-effect by numerical simulations of magnetoconvection. According to their results the α-effect is suppressed stronger than (say) β^{-3}. This calls into question whether the equilibration value of $\bar{B}^2$ is really the turbulent value $\mu_0 \rho u_{\mathrm{T}}^2$. Another possibility would be $\mu_0 \rho \eta / \tau_{\mathrm{corr}}$, with η as the microscopic magnetic diffusivity. In the high-conductivity limit this value is extremely small, so that the magnetic quenching of α is much more effective than in the above standard case (Cattaneo & Vainshtein 1991, Vainshtein & Cattaneo 1992). One always has $u_{\mathrm{T}}^2 \tau_{\mathrm{corr}} (\simeq \eta_{\mathrm{T}}) \gg \eta$. Formally we can write in this case $\alpha \propto \eta / \bar{B}^2$ or $\alpha \propto 1/(\mathrm{Rm}\bar{B}^2)$ with the (large) microscopic magnetic Reynolds number $\mathrm{Rm} = \tau_{\mathrm{corr}} u_{\mathrm{T}}^2 / \eta$.

In discussing this question it may help to consider SOCA-expressions for the α-effect in the high-conductivity limit, $\eta \to 0$. Consider the α-effect for a homogeneous, isotropic but helical turbulence. Its spectral tensor contains an antisymmetric part, namely $\hat{Q}_{ij} =$

$\cdots - \mathrm{i}\epsilon_{ijk}k_k Q_2(k,\omega)$, which leads to an α-effect with the structure $\alpha = \alpha_0 - \Gamma \bar{B}^2$, with

$$\alpha_0 = \frac{2}{3}\eta \int \frac{k^4 Q_2 \, \mathrm{d}\boldsymbol{k} \, \mathrm{d}\omega}{\omega^2 + \eta^2 k^4}, \qquad \Gamma = \frac{4}{5}\frac{\eta}{\mu\rho}\int \frac{k^6(\nu\eta k^4 - \omega^2)Q_2 \, \mathrm{d}\boldsymbol{k}\mathrm{d}\omega}{(\omega^2 + \nu^2 k^4)(\omega^2 + \eta^2 k^4)^2} \qquad (4.51)$$

(Rüdiger 1974). With the τ-approximation (3.47) also for η one finds $\alpha_0 \propto u_T$ and $\Gamma \propto (\mu_0 \rho u_T)^{-1}$, so that $\alpha \simeq u_T$ for weak fields, whereas for strong fields results[8]

$$\alpha \propto \frac{u_T^3}{\bar{B}^2/\mu_0\rho}. \qquad (4.52)$$

The situation changes completely in the high-conductivity limit. The α_0-expression remains almost unchanged. One finds that in this case (as usual $\nu \to \ell_{\mathrm{corr}}^2/\tau_{\mathrm{corr}}$) the result for Γ is $\Gamma \propto \eta^{-1}$, which is never small for high conductivity. This result does not depend on the special form of Q_2, as for $\eta \to 0$ there are Dirac delta-functions in (4.51)[9]. For strong fields one obtains

$$\alpha \propto \frac{1}{\mathrm{Rm}}\frac{u_T^3}{\bar{B}^2/\mu_0\rho}, \qquad (4.53)$$

which differs from Eq. (4.52) by the very small factor Rm^{-1} ($\sim 10^{-3}$ for stars). Equation (4.53) indicates an extremely strong quenching. A more detailed discussion of such 'catastrophic quenching' is given by Blackman & Brandenburg (2002) and Brandenburg & Subramanian (2004).

From the original expressions of Rüdiger (1974) or with Eq. (4.44) one finds that for strong magnetic fields the *cubic* quenching $\alpha \propto (\sqrt{\eta}/\bar{B})^3$ yields an even better approximation.

4.2.4   Weak-Compressible Turbulence

Turbulence in a rotating and density-stratified fluid possesses a negative (positive) kinetic helicity, Eq. (4.9), in the northern (southern) hemisphere (see Miesch et al. 2000, their Fig. 22). In this case the helicity is always associated with an α-effect of opposite sign, i.e. $\alpha \propto -\mathcal{H}_{\mathrm{kin}}$. The standard sign of the α-effect in rotating, density-stratified fluids with forced turbulence is thus positive (negative) for the northern (southern) hemisphere.

However, this antiphase relation between the α-effect and the kinetic helicity vanishes for magnetic-dominated compressible turbulence models including buoyancy. In this case it is the *current helicity* that starts to play the dominant role for the α-effect. In particular the buoyancy turns out to be very important for the generation of α and angular momentum transport (Brandenburg & Schmitt 1998).

We present results here for a compressible turbulence field driven by a (given) Lorentz force in a rotating convection zone. The numerical findings of Brandenburg & Schmitt (1998) are confirmed that in the northern (southern) hemisphere the α-effect and the kinetic helicity are positive (negative) and the current helicity is there negative (positive). For the purely hydrodynamic constellation in Sect. 3.4 the kinetic helicity had the opposite sign.

---

[8] the quenching of the α-effect has been reformulated here as $\alpha = \alpha_0/(1 + (\Gamma/\alpha_0)\bar{B}^2)$

[9] see Rüdiger (1989, p. 267) for details

Consider compressible turbulence in a medium with uniform density. Then, in the inertial system, the momentum equation for rigid rotation is

$$\frac{\partial \boldsymbol{u}'}{\partial t} + \boldsymbol{\Omega} \times \boldsymbol{u}' + s\Omega(\hat{\boldsymbol{e}}_\phi \cdot \nabla)\boldsymbol{u}' =$$

$$-\frac{1}{\rho} \nabla \left(P' + \frac{\boldsymbol{B}^{(0)} \cdot \bar{\boldsymbol{B}}}{\mu_0} \right) + \frac{\rho'}{\rho} \boldsymbol{g} + \frac{1}{\mu_0 \bar{\rho}} (\bar{\boldsymbol{B}} \cdot \nabla) \boldsymbol{B}^{(0)} + \nu \Delta \boldsymbol{u}'. \qquad (4.54)$$

We take the fluctuating magnetic field $\boldsymbol{B}^{(0)}$ (and also the mean magnetic field $\bar{\boldsymbol{B}}$) to be known. The correlations of the random magnetic fluctuations $\boldsymbol{B}^{(0)}$ may describe a homogeneous and isotropic field of magnetic fluctuations.

Mass conservation requires that the density fluctuations ρ' satisfy the continuity equation $\partial \rho'/\partial t + \bar{\rho} \nabla \cdot \boldsymbol{u}' = 0$. Notice that the mean density has been assumed as homogeneous in space and that the density fluctuations vary in time. We do *not* adopt the anelastic approximation. The acoustic term $\partial \rho'/\partial t$ is necessary in order to get the presented results. The details of the overall procedure are given by Pipin (2003).

For the energy equation the polytrope relation $P' = c_{\rm ac}^2 \rho'$ is adopted, where $c_{\rm ac}$ is the isothermal sound speed. One can now find the correlation tensor of the turbulence driven by the Lorentz force in Eq. (4.54). The total angular momentum transport with Maxwell stress included is given by

$$T_{r\phi} = \langle u_r' u_\phi' \rangle - \frac{1}{\mu_0 \bar{\rho}} \langle B_r' B_\phi' \rangle. \qquad (4.55)$$

By definition, all off-diagonal correlations of the given fluctuation $\boldsymbol{B}^{(0)}$ are zero, but this is not true for the magnetic fluctuation $\boldsymbol{B}'$, resulting from the interaction of the mean magnetic field and the flow field due to Eq. (4.54). According to the induction equation in its linearized form,

$$\frac{\partial \boldsymbol{B}'}{\partial t} + s\Omega(\hat{\boldsymbol{e}}_\phi \cdot \nabla)\boldsymbol{B}' - \eta \Delta \boldsymbol{B}' = \nabla \times (\boldsymbol{u}' \times \bar{\boldsymbol{B}}), \qquad (4.56)$$

the influence of global rotation only concerns the nonaxisymmetric field components.

The resulting magnetic fluctuations can be used to compute the current helicity

$$\mathcal{H}_{\rm curr} = \langle \boldsymbol{J}' \cdot \boldsymbol{B}' \rangle, \qquad (4.57)$$

which has the same kind of equatorial (anti-)symmetry as the dynamo-α. For *homogeneous* global magnetic fields, the dynamo-α is related to the turbulent EMF according to $\boldsymbol{\mathcal{E}} = \langle \boldsymbol{u}' \times \boldsymbol{B}' \rangle \simeq \alpha \circ \bar{\boldsymbol{B}}$ so that $\alpha_{ij} \bar{B}_i \bar{B}_j = \boldsymbol{\mathcal{E}} \cdot \bar{\boldsymbol{B}}$. Rädler & Seehafer (1990) read this equation as $\alpha_{\phi\phi} = \boldsymbol{\mathcal{E}} \cdot \bar{\boldsymbol{B}}/\bar{B}_\phi^2$, where $\alpha_{\phi\phi}$ is the dominant component of the α-tensor. We are particularly interested in checking the antiphase relation,

$$\alpha_{\phi\phi} \mathcal{H}_{\rm curr} < 0, \qquad (4.58)$$

between the α-effect and current helicity (Keinigs 1983, Rädler & Seehafer 1990), resulting from $\langle \boldsymbol{E}' \cdot \boldsymbol{B}' \rangle = 0$ with $\boldsymbol{E}$ as the electrical field, which may generally hold for homogeneous and stationary turbulence (Seehafer 1996). An increasing number of contributions present observations of the current helicity at the solar surface, all showing that it is negative (positive) in the northern (southern) hemisphere (Hale 1927, Seehafer 1990, Rust & Kumar 1994, Abramenko, Wang & Yurchishin 1996, Bao & Zhang 1998, Pevtsov, Canfield & Latushko 2001, Kleeorin et al. 2003). Convincing observations are shown in Fig. 4.27, note (i) the equatorial antisymmetry and (ii) the temporal constancy of the phenomenon. If Eq. (4.58) is correct then there is a strong empirical evidence that the α-effect is positive (negative) in the northern (southern) hemisphere of the solar convection zone.

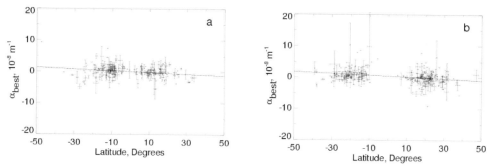

Figure 4.27: The current helicity (from magnetic-field geometries of sunspots) in cycle 22 (*left*) and cycle 23 (*right*) is negative (positive) in the northern (southern) hemisphere. Courtesy A. Pevtsov.

The Current Helicity

The (massive) algebra of the equations resulting from a Fourier transformation is described by Pipin (2003). Here only the characteristic expressions within the τ-approximation are discussed. The correlation length of the magnetic field fluctuations is denoted by ℓ_{corr}. For the current helicity, Eq. (4.57), we find

$$\mathcal{H}_{\mathrm{curr}} = \frac{2}{5} \frac{\tau_{\mathrm{corr}}^3}{\ell_{\mathrm{corr}}^2} \frac{\bar{B}_\phi^2}{\mu_0} \frac{\langle B^{(0)\,2} \rangle}{\mu_0 \bar{\rho} c_{\mathrm{ac}}^2} (\boldsymbol{g} \cdot \boldsymbol{\Omega}), \qquad (4.59)$$

with $\boldsymbol{g}$ as the inward directed gravity. The current helicity is thus negative (positive) in the northern (southern) hemisphere. This is exactly the numerical result of Brandenburg (2000) for the current helicity of the magnetic field fluctuations, and it also agrees with the observations.

Turning next to the α-effect, only the most important component, $\alpha_{\phi\phi}$, will be given, i.e.

$$\alpha_{\phi\phi} = -\frac{1}{5} \frac{\tau_{\mathrm{corr}}^2}{c_{\mathrm{ac}}^2} \frac{\langle B^{(0)\,2} \rangle}{\mu_0 \bar{\rho}} (\boldsymbol{g} \cdot \boldsymbol{\Omega}). \qquad (4.60)$$

For rigid rotation the α-effect proves to be *positive* in the northern hemisphere and negative in the southern hemisphere. Again our result agrees with the numerical simulations. The current helicity and α-effect have opposite signs, their ratio being

$$\frac{\alpha_{\phi\phi} \bar{B}_\phi^2}{\mathcal{H}_{\mathrm{curr}}} = -\frac{\mu_0}{2} \frac{\ell_{\mathrm{corr}}^2}{\tau_{\mathrm{corr}}}. \qquad (4.61)$$

For the kinetic helicity the model yields

$$\mathcal{H}_{\text{kin}} = -\frac{8}{15} \frac{\tau_{\text{corr}}^3}{\ell_{\text{corr}}^2} \frac{\bar{B}_\phi^2}{\mu_0 \bar{\rho}} \frac{\langle B^{(0)2} \rangle}{\mu_0 \bar{\rho} c_{\text{ac}}^2} \left(\boldsymbol{g} \cdot \boldsymbol{\Omega} \right), \tag{4.62}$$

which is *positive* (negative) in the northern (southern) hemisphere. If a rising eddy can expand in a density-reduced surrounding then a negative value of the kinetic helicity is expected. The magnetic-buoyancy model, however, leads to another result. Upward flow results from negative density fluctuation and downward flow results from positive density fluctuation. The fluctuating flow is thus compressed when flowing upward, and expands when flowing downward. Thus, for magnetically driven turbulence with density fluctuations there is *no* minus sign between the α-effect and kinetic helicity (as in Keinigs' relation), but nevertheless the minus sign is present in the relation between the α-effect and current helicity – and the α-effect is again positive.

We have shown that the simplest turbulence model, including magnetic buoyancy under the action of a slow global rotation, but without density stratification and shear, exactly fulfills the rules found by Brandenburg & Schmitt (1998) in a numerical simulation for the physical representation of the solar tachocline. For the northern hemisphere we find the dynamo-α and the kinetic helicity being positive and the current helicity being negative.

There is also a minus sign between the α-effect (northern hemisphere) and the angular momentum transport (the 'viscosity-α') suggested by Brandenburg (1998) with a shear flow simulation. Our model yields this negative sign also for the case of turbulence subject to rigid rotation (see Eq. (4.63)). However, the main power of such an antiphase relation between the α-effect and angular momentum transport will be developed in a theory of accretion disks, where a positive angular momentum transport is needed, so that Brandenburg's law would yield a *negative* α-effect on the upper disk plane – with strong implications to the accretion-disk dynamo and jet theory (see Sect. 5.9).

It is not yet clear whether the observations of surface values of the current helicity reflect the helicity of the fluctuating fields or the helicity of the large-scale fields. The high level of noise in the observations seems to indicate a fluctuating-field origin of the phenomenon (see Fisher et al. 1999). In any case, from Eq. (4.61) we should expect also a highly noisy α-effect. In Table 4.2 our results are summarized for both of the discussed turbulence models. In all cases the signs of the (antisymmetric) scalars are given for the northern hemisphere.

Table 4.2: The signs of turbulence quantities for the northern hemisphere for kinetic- and magnetic-driven turbulence models. From Rüdiger & Pipin (2000).

	kinetic-driven	magnetic-driven	total	observ.
$\mathcal{H}_{\text{kin}}$	negative	positive	≈ 0	
$\mathcal{C}$	negative	negative	negative	$\lesssim 0$
$\mathcal{H}_{\text{curr}}$	negative	negative	negative	negative
α-effect	positive	positive	positive	

If the real turbulence model can be considered as a mixture of both the approximations then we expect

- a rather small kinetic helicity $\mathcal{H}_{\mathrm{kin}} \simeq 0$,
- a negative current helicity $\mathcal{H}_{\mathrm{curr}} < 0$,
- a positive α-effect.

Observations and simulations fully seem to comply with these formulations.

Angular Momentum Transport

The angular momentum transport is described by the off-diagonal elements of the correlation tensor that do not vanish for rigid rotation. Our turbulence model provides the negative expression

$$\Lambda_{\mathrm{V}} = -\frac{18}{105} \frac{\tau_{\mathrm{corr}}^3 g^2}{c_{\mathrm{ac}}^2} \frac{\bar{B}^2}{\mu_0 \bar{\rho}} \frac{\langle B^{(0)^2} \rangle}{\mu_0 \bar{\rho} c_{\mathrm{ac}}^2} \tag{4.63}$$

for the Reynolds stress in the rigidly rotating plasma. Note that Λ_{V} is even in g. For the magnetic-induced angular momentum transport $\langle B_r' B_\phi' \rangle = 0$ is obtained. Now replace $g \simeq c_{\mathrm{ac}}^2/H_{\mathrm{p}}$, $\ell_{\mathrm{corr}} = \alpha_{\mathrm{MLT}} H_{\mathrm{p}}$, where H_{p} is the vertical pressure scale, i.e.

$$\Lambda_{\mathrm{V}} = -\frac{18}{105} \alpha_{\mathrm{MLT}}^2 \frac{\tau_{\mathrm{corr}}^3}{\ell_{\mathrm{corr}}^2} V_{\mathrm{A}}^2 \frac{\langle B^{(0)^2} \rangle}{\mu_0 \bar{\rho}}, \tag{4.64}$$

with the Alfvén velocity V_{A} associated with the dominant toroidal field component $\bar{B}_\phi$. According to the results presented in Fig. 3.15 (right) negative values in Eq. (4.64) lead to rotation laws decreasing outward, i.e. $\partial\Omega/\partial r < 0$. Indeed, such a regime (of slow rotation) holds in the subsurface supergranulation layers (Fig. 3.2, left).

4.3   Magnetic-Diffusivity Tensor and η-Quenching

Like the eddy viscosity tensor, the magnetic diffusivity tensor η_{ijk} defined by Eq. (4.7) also depends on the influence of global rotation and large-scale magnetic fields (Roberts & Soward 1975).

4.3.1   The Eddy Diffusivity Tensor

The magnetic field diffusivity tensor η_{ijk} for a rotating fluid reads

$$\eta_{ijk} = \epsilon_{ijm}\left(\eta_{\mathrm{T}}\delta_{km} + \eta_{\|}\frac{\Omega_k\Omega_m}{\Omega^2}\right) + (a-b)\delta_{ik}\Omega_j - b\,\delta_{ij}\Omega_k + \frac{c}{\Omega^2}\,\Omega_i\Omega_j\Omega_k. \tag{4.65}$$

This equation includes all the contributions to the mean EMF proportional to the spatial derivatives of the mean magnetic field. An important complication is that not all of these terms describe an effective field dissipation. Only the η_{T}- and $\eta_{\|}$-terms in Eq. (4.65) represent the anisotropic diffusion. The middle terms in Eq. (4.65) can be written as $(a - 2b)\delta_{ik}\Omega_j + b(\delta_{ik}\Omega_j - \delta_{ij}\Omega_k)$. Here the last term forms the $\boldsymbol{\Omega} \times \boldsymbol{J}$-effect of Rädler (1969), while the first one can be written as a gradient (for rigid rotation), the *curl* of which disappears. Hence a is not important and b is the coefficient of the $\boldsymbol{\Omega} \times \boldsymbol{J}$-effect that, however, vanishes in the frame of the τ-approximation. We shall meet this situation many times later: not all contributions to the diffusivity tensors actually describe dissipative processes (as also, for example, the Hall effect).

The coefficients in Eq. (4.65) are the spectral integrals

$$\eta_{\mathrm{T}} = \int\limits_0^\infty \int\limits_0^\infty \frac{\eta_{\mathrm{t}} k^2 E(k,\omega)}{\omega^2 + \eta_{\mathrm{t}}^2 k^4} K(k,\omega,\Omega) \, \mathrm{d}k \, \mathrm{d}\omega,$$

$$\eta_{\|} = \int\limits_0^\infty \int\limits_0^\infty \frac{\eta_{\mathrm{t}} k^2 E(k,\omega)}{\omega^2 + \eta_{\mathrm{t}}^2 k^4} K_{\|}(k,\omega,\Omega) \, \mathrm{d}k \, \mathrm{d}\omega,$$

$$b = \int\limits_0^\infty \int\limits_0^\infty \frac{\eta_{\mathrm{t}}^2 k^4 \omega^2 E(k,\omega)}{(\omega^2 + \eta_{\mathrm{t}}^2 k^4)^2 (\omega^2 + \nu_{\mathrm{t}}^2 k^4)} K_b(k,\omega,\Omega) \, \mathrm{d}k \, \mathrm{d}\omega,$$

$$c = \int\limits_0^\infty \int\limits_0^\infty \frac{\eta_{\mathrm{t}}^2 k^4 \omega^2 E(k,\omega)}{(\omega^2 + \eta_{\mathrm{t}}^2 k^4)^2 (\omega^2 + \nu_{\mathrm{t}}^2 k^4)} K_c(k,\omega,\Omega) \, \mathrm{d}k \, \mathrm{d}\omega, \qquad (4.66)$$

where the kernels K depend on the angular velocity (Kitchatinov, Pipin & Rüdiger 1994). The general expressions for the kernels are so complex that simplifications are necessary. In the slow-rotation limit ($\Omega \to 0$), $\eta_{\|} = 0$, while K tends to 1/3, reproducing the isotropic diffusion coefficient. The other kernels are $K_b = 16/15$ and $K_c = 0$, in agreement with Roberts & Soward (1975). For rapid rotation we have

$$K = K_{\|} = \frac{\pi(\omega^2 + \nu_{\mathrm{t}}^2 k^4)}{16\Omega\nu_{\mathrm{t}} k^2}. \qquad (4.67)$$

The $\eta_{\|}$-parameter is the additional diffusion coefficient along the rotation axis, i.e. the effective diffusivity for this direction is $\eta_{\mathrm{T}} + \eta_{\|}$, while normal to the rotation axis the diffusivity is η_{T}. The result of *curling* of the EMF with Eq. (4.65) is

$$\nabla \times \boldsymbol{\mathcal{E}} = \eta_{\mathrm{T}} \Delta \boldsymbol{B} + \eta_{\|} \frac{\partial^2 \boldsymbol{B}}{\partial z^2}, \qquad (4.68)$$

so that obviously the dissipation is enhanced in the direction of $\boldsymbol{\Omega}$ (Rüdiger, Elstner & Stepinski 1995). Equation (4.67) shows that in the rapid-rotation limit the diffusion along the rotation axis is stronger, by as much as a factor of 2. This diffusivity anisotropy is due to the rotation-induced anisotropy of the turbulence. The other kernels in Eq. (4.66) in this rapid-rotation limit are small, i.e. K_b and K_c scale as Ω^{-3}.

Now the τ-approximation (3.47) is introduced. One of its advantages is that the resulting simplifications are strong enough to formulate the results for arbitrary rotational velocities. Another is that the rotation rate dependencies are expressed in terms of the simple parameter Coriolis number Ω^*.

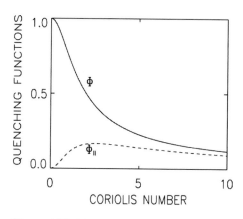
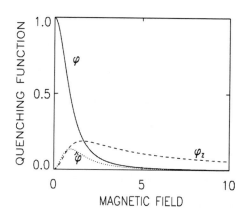

Figure 4.28: The quenching functions in the η-tensor for (*left*) the influence of rotation and (*right*) the influence of magnetic field.

Application of Eq. (3.47) to Eq. (4.66) results in $b = c = 0$, and

$$\eta_{\mathrm{T}} = \eta_0 \phi(\Omega^*), \qquad \eta_{\|} = \eta_0 \phi_{\|}(\Omega^*), \qquad\qquad (4.69)$$

where

$$\eta_0 = c_\eta \tau_{\mathrm{corr}} u_{\mathrm{T}}^2, \qquad\qquad (4.70)$$

with $c_\eta = 1/3$ (as the standard value) is the magnetic diffusivity for a nonrotating fluid[10]. The quenching functions $\phi(\Omega^*)$ and $\phi_{\|}(\Omega^*)$ are

$$\phi = \frac{3}{4\Omega^{*2}} \left(1 + \frac{\Omega^{*2} - 1}{\Omega^*} \tan^{-1} \Omega^* \right), \quad \phi_{\|} = \frac{3}{4\Omega^{*2}} \left(-3 + \frac{\Omega^{*2} + 3}{\Omega^*} \tan^{-1} \Omega^* \right),$$

with $\phi(0) = 1$ and $\phi_{\|}(0) = 0$ (see Fig. 4.28, left).

For slow rotation and weak magnetic field the eddy diffusivity tensor takes the simple and well-known form $\eta_{ijk} = \eta_{\mathrm{T}} \epsilon_{ijk}$, so that $\mathcal{E} = -\eta_{\mathrm{T}} \nabla \times \bar{B}$. The eddy diffusivity, however, is only a simple tensor without the magnetic feedback. For strong magnetic fields the η_{T}-tensor becomes much more complex, having the form

$$\eta_{ijk} = \eta_{\mathrm{T}}(\bar{B}) \, \epsilon_{ijk} + \hat{\eta}(\bar{B}) \, \epsilon_{ilk} \bar{B}_j \bar{B}_l + \dots . \qquad\qquad (4.71)$$

Its most convenient representation concerns the turbulent EMF

$$\mathcal{E} = \dots - \eta_{\mathrm{T}}(\bar{B}) \nabla \times \bar{B} + U^{\mathrm{mag}} \times \bar{B}, \qquad\qquad (4.72)$$

with the 'magnetic velocity'

$$U^{\mathrm{mag}} = \hat{\eta}(B) \, \nabla \log \bar{B}^2 + \eta_z(\bar{B}) \, \frac{\bar{J} \times \bar{B}}{\bar{B}^2}. \qquad\qquad (4.73)$$

---

[10] Reighard & Brown (2001) present an experimental realization, in liquid sodium, of the turbulence-induced conductivity reduction expressed by Eq. (4.70)

Here a similar computation to that which led to Eq. (4.68) provides the finding that again the magnetic dissipation in field direction is enhanced (Kim 1997). For a magnetic field in the z-direction one finds

$$\mathcal{E}_y = (\eta_{\mathrm{T}} + \eta_z - 2\hat{\eta})\frac{\mathrm{d}\bar{B}}{\mathrm{d}x}, \tag{4.74}$$

expressing a *reduction* of the dissipation in the orthogonal direction.

Within the τ-approximation one finds the quenching expressions $\eta_{\mathrm{T}} = \eta_0 \varphi(\beta)$, $\hat{\eta} = \eta_0 \hat{\varphi}(\beta)$ and $\eta_z = \eta_0 \varphi_z(\beta)$ with

$$\varphi = \frac{3}{2\beta^2}\left\{-\frac{1}{1+\beta^2} + \frac{1}{\beta}\tan^{-1}\beta\right\}, \qquad \hat{\varphi} = \frac{3}{8\beta^2}\left\{-\frac{5\beta^2+3}{(1+\beta^2)^2} + \frac{3}{\beta}\tan^{-1}\beta\right\},$$

$$\varphi_z = \frac{3}{8\beta^2}\left\{1 + \frac{2}{1+\beta^2} + \frac{\beta^2-3}{\beta}\tan^{-1}\beta\right\} \tag{4.75}$$

(Fig. 4.28, right). Here $\varphi \simeq 1 - 6\beta^2/5$, $\hat{\varphi} \simeq 3\beta^2/5$ and $\varphi_z \simeq 2\beta^2/5$ are valid for weak magnetic fields. The η-quenching starts to become important for magnetic fields of this order. Again, as for the α-effect, $\varphi \simeq 3\pi/4\beta^3$ yields a cubic quenching for the eddy diffusivity, but (again) Rogachevskii & Kleeorin (2001) present a weaker quenching, $O(\beta^{-1})$. So far the above η-quenching expressions have been applied to the theory of sunspot decay, to stellar activity cycles as well as the decay of (hypothetical) strong initial galactic magnetic fields.

We are confronted with the situation that in the mean-field theory both the inputs ($\alpha, \eta_{\mathrm{T}}$) are strongly influenced by magnetic deformation and suppression. This must be a delicate situation. For example, one cannot imagine that an α^2-dynamo could exist if the dissipation (η_{T}) is more strongly quenched by the resulting magnetic field than the induction (α). Tobias (1996) has shown with a special dynamo model[11] how complex the solution is. The dynamo only works normally in the regime close to the critical dynamo number. If, however, the dynamo is far in the nonlinear regime ($\beta \gg 1$), the quenching becomes more and more effective, so that details of the quenching functions become important. This is true in particular if the shear is fixed, so that for an $\alpha\Omega$-dynamo the dynamo number $\mathcal{D} = C_\alpha C_\Omega$ grows with growing magnetic field. This might also be the background for the existence of weak- and strong-field solutions and the hysteresis between them found by Tobias (1996). So far, there is no spherical dynamo model for which the interplay between α-quenching and η-quenching has been studied in detail.

Finally, the same considerations started with Eq. (4.51) can be undertaken concerning the η-quenching. The corresponding series is $\eta_{\mathrm{T}} = \eta_0 - \Gamma^* \bar{B}^2$, with

$$\eta_0 = \frac{\eta}{3}\int\frac{k^2 \hat{Q}_{ll}\mathrm{d}\boldsymbol{k}\mathrm{d}\omega}{\omega^2 + \eta^2 k^4}, \qquad \Gamma^* = \frac{2}{5}\frac{\eta}{\mu_0\rho}\int\frac{k^4(\nu\eta k^4 - \omega^2)\hat{Q}_{ll}\mathrm{d}\boldsymbol{k}\mathrm{d}\omega}{(\omega^2 + \nu^2 k^4)(\omega^2 + \eta^2 k^4)^2} \tag{4.76}$$

(Kitchatinov, Pipin & Rüdiger 1994). Once again, in the high-conductivity limit η_0 remains finite, but yields $\Gamma^* \propto 1/\eta$, describing a 'catastrophic' quenching. Such extremely strong quenching has been reported for 2D turbulence by Cattaneo & Vainshtein (1991) and Parker (1992). With their numerical simulations of magnetoconvection under the influence of an inhomogeneous large-scale magnetic field Nordlund, Galsgaard & Stein (1994) find a significant η-quenching only for 2D turbulence subject to strong fields (in the plane of the motion).

---

[11] interface dynamo: shear exists below an interface and α-effect exists above it

4.3.2   Sunspot Decay

One of the few examples where magnetic η-quenching can be observed is in the decay of sunspots. Observational studies of sunspots have a long history (see Solanki 2003), which makes them an excellent test sample for predictions of the mean-field theory. We are here mainly concerned with the time-evolution of large, long-lived spots. Figure 4.29 displays the basic observational finding that when a big recurrent spot decays, the rate of decline of the spot's area is almost constant in time (Bray & Loughhead 1964, Zwaan 1992, Skumanich et al. 1994). Martínez Pillet, Moreno-Insertis & Vázquez (1993) obtained a decay rate of

$$\dot{A} = -2 \cdot 10^{12}\ \mathrm{cm}^2/\mathrm{s}. \tag{4.77}$$

The magnetic field in the umbra of a decaying spot is also known to be almost time independent. This implies a linear law $\dot{\Phi} \simeq \mathrm{const.}$ for the decrease of the spot's magnetic flux

$$\Phi = \int \bar{B}_z\ \mathrm{d}A. \tag{4.78}$$

Following Stix (1989), a linear decay of the flux can be reproduced by the solution of a 1D diffusion equation with constant eddy diffusivity. Krause & Rüdiger (1975) noticed, however, that the solution is not consistent with the linear decline of the spot area.

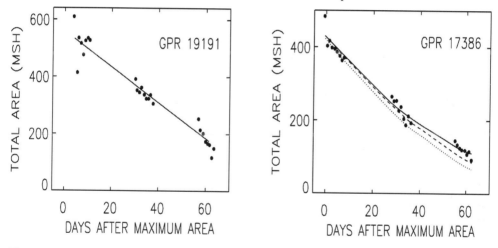

Figure 4.29: Temporal decay of the spot area of recurrent spots. The examples show linear (*left*) and slightly nonlinear (*right*) decay laws after Martínez Pillet, Moreno-Insertis & Vázquez (1993).

Here we can demonstrate that the diffusive decay law can be improved considerably by including the magnetic eddy diffusion quenching. The nonlinear diffusion model by Rüdiger & Kitchatinov (2000a) reproduces linear decay laws for both spot flux and area.

Our 2D model solves two diffusion equations for magnetic field and entropy

$$\frac{\partial \bar{B}}{\partial t} = \nabla \times \left(U^{\mathrm{mag}} \times \bar{B} - \eta_{\mathrm{T}} \nabla \times \bar{B} \right), \qquad \rho \bar{T} \frac{\partial \bar{S}}{\partial t} = -\nabla F^{\mathrm{conv}}, \tag{4.79}$$

with

$$F_i^{\mathrm{conv}} = -\rho \bar{T} \chi_{ij} \frac{\partial \bar{S}}{\partial x_j} \tag{4.80}$$

in a horizontal slab with axial symmetry. In this equation, the effective velocity (4.73) accounts for the anisotropy of the nonlinear diffusion. We started from an initial state of highly concentrated purely vertical magnetic field and an entropy distribution that is the steady solution of the heat transport equation (4.79) for the nonmagnetic case. The simulated 'spot' is defined as the circular area where the luminosity is reduced below 75% of its undisturbed value. Though there was no spot at the beginning, it develops, however, shortly after and it starts decaying after a while.

A typical example of the results is shown in Fig. 4.30. The main finding is that the decrease of both the spot area and magnetic flux with time is rather linear, which results from the nonlinear effect of magnetic quenching of the eddy diffusivities. These results are in very good agreement with the empirical arguments of Martínez Pillet (2002). By rescaling the eddy diffusivity by a factor of 10 to $\eta_T \simeq 5 \cdot 10^{11}$ cm^2/s the decay time can easily be adjusted. A much stronger magnetic quenching of the diffusivity also leads to longer lifetimes of the spots: Large-scale patterns with much weaker magnetic fields decay with a magnetic diffusion coefficient of $6 \cdot 10^{12}$ cm^2/s (Sheeley 1992, Schrijver & Zwaan 2000).

After a detailed statistical study of sunspot data, Petrovay & van Driel-Gesztelyi (1997) stress that $\dot{A}$ is slightly time dependent rather than constant, so that the decay law becomes slightly parabolic. Certainly, the character of the decay law depends on the order of the quenching law. With the *cubic* quenching used in our model the decay law of the sunspots remains linear. If it is not linear in reality then the real quenching must be (i) of much higher order (Petrovay & Moreno-Insertis 1997) or (ii) it must be stronger in the sense of Eq. (4.76).

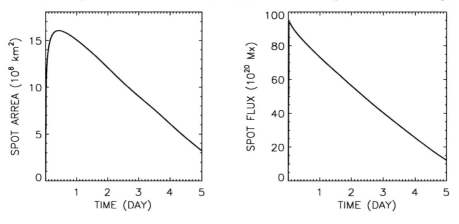

Figure 4.30: Time evolution of the spot area (*left*) and the magnetic flux (*right*) in the nonlinear diffusion model of Rüdiger & Kitchatinov (2000a). An example with $\Phi = 10^{22}$ Mx is shown. By rescaling of the eddy diffusivity the decay time changes.

The model can be further improved by taking the fluid motion into account. The barocline and magnetic forces are not in balance, leading to a meridional circulation. This flow was found to be important for the MHD-equilibrium in the sunspots (Kitchatinov & Mazur 2000). It was possible to reproduce the near-constancy of the field strength within the spot umbra with the meridional flow. The flow is convergent on the top, changing to a downdraft near the edge of the simulated spot. It helps to stabilize the highly concentrated flux against the magnetic pressure. Such flow patterns are indeed observed with local helioseismology of solar active regions (Duvall et al. 1996, Kosovichev 2002). The vertical extent of the cool sunspot proved to be only 5,000 km (Kosovichev 2002).

Sunspots can also be used to probe the properties of the magneto-acoustic-gravity ('MAG') waves traveling along the magnetic field under the influence of density stratification, radiation and compressibility. For the simplest case of an isothermal unstratified medium the equations are as simple as given in Sect. 2.5 for the hydromagnetic waves under the influence of rotation. Now the continuity equation changes from $\nabla \cdot \boldsymbol{u} = 0$ to $\partial \log \rho / \partial t + \nabla \cdot \boldsymbol{u} = 0$, which leads to the dispersion relation

$$\omega_{\pm} = \frac{1}{2}\left(k^2 V_{\mathrm{A}}^2 + \omega_{\mathrm{ac}}^2 \pm \sqrt{\left(k^2 V_{\mathrm{A}}^2 + \omega_{\mathrm{ac}}^2\right)^2 - 4\omega_{\mathrm{A}}^2 \omega_{\mathrm{ac}}^2}\,\right), \tag{4.81}$$

besides the particular Alfvén solution $\omega_0^2 = \omega_{\mathrm{A}}^2 \equiv (\boldsymbol{k} \cdot \boldsymbol{V}_{\mathrm{A}})^2$. The complication in sunspots is that they consist of regions with dominating magnetic field (top) or dominating gas pressure (bottom). The three MAG-modes, therefore, couple and mix in the layer where the magnetic Mach number approaches unity. Rosenthal et al. (2002) and Bogdan et al. (2003) have simulated the wave propagation in stratified magnetized atmospheres for an oscillation that is convectively driven with 42 mHz (~ 24 s) subject to a magnetic field of 5000 G. The field is mainly vertical over a horizontal domain of 2000 km and the adiabatic sound speed is 8.5 km/s. The wave propagation is computed with a 2D compressible MHD code.

The oscillations can also be observed. There are three bands observed of oscillation periods with peaks at 2'–3', 5' and $\gtrsim 20$'. Recent reviews are by Bogdan (2000) and Staude (2002). The 3' oscillations are mainly observed in the chromosphere above the umbra. At photospheric levels the flow amplitudes are much lower ($\lesssim 50$ m/s) and they are not always discovered. The amplitudes become larger with increasing height z. Recent observations have shown that significant signals of magnetic field oscillations exist but they are limited to much smaller regions inside the spots than the velocity oscillations that cover much larger parts of a sunspot (Fig. 4.31).

It is unclear at present whether the oscillations are eigenoscillations of the sunspot itself and/or a passive response to forcing by convection or 'normal' magnetic-deformed p-mode oscillations.

4.4   Mean-Field Stellar Dynamo Models

We shall now apply the findings about the parameterizations of the turbulence-induced EMF (4.7) to Eq. (4.5). The resulting dynamo equation can be written in the form

$$\frac{\partial \bar{\boldsymbol{B}}}{\partial t} = \nabla \times \left(\bar{\boldsymbol{u}} \times \bar{\boldsymbol{B}} + \alpha \circ \bar{\boldsymbol{B}} - \sqrt{\eta_{\mathrm{T}}}\nabla \times \left(\sqrt{\eta_{\mathrm{T}}}\bar{\boldsymbol{B}}\right)\right), \tag{4.82}$$

which includes the diagonal elements of the α-effect and the diamagnetism due to nonuniform turbulence, Eq. (4.22). Note that after differentiation the factor $\sqrt{\eta_{\mathrm{T}}}$ exactly produces the advection expression $-1/2 \cdot \nabla \eta_{\mathrm{T}}$. If there are strong gradients of turbulence intensity, this pumping term will dominate the transport of the mean magnetic fields. The tensorial nature of η is ignored here, as its structure is not known to have any dramatic consequences for the

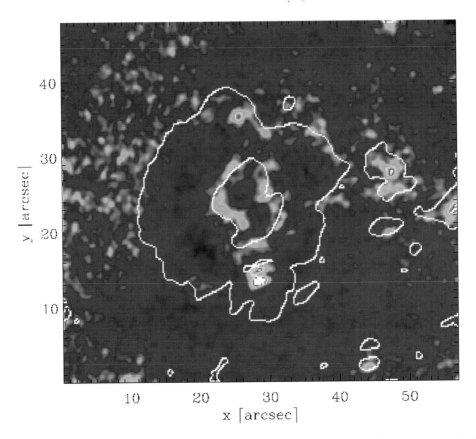

Figure 4.31: Map of oscillatory power of the magnetic field strength in the 3' period range. Dark and white indicate increasing power. The white lines indicate the boundary between umbra and penumbra. From Staude (2002).

dynamo theory. However, Kitchatinov (2002) found a moderate influence of the diffusion anisotropy on the resulting butterfly diagram.

Only the exact balance of the inducing and dissipating processes allows the dynamo to work. In order to demonstrate the variety of the solutions of the linear dynamo equation (4.82) with $\eta = \text{const.}$ and $\nabla \cdot \boldsymbol{B} = 0$, it is applied to the simplest case of a 1D dynamo in Cartesian coordinates (Parker 1971). We put $\boldsymbol{u} = (0, u_y(x), 0)$ and $\boldsymbol{B} = (B_x(z), B_y(z), 0)$, $\mathrm{d}u_y/\mathrm{d}x$ describing a differential rotation such as $\partial\Omega/\partial r$. One finds

$$\frac{\partial \bar{B}_x}{\partial t} = -\alpha\frac{\partial \bar{B}_z}{\partial z} + \eta_\mathrm{T}\frac{\partial^2 \bar{B}_x}{\partial z^2}, \qquad \frac{\partial \bar{B}_y}{\partial t} = \alpha\frac{\partial \bar{B}_x}{\partial z} + \bar{B}_x\frac{\mathrm{d}u_y}{\mathrm{d}x} + \eta\frac{\partial^2 \bar{B}_y}{\partial z^2}, \qquad (4.83)$$

which might be solved with $B \propto \mathrm{e}^{\mathrm{i}(kz-\omega t)}$ so that a dispersion relation

$$\left(-\mathrm{i}\omega + \eta k^2\right)^2 = \alpha^2 k^2 + \mathrm{i}\alpha k\frac{\mathrm{d}u_y}{\mathrm{d}x} \qquad (4.84)$$

results. Without the shear term $-\mathrm{i}\omega = \pm\alpha k - \eta k^2$, so that all modes with $k \leq |\alpha|/\eta$ are dynamo-unstable. The maximum growth rate is $\alpha^2/2\eta$ for $k = \alpha/2\eta$. The pattern of this 'α^2-dynamo' is stationary. If, on the other hand, the α^2-term in Eq. (4.84) is neglected, one finds $-\mathrm{i}\omega = (1 + \mathrm{i})\sqrt{\alpha k u_{y,x}/2} - \eta k^2$, which indicates marginal instability for modes with $k = (\alpha u_{y,x}/2\eta^2)^{1/3}$, traveling with a frequency $\omega_{\mathrm{cyc}} = (\alpha^2 u_{y,x}^2/4\eta)^{1/3}$ which runs with $\Omega^{4/3}$ if both α-effect and shear are linear Ω (Tuominen, Rüdiger & Brandenburg 1988). The maximum linear growth rate $0.30(\alpha^2 u_{y,x}^2/\eta)^{1/3}$ is of the same order of magnitude. This oscillating dynamo is called the '$\alpha\Omega$-dynamo'. The transition between the regimes of α^2-dynamos and $\alpha\Omega$-dynamos is rather complicated, and will be discussed below for a spherical model (see Fig. 4.38, right).

4.4.1   The α^2-Dynamo

The model consists of a turbulent fluid in a spherical shell of inner radius x_{in} and outer radius 1. The induction equation is (4.82). A magnetic field is generated in the shell by the α-effect (Steenbeck & Krause 1966, Roberts 1972). The turbulent magnetic diffusivity η_0 is constant in the shell. For $x > 1$ there is assumed to be a conductor with large magnetic diffusivity η_{out}, and for $x < x_{\mathrm{in}}$ a conductor with high electrical conductivity, i.e. with small magnetic diffusivity η_{in}. Here we are interested in the structure of the solution of an α^2-dynamo. However, there is no case known in which the α-tensor has a simple structure. The general structure of the α-tensor is given by Eq. (4.30). It is reduced to its first term in almost all models of α-effect dynamos.

In the models investigated the critical eigenvalues for dipolar and quadrupolar fields are always rather close together[12]. However, only the inclusion of all the remaining symmetric parts of the α-tensor reveals the variety of the solutions of the α^2-dynamo. In any case a dimensionless value

$$C_\alpha = \frac{|\alpha_1| \cdot R}{\eta_{\mathrm{T}}} \tag{4.85}$$

can be defined as the 'dynamo number'. The material within the inner boundary may be considered as a perfect conductor. The numerical outer boundary of the sphere is fixed at 1.5 stellar radii, where the standard conditions for a pseudovacuum are used. Between $x = 1$ and $x = 1.5$ the value for the magnetic diffusivity is increased by a factor 100 in order to mimic vacuum boundary conditions at the stellar surface. In the inner perfect-conducting part the magnetic diffusivity is reduced by a factor of 10^{-8}.

The following models are computed with a grid-point method rather than with the standard method of spectral development (see Sect. 6.5 for details). The regularity conditions of the magnetic field and its derivatives on the rotation axis form an essential point in this concept. They are almost trivial for the axisymmetric solutions ($B_\theta = B_\phi = \partial B_r/\partial\theta = 0$). For the modes with $m = 1$ they read $B_r = \partial B_\theta/\partial\theta = \partial B_\phi/\partial\theta = 0$, while for all higher m all the magnetic field components must vanish at the axis[13].

---

[12] for perfectly conducting boundaries Proctor (1977) even showed that they are identical

[13] these conditions are automatically fulfilled by the Legendre polynomials in the spectral codes

The Symmetric Part of the α-Tensor

There are also studies in the literature where the natural antisymmetry of the α-effect with respect to the equator has been neglected (Rädler & Bräuer 1987, Schubert & Zhang 2000, Stefani & Gerbeth 2003). The realization of such models can be imagined in technical installations like the Karlsruhe dynamo experiment, with a fixed helicity and a uniform flow field in the vertical (z) direction (Stieglitz & Müller 2001, Fig. 4.32), so that $\alpha_{zz} = 0$ is obvious (Rädler et al. 2002). In order to demonstrate the differences between isotropic and anisotropic α-tensors also in the case of a *homogeneous* α-effect (i.e. $\cos\theta$ ignored), the following calculations are presented for the cases (i) $\alpha_{zz} = \alpha_{\phi\phi}$ and (ii) $\alpha_{zz} = 0$. While the first case seems to be very academic the second one fits the situation in the dynamo experiment mentioned.

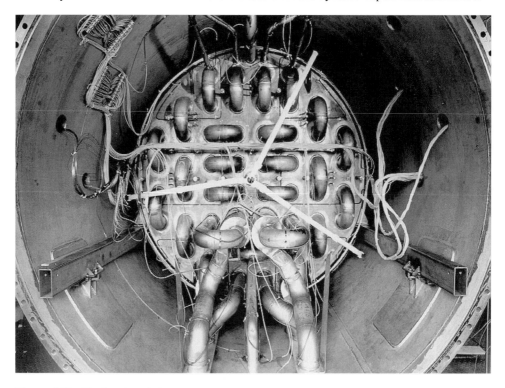

Figure 4.32: The interior of the Karlsruhe dynamo experiment. The sodium flows in the pipes with prescribed helicity in the z-direction (perpendicular to the paper plane) that leads to $\alpha_{zz} = 0$. Courtesy Forschungszentrum Karlsruhe GmbH.

Any mode has its own dynamo number, the mode with the lowest dynamo number is the preferred stable mode (Krause & Meinel 1988). In Fig. 4.33 the resulting minimum numbers C_α and the associated latitudinal mode numbers are given for various inner shell radii x_{in}. The dipole mode only dominates for the thick α-layers (see Krause & Rädler 1980, p. 177). For $x_{in} \gtrsim 0.5$, however, the quadrupoles are favored, while even higher modes appear for thinner and thinner layers. All the solutions are steady. We do not find any oscillatory α^2-dynamo

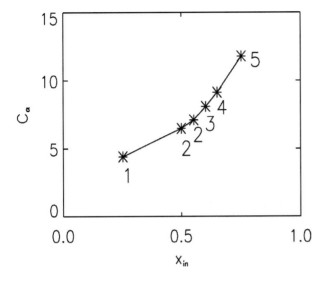

Figure 4.33: The simplest dynamo model (α = const.) The lowest critical dynamo numbers C_α for models with various x_{in}. The curve is marked with the latitudinal mode number for which the dynamo number is minimum. All solutions are steady, oscillating modes are located *above* this line (Rüdiger, Elstner & Ossendrijver 2003).

unless its α-amplitude C_α is not the lowest one. Oscillatory modes with low n do exist, but their C_α are not the lowest ones.

More important are the solutions with homogeneous but anisotropic α-tensor. The α-tensor may not have any zz-component, i.e. $\alpha_{zz} = 0$ in cylindrical coordinates. Now a rotation axis is clearly defined so that it makes sense to ask for the axisymmetry of the solutions. Table 4.3 gives both the results for uniform α-effect and the results if the α-effect is $\alpha \sim \cos\theta$. The notation is the standard one, i.e. Am (Sm) denotes a solution with antisymmetry (symmetry) with respect to the equator and with the azimuthal quantum number m (Fig. 4.34). It was indeed important to include the nonaxisymmetric modes, as they always possess the lowest C_α. The axisymmetric solutions (mostly oscillating) that have been found by Busse & Miin (1979), Weisshaar (1982) and Olson & Hagee (1990) are probably not stable.

There are no basic differences for $\alpha \sim$ const. and $\alpha \sim \cos\theta$ in Table 4.3. The dominance of the modes with $m = 1$ for anisotropic α-effect has been discussed by Rüdiger (1980),

Table 4.3: Marginal dynamo numbers C_α for axisymmetric and nonaxisymmetric magnetic field modes for anisotropic ($\alpha_{zz} = 0$) but uniform α-effect (*left*, see Karlsruhe dynamo experiment) and for $\alpha \propto \cos\theta$ (*right*). The minimum values are marked in bold and oscillating solutions are marked with $\sim$.

	$\alpha =$ const.				$\alpha \propto \cos\theta$			
x_{in}	A0	S0	A1	S1	A0	S0	A1	S1
0.25	9.06	8.40	**6.72**	**6.72**	15.0 ($\sim$)	14.9 ($\sim$)	10.6	**9.97**
0.50	10.8 ($\sim$)	10.6 ($\sim$)	**9.50**	**9.50**	15.7 ($\sim$)	15.5 ($\sim$)	11.8	**11.7**
0.75	17.3 ($\sim$)	17.3 ($\sim$)	**16.2**	**16.2**	21.6 ($\sim$)	21.58 ($\sim$)	**18.4**	**18.4**

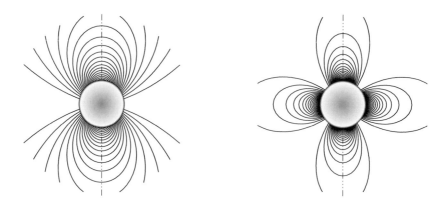

Figure 4.34: *Left*: The S1 mode as it is standard for the resulting magnetic geometry in the theory of the α^2-dynamo. *Right*: The A1 mode.

Rädler et al. (1990)[14] and by Rüdiger & Elstner (1994). In light of these calculations it is thus not a surprise that the Karlsruhe dynamo experiment provides the nonaxisymmetric mode with $m = 1$.

Now the models with equatorial antisymmetry of the α-effect (with $\cos\theta$, as induced by the global rotation) are developed in more detail. Different α-profiles in latitude are used in order to simulate a possible concentration of the α-effect to the equator. The latitudinal profile is fixed such that $\alpha \propto \sin^{2\lambda}\theta \cos\theta$, where λ is a free parameter describing the latitudinal profile of the α-effect. For $\lambda > 0$ the α-effect at the poles vanishes[15]. For increasing values of λ the α-effect is more and more concentrated at lower latitudes. Dynamo models with $\lambda > 0$ seem to be of academic interest only, but it is necessary to know that only for such models do oscillatory solutions appear with the lowest dynamo numbers. The results are summarized in Tabs. 4.4 and 4.5.

Table 4.4 gives the results for *isotropic* α-effect with $\alpha_{zz} = \alpha_{\phi\phi}$. For the standard case with $\lambda = 0$ the axisymmetric dipole A0 is the preferred mode. This result, however, strongly depends on the latitudinal profile of the α-effect. Already for $\lambda = 1$ the preferred mode is nonaxisymmetric and for $\lambda \geq 2$ we always find A1 as the preferred mode. Oscillatory modes also appear, but they never have the lowest dynamo numbers.

For a strongly anisotropic α-tensor ($\alpha_{zz} = 0$) we have an even clearer situation. The solutions with the lowest C_α are always nonaxisymmetric (see Table 4.4). Oscillating axisymmetric solutions occur for the same α-anisotropy, but only for very thin convection zones with $x_{\mathrm{in}} = 0.8$ and $\lambda > 0$ (Table 4.5). For such a model with vanishing α-effect in the polar regions we find that the mode with the lowest C_α yields oscillating axisymmetric magnetic

---

[14] another sort of anisotropy has been considered, see their Eq. (24)

[15] which has never been used so far in numerical simulations

Table 4.4: Dynamo numbers C_α for both isotropic α-effect ($\alpha_{zz} = \alpha_{\phi\phi}$) and for anisotropic α-effect ($\alpha_{zz} = 0$). The bottom of the convection zone is at $x_{\rm in} = 0.5$. The boldface numbers mark the lowest value of the dynamo number.

λ	$\alpha_{zz} = \alpha_{\phi\phi}$				$\alpha_{zz} = 0$			
	A0	S0	A1	S1	A0	S0	A1	S1
0	**9.41**	9.42	9.75	9.76	15.7 ($\sim$)	15.5 ($\sim$)	11.8	**11.7**
1	28.8 ($\sim$)	28.7 ($\sim$)	26.7	**26.7**	35.0 ($\sim$)	33.8 ($\sim$)	32.7	**31.3**
2	41.1 ($\sim$)	41.2 ($\sim$)	**38.8**	39.2	51.7 ($\sim$)	49.1 ($\sim$)	49.3	**47.0**
3	51.9 ($\sim$)	52.0 ($\sim$)	**49.8**	50.3	66.6 ($\sim$)	63.3 ($\sim$)	64.3	**61.4**

Table 4.5: The same as in Table 4.4 but for thin shells ($x_{\rm in} = 0.8$) and $\alpha_{zz} = 0$.

λ	A0	S0	A1	S1
0	26.3 ($\sim$)	26.3 ($\sim$)	**23.2**	**23.2**
1	63.0 ($\sim$)	**62.9** ($\sim$)	63.0	**62.9**
2	89.8 ($\sim$)	**89.4** ($\sim$)	90.2	89.9
3	112.9 ($\sim$)	**111.8** ($\sim$)	113.4	112.7

fields. After Rüdiger, Elstner & Ossendrijver (2003) such a cyclic behavior seems to be a rather exceptional case, as it only appears if three conditions are simultaneously fulfilled, i.e.

- the α-tensor must be highly anisotropic,
- the α-effect must be concentrated to the equator and
- the convection zone must be thin.

The Antisymmetric Part of the α-Tensor

The α_3-component in the tensor formulation (4.30) formally acts as a (differential) rotation – so that, if α_3 is strong enough – all α^2-dynamos can operate as (pseudo) $\alpha\Omega$-dynamos, and could thus be oscillatory. We shall denote this 'virtual' angular velocity by $\Omega_{\rm T}$, where $\Omega_{\rm T} = -\alpha_3/x$. The ratio α_3/α_1 will determine the ability of the α^2-dynamo to operate in an oscillating regime. It transforms poloidal magnetic fields to toroidal magnetic fields with a phase relation depending on the sign of $\partial\Omega_{\rm T}/\partial r$. With the notation in Ossendrijver et al. (2002) $\alpha_3 = -\gamma_\phi/\sin\theta$, i.e. $\alpha_3 = -\gamma_\phi$(equator) is found. This quantity is given in Fig. 4.35. The influence of this effect for real turbulence fields is still an open question.

Stefani–Gerbeth Effect

If in a turbulent shell the α-effect changes its sign radially then it can happen that the resulting α^2-dynamo oscillates even for the simplest case of uniform α. This surprising phenomenon

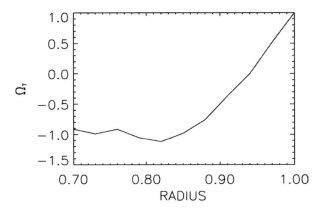

Figure 4.35: The azimuthal advection rate Ω_T as a function of depth (Ossendrijver et al. 2002, their Fig. 4). The tensor quantity γ_ϕ at the equator equals $x\Omega_T$.

only exists for some rather limited α-profiles (Fig. 4.36, see Stefani & Gerbeth 2003). This bistability behavior also exists for more realistic α-values, e.g. also including the $\cos\theta$. One can also ask whether for a given α-profile with changing signs a radial profile for η_T exists leading to oscillating solutions. The answer is yes. For example, for $\alpha \propto \sin(2\pi x)\cos\theta$ the solution oscillates if the ratio of the inner and the outer (uniform) η_T is chosen as between 0.7 and 0.8.

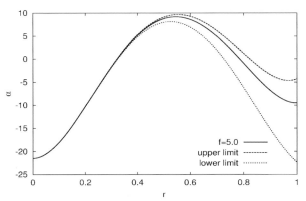

Figure 4.36: Possible radial α-profiles for which Stefani & Gerbeth (2003) find that even the simplest α^2-dynamo oscillates. Here the α-effect does not vanish for $r = 0$.

In the light of Fig. 4.23 (left), with its clear separation of positive and negative α-effect in the shell without density stratification, the Stefani–Gerbeth effect may be of interest for the reversal theory of the geodynamo. For a rather consistent model given in Krause & Rädler (1980), however, working with a distributed turbulence intensity (their Fig. 16.1), the oscillations did not appear.

4.4.2   The $\alpha\Omega$-Dynamo for Slow Rotation

We now turn to the solution of the induction equation with differential rotation included. Roberts (1972) discussed the excitation conditions of distributed $\alpha\Omega$-shell dynamos. The α-effect was located in an outer shell, the dynamo was embedded in vacuum and the differential rotation $d\Omega/dx$ was considered as uniform throughout the shell. For positive dynamo number $(\alpha_{\mathrm{north}}\cdot d\Omega/dx)$ dipoles are excited slightly easier than quadrupoles for thin convection zones, but for deep zones the quadrupoles clearly dominate (Table 4.6). At $x_{in} = 0.65$ both the values are equal (see Proctor 1977).

Table 4.6: The dynamo numbers for dipolar (A) to equatorially quadrupolar (S) solutions for $\alpha\Omega$-shell dynamos with $d\Omega/dr > 0$. At $x_{in} = 0.65$ both eigenvalues are equal. From Roberts (1972).

	$\alpha > 0$		$\alpha < 0$	
x_{in}	dipole	quadrupole	dipole	quadrupole
0	87	**76**	**74**	95
0.4	90	**84**	**82**	88
0.65	**136**	**136**	**133**	**133**
0.75	**200**	203	200	**197**

Roberts & Stix (1972) computed $\alpha^2\Omega$-dynamos with a rotation law close to the solar rotation law. The results are given in their Figs. 3 and 4. Without latitudinal differential rotation but with positive (negative) $\partial\Omega/\partial x\big|_{eq}$ the solution with quadrupolar (dipolar) parity has a lower eigenvalue than the solution with the dipolar (quadrupolar) symmetry – similar to the Steenbeck-Krause (1969) model with the α-effect being unchanged. The conclusion arises that the solar dynamo should work with Ω increasing with depth (which has not been confirmed later by helioseismology). The inclusion of the *latitudinal* shear produces still higher dynamo numbers, and the difference between quadrupolar and dipolar parity always grows.

Köhler (1973) started to consider only the excitation of modes with prescribed dipolar parity. For α-effect positive in the northern hemisphere he found for positive $\partial\Omega/\partial x$ a poleward drift of the toroidal magnetic field belts, in opposition to the observations, a situation that later has been called the 'dynamo dilemma' (Parker 1987).

Moss & Brooke (2000), in order to produce the observed equatorward migration of the toroidal fields, worked with the solar rotation law and with *negative* northern α-effect in the bulk of the convection zone. The dipolar solutions are only slightly easier to excite than the quadrupolar ones (dipole: $C_\alpha = -3.20$, quadrupole $C_\alpha = -3.25$). The parity problem does not seem to exist for negative α-effect (see also Fig. 3 in Roberts & Stix 1972). The situation, however, changes for positive northern α-effect (Roberts & Stix 1972, Fig. 4; Moss 1999). In this case one is confronted for the solar dynamo with an enormous parity problem. The basic solution in this case is of quadrupolar symmetry (see Dikpati & Gilman 2001).

For axisymmetry the mean flow and magnetic field in spherical coordinates are given by

$$\bar{u} = (0, 0, r\sin\theta\,\Omega)\,, \qquad \bar{B} = \left(\frac{1}{r^2\sin\theta}\frac{\partial A}{\partial\theta}, -\frac{1}{r\sin\theta}\frac{\partial A}{\partial r}, B\right), \qquad (4.86)$$

with A as the stream function of the poloidal field[16] and B as the toroidal field. Their evolution

---

[16] $A = $ const. are the field lines

is described by

$$\frac{\partial A}{\partial t} = \alpha s B + \eta_{\mathrm{T}} \left(\frac{\partial^2 A}{\partial r^2} + \frac{\sin\theta}{r^2} \frac{\partial}{\partial\theta} \left(\frac{1}{\sin\theta} \frac{\partial A}{\partial\theta} \right) \right),$$

$$\frac{\partial B}{\partial t} = \frac{1}{r} \frac{\partial\Omega}{\partial r} \frac{\partial A}{\partial\theta} - \frac{1}{s} \frac{\partial}{\partial r} \left(\alpha \frac{\partial A}{\partial r} \right) -$$

$$- \frac{1}{r^3} \frac{\partial}{\partial\theta} \left(\frac{\alpha}{\sin\theta} \frac{\partial A}{\partial\theta} \right) + \frac{\eta_{\mathrm{T}}}{s} \left(\frac{\partial^2 (sB)}{\partial r^2} + \frac{\sin\theta}{r} \frac{\partial}{\partial\theta} \left(\frac{1}{s} \frac{\partial (sB)}{\partial\theta} \right) \right). \quad (4.87)$$

These equations are solved with a finite-difference scheme for the radial dependence and a polynomial expansion for the angular dependence. The expansions

$$A = \mathrm{e}^{-\mathrm{i}\omega t} \sum_j a_j(x) P_j^1(\cos\theta) \sin\theta, \quad B = \mathrm{e}^{-\mathrm{i}\omega t} \sum_k b_k(x) P_k^1(\cos\theta) \quad (4.88)$$

are used, where ω is the real (for marginal stability) eigenvalue, j is odd and k is even for equatorially antisymmetric modes, and vice versa for equatorially symmetric modes. Vacuum boundary conditions at $x = 1$ yield

$$\frac{\mathrm{d}a_n}{\mathrm{d}x} + na_n = b_n = 0, \quad (4.89)$$

whereas at $x = x_{\mathrm{in}}$ we take

$$x \frac{\mathrm{d}b_n}{\mathrm{d}x} + b_n = a_n = 0 \quad (4.90)$$

corresponding to perfectly conducting boundary conditions[17]. Then the dimensionless turbulence numbers are

$$C_\Omega = \frac{V^{(0)}\Omega_\odot R^2}{\eta_{\mathrm{T}}}, \qquad C_\alpha = \frac{\alpha R}{\eta_{\mathrm{T}}}, \qquad C_\omega = \frac{\omega R^2}{\eta_{\mathrm{T}}}. \quad (4.91)$$

Here C_Ω is the normalized radial shear, which is positive for superrotation ($\partial\Omega/\partial x > 0$) and negative for subrotation ($\partial\Omega/\partial x < 0$). For oscillatory solutions C_ω is the normalized cycle frequency. We always consider C_Ω as given and compute for marginal instability (ω real) the resulting eigenvalues C_α and C_ω.

Let the latitudinal dependence of the rotation law be neglected. This is allowed for slowly rotating dynamos, with Ω^* so small that the radial Λ-effect only exists with $V^{(1)} = H^{(1)} = 0$. The radial rotation law simply results from $x\mathrm{d}\Omega/\mathrm{d}x = V^{(0)}\Omega$, which might be approached by the linear rotation law $\Omega \simeq (1 + V^{(0)}x)\Omega_\odot$. Figure 4.37 gives the results for a thick ($x_{\mathrm{in}} = 0.1$) and a thin ($x_{\mathrm{in}} = 0.8$) outer convection zone that – in the $\alpha\Omega$-regime – might be compared with the results in Table 4.6 by Roberts. Again, thick shells with superrotation excite oscillating quadrupolar solutions, while for thin shells dipoles and quadrupoles are excited approximately equally easily. Hence, the distributed dynamo models with weak or strong superrotation below the equatorial region can not explain the dominant dipolar geometry of the solar magnetic field. Moreover, the dynamo computations for positive C_Ω always lead to positive values of $\bar{B}_r \cdot \bar{B}_\phi$ in contrast to the observations. The cycletime statistics (Fig. 4.2)

---

[17] provided $\alpha(x_{\mathrm{in}}) = 0$

also do not confirm the strong differences in C_ω for thin and thick convection zones, which are summarized in Fig. 4.38 (left). For simple shell dynamos

$$\tau_{\mathrm{cyc}} \simeq 0.26 \, \frac{DR}{\eta_{\mathrm{T}}} \simeq \frac{10 \, \mathrm{yr}}{\eta_{\mathrm{T}} / \left(10^{12} \, \mathrm{cm}^2/\mathrm{s}\right)} \tag{4.92}$$

results, the latter with solar values adopted.

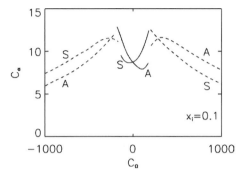
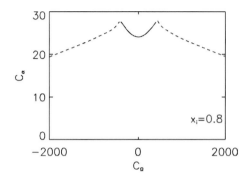

Figure 4.37: The excitation conditions for $\alpha^2 \Omega$-dynamos for a thick (*left*) and a thin (*right*) outer convection zone. Stationary solutions are represented by solid lines, oscillatory solutions by dashed lines. Dipolar symmetry: A, quadrupolar symmetry: S.

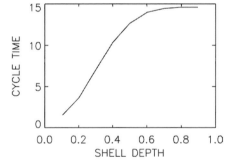
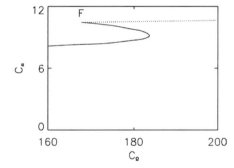

Figure 4.38: *Left*: The cycletime for oscillating outer-shell $\alpha\Omega$-dynamos with increasing shell thickness and with $\eta_{\mathrm{T}} = 1.4 \cdot 10^{12} \, \mathrm{cm}^2/\mathrm{s}$. Note the linear growth with the shell depth. *Right*: The transition of stationary dipolar solutions to oscillatory dipolar solutions close to the bifurcation point F. Courtesy A. Bonanno.

Note that in Fig. 4.37 there is no continuous transition between stationary and oscillatory solutions. Surprisingly, when at small C_Ω the first oscillatory solution appears, it needs a stronger α-effect than the stationary α^2-dynamos do (Charbonneau & MacGregor 2001). Only for rather strong C_Ω does the necessary C_α become smaller than for the α^2-dynamos.

The question arises how the transition between the two dynamo regimes happens in detail. In Fig. 4.38 (right) details are presented. There is a jump in the lowest C_α between stationary and oscillatory fields. The oscillating solution bifurcates from a second branch of the α^2-dynamos rather than from the lowest one. The α^2-dynamo only possesses growing modes between the first and the second solid line. There are no growing modes between the second solid line and the dashed line. A simultaneous excitation of stationary and oscillating modes in the linear regime is therefore excluded. A stable flip-flop phenomenon (here for the poles) cannot exist.

4.4.3   Meridional Flow Influence

The influence of a meridional flow on the dynamo might also be important. Roberts & Stix (1972) found that a (slow) meridional flow of either sense inhibits the dynamo action (their Fig. 6). For clockwise flow (toward the pole at the bottom of the convection zone) the oscillation frequency grows for growing drift amplitude, i.e. cycletime and drift amplitude become anticorrelated. These authors also already used the meridional flow to obtain corrections to the form of the butterfly diagram.

Figure 4.39 shows the influence of a slow meridional flow on the excitation and cycle period of a nonlinear dynamo, which works with superrotation and a negative surface α-effect. The butterfly diagrams are given with and without the meridional flow. For positive bottom flow (counterclockwise flow) the cycle period and flow amplitude are positively correlated. The faster the flow the longer the cycle – in opposition to the observations (Fig. 3.10). Note that the dynamo does not survive if the flow is too fast.

4.5   The Solar Dynamo

A final, properly working mean-field model of the solar dynamo does not yet exist. For its construction we would certainly need more information about the dynamics within the convection zone, and also about the solar-stellar connections (internal rotation, cycle statistics). In the following we present the two main models. The interior rotation law of the Sun, which is known from helioseismology (except in the polar region), is always used in the calculations. The first model works with a negative α-effect, which is assumed to exist in the (thin) solar overshoot region. Here the positive α-values that are expected in the bulk of the convection zone are simply neglected. The second model works with the meridional flow that results from the mean-field models presented in Sect. 3.3. The overall problems with the solar dynamo are discussed in considerable detail by Ossendrijver (2003). Here only the two basic concepts are presented.

4.5.1   The Overshoot Dynamo

The spatial location of the dynamo action is still unknown, until helioseismology can reveal the exact position of the magnetic toroidal belts beneath the solar surface (see Dziembowski & Goode 1991, Antia, Basu & Chitre 1998). There are arguments in favor of locating it deep within or below the convection zone, namely

- Hale's law of sunspot parities can only be fulfilled with strong toroidal magnetic field belts (10^5 G, see Moreno-Insertis 1983, Choudhuri 1989, 1990, Fan, Fisher & DeLuca 1993, Caligari, Moreno-Insertis & Schüssler 1995).

- The dynamo field strength approaches the equipartition value $B_{eq} = \left(\mu_0 \rho u_T^2\right)^{1/2}$. Using mixing-length theory arguments $\rho u_T^3 \simeq$ const., hence B_{eq} gradually increases inward. At the bottom of the convection zone it is of order 10 kG.
- The radial gradient of Ω is maximal below the convection zone.

High field amplitudes might thus be generated only in the layer between the convection zone and the radiative interior (van Ballegooijen 1982). On the other hand, if there is some form of turbulence in this layer, it would be hard to understand the present-day Li concentration in the solar convection zone. Lithium burning starts only 40,000 km below the bottom of the convection zone. Any turbulence in this domain would lead to a rapid and complete depletion of the Li in the convection zone.

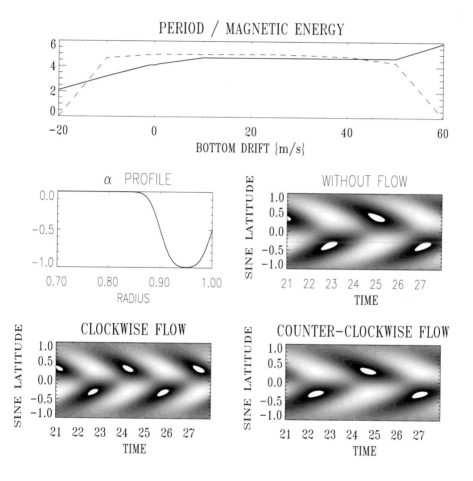

Figure 4.39: A nonlinear dynamo with prescribed superrotation and with $\partial\Omega/\partial\theta = 0$, negative $\alpha_{\phi\phi}$ and meridional flow of low magnetic Reynolds number ($\eta_T = 10^{12}$ cm^2/s). Positive bottom drift means counterclockwise flow and vice versa. *Top*: Cycle period in years (solid) and magnetic energy (dashed). *Bottom*: Butterfly diagrams for clockwise flow and counterclockwise flow. From Rüdiger & Arlt (2003).

The negativity of α in the overshoot region (Fig. 4.22, right) can only be relevant for the dynamo if the bulk of the convection zone is free from positive α. It has been argued that the short rise-times of the flux tubes in the convection zone prevent the formation of the α-effect (Spiegel & Weiss 1980, Schüssler 1987, Stix 1991). The rise-times may be much longer in the overshoot region (Ferriz-Mas & Schüssler 1995, van Ballegooijen 1998).

There is a serious shortcoming in the concept of overshoot dynamos as the characteristic scales of the mean magnetic fields cannot exceed the scales of the turbulence. The validity of the local formulations of the mean-field electrodynamics is not ensured for such thin layers. Nevertheless, a number of quantitative models exist (Choudhuri 1990, Belvedere, Lanzafame & Proctor 1991, Markiel & Thomas 1999). For a demonstration of the abilities of such a model Rüdiger & Brandenburg (1995) worked under the assumptions that

- α exists only in the overshoot region, η_{T} also in the convection zone,
- a possible vanishing of the α-effect in the polar regions is parameterized with $\alpha \sim (1 - \alpha_{\mathrm{u}} \cos^2 \theta) \cos \theta$,
- the correlation time is about 10^6 s so that $\Omega^* \simeq 5$ results,
- the tensors α and η_{T} are computed for a velocity profile by Stix (1991).

The main results from the model are that

- for $\alpha_{\mathrm{u}} \simeq 0$ the magnetic activity is concentrated near the poles, for $\alpha_{\mathrm{u}} \simeq 1$ it moves to the equator (Figs. 4.40 and 4.41),

- for too thin α-layers there are too many toroidal magnetic belts in each hemisphere (Fig. 4.41).

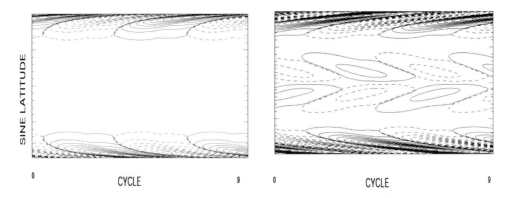

Figure 4.40: Butterfly diagram for an overshoot dynamo for negative $\alpha_{\phi\phi}$ in a (thick) layer of 70 Mm. *Left*: $\alpha_{\mathrm{u}} = 0$, α-effect maximum at the poles. Only polar spots can be expected. *Right*: $\alpha_{\mathrm{u}} = 1$, α-effect vanishing at the poles. A polar branch migrating polewards and an equatorial branch migrating equatorwards result.

The linear solutions have the correct cycle period, which has just the same sensitivity to the overshoot depth as the shell dynamo, i.e. $\tau_{\mathrm{cyc}} \sim D$ for τ_{cyc} in years and D in Mm ($D \simeq$ 15–35 Mm). Its increase with D is in agreement with the linear relation for spherical shell dynamos (see Fig. 4.38). For too thin boundary layers the cycletime becomes too short.

Note that the overshoot dynamo with the standard $\cos\theta$-profile of the α-effect only produces strong toroidal magnetic belts in the polar region (Fig. 4.40, left). Such spot geometry does indeed exist for rapidly rotating stars (see Figs. 4.9 and 4.11). No simulation so far has provided a vanishing α-effect at the poles (see Ossendrijver, Stix & Brandenburg 2001).

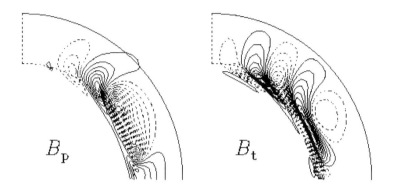

Figure 4.41: The magnetic field geometry for the nonlinear overshoot dynamo in a layer of only 35 Mm with buoyancy. An obvious problem of the dynamo is the large number of toroidal field belts.

4.5.2 The Advection-Dominated Dynamo

Already Spörer (1894, refering to Wolf) speculated that the equatorward drift of the sunspots ('Spörer's law') might be due to the action of a meridional flow toward the equator. Indeed, the meridional flow $\boldsymbol{u}^{\mathrm{m}}$ also influences the mean-field dynamo. This influence can be expected to be small if its characteristic timescale τ_{drift} exceeds the cycletime τ_{cyc}. The equations for the fields with meridional flow included are

$$\frac{\partial A}{\partial t} + (\boldsymbol{u}^{\mathrm{m}} \cdot \boldsymbol{\nabla})A = \alpha s B + \eta_{\mathrm{T}}\left(\frac{\partial^2 A}{\partial r^2} + \frac{\sin\theta}{r^2}\frac{\partial}{\partial\theta}\left(\frac{1}{\sin\theta}\frac{\partial A}{\partial\theta}\right)\right),$$

$$\frac{\partial B}{\partial t} + s\rho(\boldsymbol{u}^{\mathrm{m}} \cdot \boldsymbol{\nabla})\frac{B}{s\rho} = \frac{1}{r}\left(\frac{\partial\Omega}{\partial r}\frac{\partial A}{\partial\theta} - \frac{\partial\Omega}{\partial\theta}\frac{\partial A}{\partial r}\right) - \frac{1}{s}\frac{\partial}{\partial r}\left(\alpha\frac{\partial A}{\partial r}\right) -$$

$$-\frac{1}{r^3}\frac{\partial}{\partial\theta}\left(\frac{\alpha}{\sin\theta}\frac{\partial A}{\partial\theta}\right) + \frac{\eta_{\mathrm{T}}}{s}\left(\frac{\partial^2(sB)}{\partial r^2} + \frac{\sin\theta}{r}\frac{\partial}{\partial\theta}\left(\frac{1}{s}\frac{\partial(sB)}{\partial\theta}\right)\right) \qquad (4.93)$$

with $\boldsymbol{u}^{\mathrm{m}} = (\bar{u}_r, \bar{u}_\theta, 0)$. The dynamo problem is studied for dynamos with small eddy diffusivities so that the diffusion time exceeds the advection time and the meridional flow advects the toroidal magnetic field belts (Wang, Sheeley & Nash 1991[18], Nordlund, Galsgaard & Stein 1994, Durney 1995, Choudhuri, Schüssler & Dikpati 1995). Models exist with positive α-effect mainly at the top of the convection zone, and also with positive α-effect mainly at the

---

[18] an extremely anisotropic eddy diffusivity is used

bottom of the convection zone. Distinct differences in the parity characteristics of the two models appear (Dikpati & Gilman 2001).

The meridional flow only has a strong impact in the mean-field dynamo for low eddy diffusivity. For $\eta_T = 10^{11} \mathrm{cm}^2/\mathrm{s}$ (known from the sunspot decay rate) the magnetic Reynolds number $\mathrm{Rm} = u^m R/\eta_T$ reaches values of the order of 10^3 for flows of 10 m/s, and strong modifications of magnetic field geometry and cycle period can be expected. This possibility has been the subject of intense numerical investigations (Dikpati & Charbonneau 1999, Dikpati & Gilman 2001, Küker, Rüdiger & Schultz 2001, Nandy & Choudhuri 2002), where it has been shown that solutions with high magnetic Reynolds number yield the correct cycle period, butterfly diagrams *and* magnetic phase relations with a *positive* α-effect in the northern hemisphere. Quadrupolar field configurations are more easily excited than dipolar ones, however, if there is no α-effect below (say) $x = 0.8$. Dipolar solutions only dominate if the α-effect is concentrated at the bottom layer of the convection zone (Bonanno et al. 2002).

With the following models we shall demonstrate the situation. The $\alpha_{\phi\phi}$ is assumed never to change its sign at a certain radius, or even in the overshoot layer. Below the convection zone the magnetic diffusivity is 10 times smaller than within the convection zone. The known rotation law $\Omega = \Omega(r, \theta)$ within the convection zone is always used in the calculations.

α-Effect in the Entire Convection Zone

Let a (positive) α-effect exist throughout the whole convection zone. In Table 4.7 the results are given. The drift amplitude at the bottom of the convection zone varies between 2 m/s and 6 m/s and the dipole-solution only for slow flow occurs with the lowest α-effect amplitude. With the small values of eddy diffusivity the cycle period becomes much too long compared with the solar value. Figure 4.42 shows the magnetic geometry of the dynamo with 6 m/s drift amplitude. The toroidal field belts are concentrated at the bottom of the convection zone. They migrate toward the equator, but the maximal field strengths occur in the polar region. $\bar{B}_r$ and $\bar{B}_\phi$ are mainly out of phase.

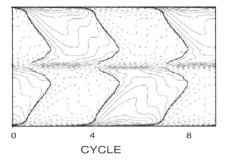
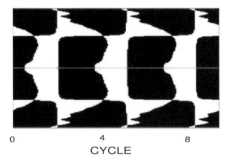

Figure 4.42: α-effect in the entire convection zone: Butterfly diagram (*left*) and field phase relation (*right*) of a dynamo with critical turbulence $\alpha_0 = 2.5$ cm/s, $\eta_T = 10^{11}$ cm^2/s, $u^m = 6$ m/s. Black means negative $\bar{B}_r \cdot \bar{B}_\phi$.

Table 4.7: Critical α-values and cycle periods for models with α-effect in the entire convection zone and $\eta_T = 10^{11}$ cm^2/s. Bold is used for the solution with the smallest α-amplitude. The dipolar symmetry is denoted by A, the quadrupolar symmetry is denoted by S.

u^m [m/s]	α^A [cm/s]	τ^A_{cyc} [yr]	α^S [cm/s]	τ^S_{cyc} [yr]
2	**0.90**	∞	1.28	131
3	1.83	82	**1.70**	83
6	2.46	51	**2.17**	54

α-Effect at the Top

Models with the α-effect located only at the top of the convection zone are now considered. The results of the simulations confirm the basic features of the advection-dominated dynamo, namely that for a flow of a few m/s and for low diffusivity the butterfly diagram shows the correct equatorward migration of the toroidal field and the phase relation of the magnetic fields is mostly negative.

The results are summarized in Table 4.8. As far as the parity selection is concerned, variations of the ratio η_T/η_c are not significant. For slow flow the solutions are of quadrupolar

Table 4.8: α-effect at the top (thin layer): Critical α-values and cycle periods for various values of the flow with $\eta_T = 10^{11}$ cm^2/s.

u^m [m/s]	α^A [cm/s]	τ^A_{cyc} [yr]	α^S [cm/s]	τ^S_{cyc} [yr]
2	7.11	∞	**5.11**	∞
3	**7.24**	253	7.93	∞
5	9.88	176	**6.59**	137

symmetry and even stationary. Only for intermediate values of the flow dipole solutions have a smaller critical α-value. For fast flow quadrupolar fields are again more easily excited and the cycletimes remain much too long (Dikpati & Charbonneau 1999, Dikpati & Gilman 2001).

α-Effect at the Bottom

The situation is changed if the α-effect is located at the bottom of the convection zone. We have worked with a thin α-layer where the α-effect only exists between 0.7 and 0.8. This constellation is rather close to the 'interface dynamo' by Parker (1993, in Cartesian geometry) and by Charbonneau & MacGregor (1997, in spherical geometry). Figure 4.43 presents an example. Once again, the toroidal field is concentrated at the bottom of the convection zone, but the highest field amplitudes occur in the polar regions. The diagram for $\bar{B}_r \cdot \bar{B}_\phi$ shows dominance of the negative sign.

For the models with a thin α-layer at the bottom of the convection zone the solution with the dipolar symmetry always possesses the lowest α-value. There is thus no parity problem

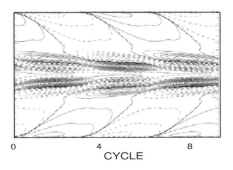
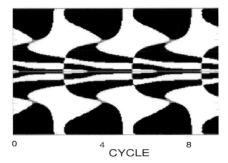

Figure 4.43: The same as in Fig. 4.42 but for the α-effect at the bottom ($\alpha = 4$ cm/s, $\eta_T = 10^{11}$ cm²/s, $u^m = 3$ m/s). Cycle period is 74 yr. The α-effect is located between 0.7 and 0.75. From Bonanno et al. (2002).

if the α-layer is located at the base of the convection zone. However, we find from Fig. 4.44 that for small η_T the cycletime is too large, but it is reduced by increasing the amplitude of the meridional flow at the bottom of the convection zone. Even in this case though the dipole solutions do not perfectly match the 11-year cycle period of the Sun. The overall result for such models is that the dipolar solutions are always more easily excited, the butterfly diagram shows the right characteristics, and the observed anticorrelation of $\bar{B}_r$ and $\bar{B}_\phi$ is also realized.

For too slow flow, however, the cyclic behavior of the dynamo disappears so that only meridional flows with amplitudes exceeding 3 m/s are here relevant. On the other hand, from Fig. 4.44 (left) one can take the finding that $\omega_{\rm cyc}$ grows with growing meridional flow (at the bottom). Hence, with the relation $u^m \propto \Omega$ taken from Fig. 3.25 (right), one finds a weak dependence (4.2) with about $n \simeq 0.30$, rather than a dependence on Ω^* (see Fig. 4.44 (right) and Messina & Guinan 2003, their Fig. 5).

Two Cells

It is very tempting to construct an advection-dominated dynamo with a two-cell flow system where at the bottom the polar cell flows poleward and the equatorial cell flows equatorward (see Fig. 3.8). In Fig. (4.45) one example is given with a resulting dipolar magnetic field of the correct cycle period of 22.6 yr. Without flow the butterfly diagram given at the top of the figure is of the type presented by Köhler (1973). The polar cell of the used circulation pattern is slightly smaller than the equatorial cell. The differences of the flow-dominated dynamo model to the flow-free model are obvious. Systematic studies for such dynamos remain to be done.

4.6 Dynamos with Random α

All the averages in the foregoing equations are imagined as taken over an 'ensemble', i.e. over a great number of identical examples. The other possibility to explain the temporal irregularities is to consider the characteristic turbulence values as a time series. The idea is that the averaging procedure concerns only a periodic spatial coordinate, e.g. the azimuthal

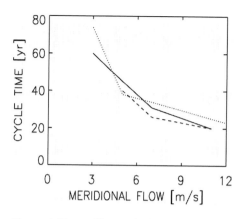 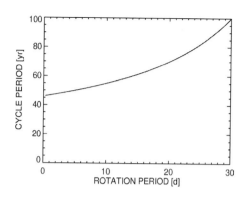

Figure 4.44: α-*effect at the bottom. Left:* The cycle periods for the dipolar modes from the various models of Dikpati & Charbonneau, Küker et al. and Bonanno et al. There is a clear anticorrelation of the drift velocity and the cycletime (see Hathaway et al. 2003). *Right:* The cycletimes grow slowly with the rotation periods if the models of Fig. 3.25 (right) are included. This result corresponds to small exponents n in Eq. (4.2).

angle ϕ (Braginsky 1964, Hoyng 1993). In other words, when expanding in Fourier series such as $e^{im\phi}$ the mode $m = 0$ is considered as the mean value. If the timescale of this mode does not vary significantly during the correlation time, local formulations such as Eq. (4.95) below are reasonable. Nevertheless, the turbulence intensity, the α-effect and the eddy diffusivity become time-dependent quantities (Hoyng 1988, Choudhuri 1992, Moss et al. 1992, Hoyng 1993, Hoyng, Schmitt & Teuben 1994, Vishniac & Brandenburg 1997, Otmianowska-Mazur et al. 1997, Mininni & Gómez 2002).

The turbulent EMF for a given position forms a time series with the correlation time τ_{corr} as a characteristic scale. The peak-to-peak variations in the time series should depend on the number of cells. They remain finite if the number of cells is restricted. For an infinite number of turbulence cells the peak-to-peak variation in the time series goes to zero.

The tensors constituting the local mean-field EMF must be calculated from one and the same turbulence field. We define a helical turbulence and compute simultaneously the EMF coefficients. We restrict ourselves to the high-conductivity limit. Then the SOCA yields

$$\mathcal{E} = \int_0^\infty \left\langle \boldsymbol{u}'(\boldsymbol{x},t) \times \nabla \times \left(\boldsymbol{u}'(\boldsymbol{x},t-\tau) \times \bar{\boldsymbol{B}}(\boldsymbol{x},t) \right) \right\rangle \, \mathrm{d}\tau, \tag{4.94}$$

which for short correlation times can be written in the form of Eq. (4.7). For a simple dynamo only the components $\mathcal{E}_x$ and $\mathcal{E}_y$ are relevant[19]. In components it reads

$$\mathcal{E}_x = \alpha_{xx}\bar{B}_x + \eta_{\mathrm{T}}\frac{\partial \bar{B}_y}{\partial z}, \qquad\qquad \mathcal{E}_y = \alpha_{yy}\bar{B}_y - \eta_{\mathrm{T}}\frac{\partial \bar{B}_x}{\partial z}, \tag{4.95}$$

---

[19] it would be tempting to apply Eq. (4.94) as it stands

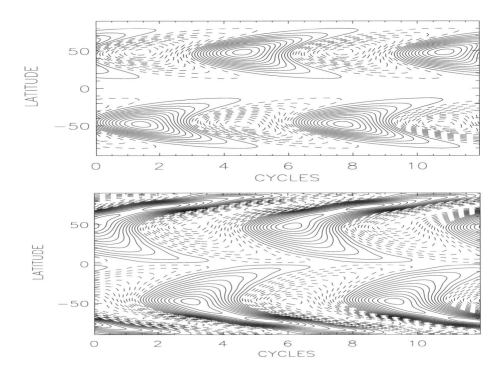

Figure 4.45: Butterfly diagram of a dynamo without meridional flow (*top*) and with a two-cell flow (*bottom*) with α in the entire convection zone (a dipole!) with maximally 5.4 m/s flow amplitude. Courtesy A. Bonanno.

with η_T as an eddy diffusivity. From Eq. (4.94) one finds

$$\alpha_{xx} = \int\limits_0^\infty \left\langle u_y'(t)\, \frac{\partial u_z'(t-\tau)}{\partial x} - u_z'(t)\, \frac{\partial u_y'(t-\tau)}{\partial x} \right\rangle \, \mathrm{d}\tau,$$

$$\alpha_{yy} = \int\limits_0^\infty \left\langle u_z'(t)\, \frac{\partial u_x'(t-\tau)}{\partial y} - u_x'(t)\, \frac{\partial u_z'(t-\tau)}{\partial y} \right\rangle \, \mathrm{d}\tau, \tag{4.96}$$

and

$$\eta_\mathrm{T} = \int\limits_0^\infty \langle u_z'(t)\, u_z'(t-\tau) \rangle \, \mathrm{d}\tau \tag{4.97}$$

(Krause & Rädler 1980). While the α-effect comes from helicity formations, the magnetic diffusivity η_T is a much simpler integral over a two-point correlation.

4.6.1 A Turbulence Model

The time evolution of the dynamo coefficients is considered. The x-axis is parallel to the solar radius and the z-axis is directed to the south pole. The parcel is permanently perturbed by vortices of maximum helicity (Otmianowska-Mazur, Urbanik & Terech 1992). All vortices are oriented like right-handed screws. The initial state is a number N of moving turbulent cells with random inclinations and positions in the xy-plane. After a specified period of time, a fraction of them is replaced by new ones with random positions, inclination, and with lifetime starting from zero.

Table 4.9: Input and output for the turbulence models $\mathcal{A}$ and $\mathcal{D}$. N is the eddy population of the equator. Time, velocity and diffusivity result after multiplication with $2.5 \cdot 10^4$ s, 10^5 cm/s and $2.5 \cdot 10^{14}$ cm²/s.

	ℓ_{corr}	τ_{corr}	N	η_{T}	$S_{\eta_{\mathrm{T}}}$	α_{xx}	$S_{\alpha_{xx}}$	α_{yy}	$S_{\alpha_{yy}}$
$\mathcal{A}$	1	10	200	0.191	0.45	−0.046	1.03	−0.077	0.72
$\mathcal{D}$	8	80	25	1.793	1.27	−0.015	8.94	−0.017	6.62

The vortex radius ℓ_{corr} as well as the decay time τ_{corr} are varied for the cases given in Table 4.9 in normalized units.

The simulations deliver time series of the turbulence intensity, the eddy diffusivity, the α-coefficients and a standard deviation S from their time averages.

The results for a model $\mathcal{A}$ with $N = 200$, which is the sample of shortest scales of individual vortices, are given in Fig. 4.46. The diffusion coefficient η_{T} possesses positive values during most of the time. The ratio $S_{\eta_{\mathrm{T}}}$ is rather small, only 0.45. In contrast to η_{T}, the coefficient α_{yy} is negative for most of the time, although positive values are also present for short periods. The α-sign results from the assumed right-handed helicity of the vortices. The fluctuations of the α-effect dominate the fluctuations of the eddy diffusivity.

Model $\mathcal{D}$ uses much larger correlation lengths and times. The resulting fluctuations of η_{T} and α-coefficients are much higher than in the case $\mathcal{A}$. The ratios S also increase. The fluctuations of α are much higher for larger eddies with longer lifetimes. The fluctuations in the time series become more and more dominant with decreasing number of eddies. The fluctuations of both the eddy diffusivity as well as the α-effect are *much* higher than the averages. The α-effect fluctuations exceed those of the eddy diffusivity in all our models. The latter proves to be more stable than the α-effect against dilution of the turbulence (Fig. 4.46, right).

4.6.2 Dynamo Models with Fluctuating α-Effect

Choudhuri (1992) presented a plane-wave dynamo in the linear regime. The fluctuations did not exceed the 10% level. In the $\alpha\Omega$-regime the oscillations are hardly influenced; the opposite is true for the α^2-dynamo. In the latter regime the solution suffers dramatic and chaotic changes even for rather weak disturbances.

We present here a nonlinear plane 1D $\alpha^2\Omega$-model. It is *not* an overshoot dynamo (see Ossendrijver, Hoyng & Schmitt 1996). The plane has infinite extent in the radial and the

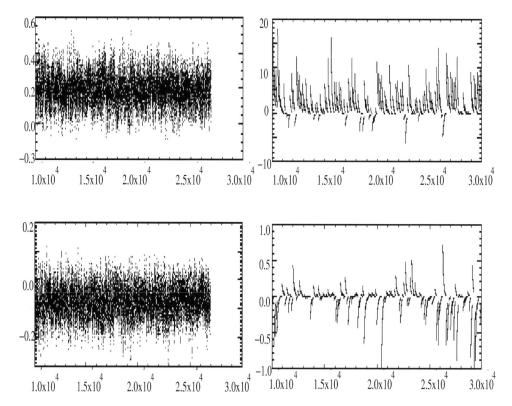

Figure 4.46: Time series for eddy diffusivity η_T (*top*) and the α-tensor component α_{yy} (*bottom*) for turbulence model $\mathcal{A}$ (*left*) and $\mathcal{D}$ (*right*). There is a characteristic difference in the behavior of eddy diffusivity and α-effect.

azimuthal direction; the boundaries are in the z-direction. We assume that the fields depend on z only.

The α-tensor has only one component, vanishing at the equator and also at the poles. The magnetic feedback is considered to be conventional α-quenching. The diffusivity is fluctuating but spatially uniform. The equations are given in Sect. 6.4.1 together with the boundary conditions.

Our dynamo numbers are $C_\alpha = 5$ and $C_\Omega = 200$. The turbulence models $\mathcal{A}$ and $\mathcal{D}$ are applied, flow patterns with small ($\mathcal{A}$) and with very large ($\mathcal{D}$) eddies are used.

A magnetic dipole field oscillating with a fixed period (Fig. 4.47, top) is produced by a model without any EMF-fluctuations. The period corresponds to an activity cycle of 8 yr in physical units. The turbulence model $\mathcal{A}$ also produces an oscillating dipole, but with a more complicated temporal behavior (Fig. 4.47). It is not a single oscillation; the power spectrum forms a broad line with substructures. The 'quality' $Q = \omega_{\mathrm{cyc}}/\Delta\omega_{\mathrm{cyc}}$ of this line (with $\Delta\omega_{\mathrm{cyc}}$ as its half-width) close to the observed quality of the solar cycle (Fig. 4.5) is produced here by a turbulence model with about 100 eddies along the equator. Variations of the cycle amplitude and the parity also exist.

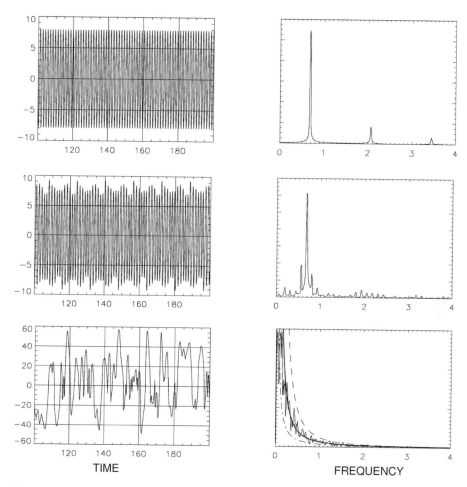

Figure 4.47: Dynamo-induced magnetic toroidal field amplitudes for the turbulence models $\mathcal{A}$ (*middle*) and $\mathcal{D}$ (*bottom*). *Left*: Time series. *Right*: Power spectra. The time unit is 2.7 yr. *Top*: a reference model without any EMF fluctuations. From Otmianowska-Mazur et al. (1997).

The turbulence field $\mathcal{D}$ leads to a highly irregular temporal behavior in the magnetic quantities. Its power spectrum peaks at several periods. The shape of the spectrum, however, no longer suggests any oscillations. The power of the lower frequencies is strongly increased; the high-frequency power decreases as $\omega_{\mathrm{cyc}}^{-5/3}$, like a Kolmogorov spectrum, indicating the existence of chaos.

Implications for the temporal evolution of the solar rotation law ('fluctuating Λ-effect') should be another output of such a cell number statistics. The consequences for the rotation law form an independent test of the theory. For only 'occasional' turbulence a nontrivial time series for the turbulence EMF-coefficients, the magnetic field, the turbulent angular momentum transport and the differential rotation are unavoidable consequences.

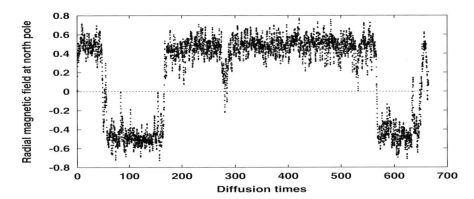

Figure 4.48: Time series of the magnetic amplitude at the north pole of a spherical model. Sometimes the α-fluctuations lead to subcritical conditions and the dynamo collapses. The recreated field is of opposite polarity. Courtesy R. Arlt.

With a 3D spherical dynamo model Arlt (2004) has probed the influence of strong α-fluctuations in a shell-dynamo located between $x = 0.5$ and the surface. The boundary conditions are those for vacuum outside the shell. The α-tensor is perturbed in space and time by smooth 'blobs' of one diffusion time duration and a rather large size. The α-tensor has been assumed as (slightly) anisotropic between the vertical and azimuthal components. The time series of the polar magnetic amplitude is shown in Fig. 4.48. During the reversals (which here are consequences of temporarily subcritical conditions) a nonaxisymmetric magnetic mode appears. The latter dominates at times of reversals or excursions. In this model the existence of reversals in the geodynamo is considered as an indication of an only slightly supercritical α-effect with slightly anisotropic components. If the lifetime of the fluctuations is reduced even just by a factor of 2 the reversals disappear.

4.7   Nonlinear Dynamo Models

A dynamo is an instability of the solution $B = 0$ if a threshold value C_{crit} of the dynamo number is reached. If the dynamo number exceeds this critical value then the magnetic field grows exponentially, until it saturates through its feedback on the original flow field. This feedback is due to the Lorentz force, as the magnetic field is accompanied by electrical currents. The question arises whether these currents can be observed by their influence on the structure of the stellar atmosphere, with consequences for the spectral lines (Stępień 1978, Landstreet 1987). The electrical currents that occur during the decay of a primordial field have already been computed by Wrubel (1952) and more recently by Moss (2003).

Krause & Meinel (1988) and Schmitt & Schüssler (1989) initiated consideration of the stability of dynamo-generated magnetic fields. According to Krause & Meinel for simple α^2-dynamos "the only stable nonlinear steady solution is that which bifurcates from the trivial solution at the marginal dynamo number of the most easily excitable linear mode". A simple example is described for a 1D α^2-dynamo with a global α-quenching $\alpha \propto \Psi(E)$ with $E =$

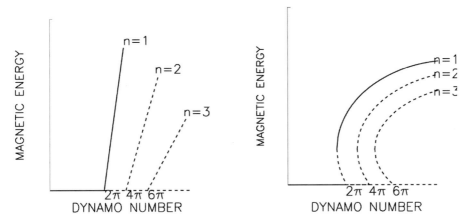

Figure 4.49: Stability maps for simple nonlinear dynamo models. Broken lines indicate unstable solution branches. Subcritical solutions may also occur (*right*). Meinel (1991).

$\int \bar{B}^2 \mathrm{d}x$. The dynamo equations are those of Eq. (4.83) with the same boundary conditions but written in the dimensionless quantities ζ and τ. The solutions are written as $\sum B_n(\zeta) \exp \omega_n \tau$ with the results

$$B_n = \sin n\pi\zeta \, \mathrm{e}^{-\mathrm{i}C_\alpha\zeta/2}, \qquad \omega_n = \frac{C_\alpha^2}{4} - n^2\pi^2. \qquad (4.98)$$

The mode B_n grows for $\omega_n > 0$, i.e. $C_\alpha > 2n\pi$. The solution $B = 0$ is not stable for $C_\alpha > 2\pi$. For the solutions the necessary stability conditions are (i) $\mathrm{d}\Psi/\mathrm{d}E < 0$ and (ii) $n = 1$. Only the first mode can be nonlinearly stable for monotonically decreasing Ψ (Fig. 4.49, left). If Ψ has a local maximum[20] then subcritical excitation can occur, but the solutions close to the bifurcation point are unstable (Fig. 4.49, right). This, in short, is the general outline of the paper by Krause & Meinel (1988), which has been confirmed by many subsequent calculations with various concepts for the nonlinear back-reaction. Some of them are presented in the following.

Of course, at a certain finite distance from the bifurcation point the stability of a solution may change, and further bifurcations can occur. It may also happen that such solutions are hardly in close connection to the kinematic modes. However, for the extrapolation of the kinematic modes into the nonlinear regime only the fundamental mode (i.e. the 'lowest' mode) is interesting. This mode is defined in the sense that is given in Fig. 4.49 as the solid line. This mode is already marginal if all the higher modes are still decaying (Meinel 1991).

4.7.1 Malkus-Proctor Mechanism

Malkus & Proctor (1975) analytically considered the nonlinear saturation of an α^2-dynamo (with uniform α-effect) by a Lorentz force-induced large-scale flow ('macro-quenching'). The resulting flow consists of differential rotation and a meridional flow crossing the equator. The

---

[20] which is not excluded, as shown by Rüdiger (1974)

latter might be a consequence of the fact that in models with $\alpha = \text{const.}$ no real equator is defined. The large-scale flows induced by the global magnetic fields play a key role in almost all studies of the solar torsional oscillations (starting with Schüssler 1979, 1981 and Yoshimura 1981). A detailed discussion of this theory is difficult because of the lack of a properly working model of the solar dynamo. However, it is commonly assumed that torsional oscillations arise from the feedback of the cyclic magnetic field on the mean flow.

Another open question with the Malkus-Proctor concept is the influence of the turbulence-induced Maxwell stresses. Even the sign of the total Lorentz force in the momentum equation, i.e. $\langle J \rangle \times \langle B \rangle + \langle J' \times B' \rangle$, is not known (Rüdiger et al. 1986). The SOCA-expressions for the turbulent Maxwell stress (4.55) by Rüdiger & Kitchatinov (1990) even weaken the Reynolds stress components for reasonable turbulence models. The remaining magnetic pressure is not relevant for the excitation of torsional oscillations.

Moss et al. (1995) studied the Lorentz force feedback in spherical $\alpha^2 \Omega$-dynamo models where the differential rotation is maintained by the (radial) Λ-effect with positive $V^{(0)}$. The resulting differential rotation produces meridional flow and induces magnetic fields. The resulting magnetic field feeds back to the meridional flow and the differential rotation. The situation is more complicated still as the combination of differential rotation *and* meridional flow can itself act as a kinematic dynamo, where according to Cowling's theorem nonaxisymmetric magnetic fields must play a role (see Dudley & James 1989). No wonder that the complete system provides complex solutions. The general result of Moss et al. (1995) is that for small Taylor number Ta (i.e. low angular momentum) the nonaxisymmetric mode S1 is the only stable one while for larger Ta the axisymmetric solution A0 dominates. For $\text{Ta} > 10^4$ the system reaches the $\alpha \Omega$-regime and the dynamos start to oscillate.

For negative $V^{(0)}$ and moderate Ta there is a tendency that after an extremely long computation time the final state is again S1. For higher Ta Barker (1993) finds an axisymmetric solution with quadrupolar symmetry.

The reason for the exclusive stability of the S1 modes in the α^2-regime is not entirely clear. The resulting Ω-isocontours are disk-like with $d\Omega/dz > 0$. As pointed out by Rädler (1986a), a *weak* differential rotation may lead to stable nonaxisymmetric fields even for an isotropic α-effect. In the presented calculations, however, the nonaxisymmetry of the magnetic fields exists for both signs of $V^{(0)}$, so that it seems to be rather robust against the modification of the rotation law.

4.7.2 α-Quenching

Another nonlinear approach is based on the idea that the induced large-scale magnetic field suppresses the turbulence, so that in particular the α-effect is quenched. If this quenching happens locally with $\bar{B}^2$ then one finds for α^2-dynamos that the induced magnetic field energy follows the law $\bar{B}^2 \propto C - C_{\text{crit}}$ where C_{crit} is the threshold value of the dynamo. In Fig. 4.50 (left) the magnetic energy of the induced polar field is given for outer-shell dynamos ($x_{\text{in}} = 0.5$) for the α-quenching law

$$\alpha \propto \frac{1}{1 + \left(\bar{B}/B_{\text{eq}}\right)^p},\tag{4.99}$$

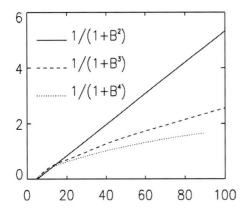 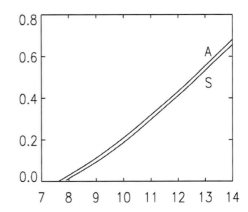

Figure 4.50: *Left*: The influence of the magnetic quenching law on the amplitude of the polar field strength in a spherical α^2-dynamo embedded in vacuum for supercritical dynamo number C_α. *Right*: The magnetic energy for $p = 2$ for growing C_α. The energy of the dipolar mode always (slightly) exceeds the energy of the quadrupolar mode.

for various p. One finds that $\bar{B} \propto (C - C_{\mathrm{crit}})^{1/p} B_{\mathrm{eq}}$. Brandenburg, Tuominen & Moss (1989) have probed the stability of α^2-dynamos with a time-dependent code. Their main question concerned the role of the growth rates, which are very similar for dipolar and quadrupolar solutions (see Fig. 4.50, right). It has been shown that the only stable solution is that with the smallest threshold dynamo number. The problem remained, however, that depending on the initial conditions for spherical models both dipolar and quadrupolar solutions are stable. This ('watershed') phenomenon can be observed with the temporal evolution of the magnetic fields. Depending on the excess of the dynamo number one finds only dipoles as the resulting stable solution or – for higher C_α – also the possibility of stable quadrupoles. This would be in contradiction to the concept of Krause & Meinel (1988) that only the first bifurcation should be stable. However, Rädler et al. (1990) showed that if one allows for nonaxisymmetric solutions, the final solution is always the axisymmetric dipole – but the time after which this final solution is reached may be very long (~ 50 (!) diffusive times).

For cyclic dynamos with differential rotation both the 2D (Brandenburg, Tuominen & Moss 1989) and 3D (Rädler et al. 1990) codes provided the same results. The model is defined by a linear radial rotation law with a fixed normalized surface rotation rate and a radially uniform α-effect. For slightly supercritical C_α a cyclic dipolar solution is stable. If, however, the α-effect is increased then both the solutions symmetric and antisymmetric with respect to the equator become unstable. The system starts to oscillate with a longer period between the even and the odd parity (Fig. 4.51, left). Figure 4.51 (right) also shows the geometry of the toroidal magnetic field, i.e. the butterfly diagram. One observes a long-term oscillation of equatorial symmetry and antisymmetry. The cycletimes at the northern and the southern hemisphere systematically differ by a rather small amount, which changes its sign after a longer period. Similar models have been proposed as one of the mechanisms that can explain the existence of grand minima like the Maunder minimum (see Brandenburg, Tuominen & Moss 1989) but the ratio of the beat period and the basic cycle period seems to be too high.

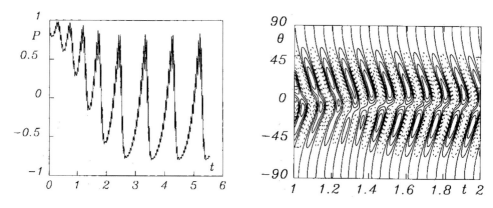

Figure 4.51: *Left*: Both the odd ($P = -1$) and even ($P = 1$) parity solutions are unstable in the $\alpha^2\Omega$-dynamo model of Brandenburg, Tuominen & Moss (1989) for $C_\alpha = 0.9$, oscillating with a period about 10 times longer than the basic cycle. *Right*: The butterfly diagram for the same model.

4.7.3   Magnetic Saturation by Turbulent Pumping

In a seminal paper Noyes, Weiss & Vaughan (1984) formulated the question of how the cycle periods of $\alpha\Omega$-dynamos depend on the nonlinear quenching mechanism. For pure α-quenching with $\alpha \propto 1/(1 + \bar{B}^2)$ the resulting cycle period does not depend on the actual value of the dynamo number $\mathcal{D}$. This is now a well-established effect (Fig. 4.52, Jennings & Weiss 1991, Rüdiger & Arlt 1996, Tobias 1998). For spherical nonlinear models the solution changes slightly. Brandenburg, Tuominen & Moss (1989), Moss, Tuominen & Brandenburg (1990) and Rüdiger et al. (1994) find a small increase of ω_{cyc} for increasing Ω. Tobias (1998) reports a stronger increase in a Cartesian model. This is surprising, as a relation $\omega_{\mathrm{cyc}} \propto \sqrt{\mathcal{D}_{\mathrm{crit}}}$ exists for the *linear* $\alpha\Omega$-dynamos. This is only valid for the initial onset though. Noyes et al. also consider the influence of an η-antiquenching, i.e. $\eta_{\mathrm{T}} \propto (1 + \bar{B}^2)$, which is assumed as an enhancement of the losses due to magnetic buoyancy. The resulting consequences are $\omega_{\mathrm{cyc}} \propto \sqrt{\mathcal{D}}$, so that there is a clear positive correlation of cycle frequency and dynamo number (see Tobias 1998, his Fig. 3).

Very similar results have been found by Schmitt & Schüssler (1989) with a 1D model (along the latitude). They started to consider nonlinear dynamo saturation by a loss of toroidal flux due to magnetic buoyancy described by a loss term in the equation of the toroidal field, such as suggested by Leighton (1969). The flux loss scales as $\bar{B}^2$, which has to exceed some threshold value that serves as the scale of the magnetic field. A summary of the results is given in Fig. 4.52. Once again the cycle frequency does not depend on the dynamo number for simple α-quenching, but it grows approximately as $\sqrt{\mathcal{D}}$ for the model with flux loss.

In the 2D models of Moss, Tuominen & Brandenburg (1990) the magnetic-buoyancy effect is included as a radial velocity scaling as $|\bar{B}|$ or $\bar{B}^2$. The turbulent advection effects are part of the α-tensor, which can be written as a large-scale velocity, which for slow rotation is reduced to a radial mean flow. If the variation of the cycle period with C_α is considered, these authors also found a relation $\omega_{\mathrm{cyc}} \propto \sqrt{\mathcal{D}}$ (their Table 2).

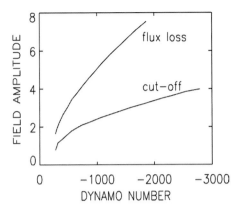 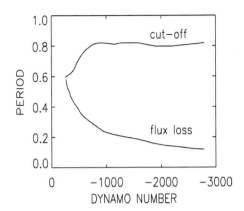

Figure 4.52: Nonlinear dynamo by Schmitt & Schüssler (1989). *Left*: Field amplitude vs. dynamo number. *Right*: Cycletime vs. dynamo number. Note that for α-quenching the cycletime does not depend on the dynamo number.

4.7.4 η-Quenching

One finds a finite value for n in Eq. (4.2) if the magnetic feedback is considered to act not only on the α-effect, but also on the eddy diffusivity tensor. What we assume here is that the magnetic field always suppresses and deforms the turbulence field and that this has consequences for both the α-effect *and* the eddy diffusivity. Such a SOCA-theory is given by Roberts & Soward (1975), Kitchatinov, Pipin & Rüdiger (1994); applications are summarized by Rüdiger & Arlt (2003).

The turbulent-diffusivity quenching concept was already described by Noyes, Weiss & Vaughan (1984). Tobias (1998), with his 2D global model in Cartesian coordinates finds the strongest scaling of the cycletimes with the dynamo number with increasing effect of η-quenching ('subcritical excitation'). The results for the Malkus-Proctor effect alone yield the weakest influences.

4.8 Λ-Quenching and Maunder Minimum

The explanation of grand minima in the magnetic activity cycle has been approached in two ways. The first approach considers the stochastic character of the turbulence, and studies its consequences for the variations with time of the α-effect and all related phenomena (see Sect. 4.6). The alternative concept includes the magnetic feedback on the internal solar rotation (Weiss, Cattaneo & Jones 1984, Jennings & Weiss 1991). Kitchatinov, Rüdiger & Küker (1994) and Tobias (1996, 1997) even introduced the conservation law of angular momentum in the convection zone including magnetic feedback in order to simulate the intermittency of the dynamo cycle and to explain the existence of grand minima (see Beer, Tobias & Weiss 1998).

A theory of differential rotation based on the Λ-effect is coupled with the induction equation in a spherical 2D mean-field model. The mean-field equations for the convection zone

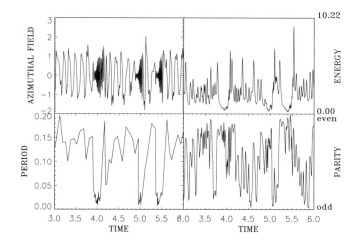

Figure 4.53: The time dependence of the dynamo for the large-scale Lorentz force feedback only ($Pm = 0.1$, $\Lambda = 1$). *Top*: Toroidal magnetic field (*left*) and magnetic energy (*right*). *Bottom*: Cycletime (*left*) and magnetic parity (*right*).

include the effects of diffusion, α-effect, toroidal field production by differential rotation and the Lorentz force. The conservation law of angular momentum is

$$\rho r \sin\theta \frac{\partial \Omega}{\partial t} = -\frac{1}{r^3}\frac{\partial}{\partial r}\left(r^3 \rho\, Q_{r\phi}\right) - \frac{1}{r\sin^2\theta}\frac{\partial}{\partial\theta}\left(\sin^2\theta \rho\, Q_{\theta\phi}\right) +$$

$$+ \frac{1}{\mu_0 r^2 \sin\theta}\left(\frac{1}{r}\frac{\partial A}{\partial\theta}\frac{\partial(Br)}{\partial r} - \frac{1}{\sin\theta}\frac{\partial A}{\partial r}\frac{\partial(B\sin\theta)}{\partial\theta}\right) \tag{4.100}$$

(see Eq. $(3.92)_2$). The computational domain is a spherical outer shell down to $x = 0.5$. The convection zone extends from $x = 0.7$ to $x = 1$. The α-effect exists only in the lower part between 0.7 and 0.8, while turbulent diffusion of the magnetic field, eddy viscosity and the Λ-effect are present in the entire convection zone. Below $x = 0.7$ both the magnetic diffusivity and the viscosity are two orders of magnitude smaller than in the convection zone. The boundaries are assumed to be stress-free.

The model is described by the magnetic Reynolds numbers of the differential rotation and the α-effect, the magnetic Prandtl number, the Elsasser number[21] Λ, and the Λ-effect amplitude $V^{(0)}$. In the α-effect the factor $\sin^2\theta$ has been used to concentrate the magnetic activity at low latitudes. The dynamo works with $C_\alpha = -10$ and $C_\Omega = 10^5$. $V^{(0)}$ ($= 0.37$) is positive in order to produce the required superrotation. With the standard eddy diffusivity expression the Elsasser number reads $\Lambda = 2/c_\eta \Omega^*$ and is thus set to unity here.

Figures 4.53 and 4.54 demonstrate the action of different effects and show the variation of the toroidal magnetic field at a fixed point ($x = 0.75$, $\theta = 30°$), the total magnetic energy, the variation of the cycle period and the parity

$$P = \frac{E_S - E_A}{E_S + E_A} \tag{4.101}$$

derived from the decomposition of the magnetic energy into symmetric and antisymmetric components. E_A and E_S are the energy of the even and odd field components. All times are given in units of a diffusion time R^2/η_T.

---

[21] Λ (see Eq. (2.4)) and Λ should not be confused here

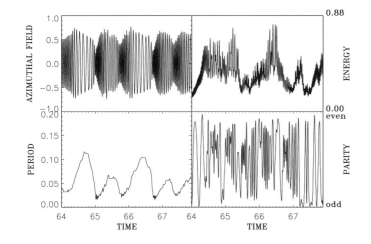

Figure 4.54: The same as in Fig. 4.53 but for strong Λ-quenching ($\lambda = 25$). *Top*: Toroidal magnetic field (*left*) and magnetic energy (*right*). *Bottom*: Cycletime (*left*) and magnetic parity (*right*). Note the strong variations of the parity. From Küker, Arlt & Rüdiger (1999).

If the large-scale Lorentz force is the only feedback on the rotation, the chaotic time series given in Fig. 4.53 may be compared with the results in Tobias (1996). A quasiperiodic behavior is shown with activity interruptions like grand minima. This model, however, neglects the feedback of strong magnetic fields on the α-effect and the differential rotation. The variations of the parity are rather strong. Deep energy minima seem to be connected with quadrupolar equator symmetry (see Knobloch, Tobias & Weiss 1998).

The same model but with a local standard α-quenching turns into the well-known simple solution with only one period and with odd parity (not shown). Similar to the suppression of dynamo action, a quenching of the Λ-effect is now introduced according to

$$V^{(0)} \propto \frac{1}{1 + \lambda(\bar{B}/B_{\mathrm{eq}})^2}. \tag{4.102}$$

If λ is near unity, the maximum field strength and total magnetic energy decrease slightly but the periodic behavior remains the same, i.e. the effect of the Λ-quenching is too small to alter the differential rotation significantly. However, an increase of λ leads to grand minima – an example for $\lambda = 25$ is given in Fig. 4.54. Minima in cycle period occur *shortly after* a grand activity minimum, in agreement with the analysis of sunspot data by Frick et al. (1997). The amplitude of the period fluctuations is much lower than in the Malkus-Proctor model but is still stronger than that observed.

Spectra of long time series of the toroidal magnetic field are given in Fig. 4.55 for both the Malkus-Proctor model and the model with the strong Λ-quenching. The long-term variations of the field will be represented by a set of close frequencies whose difference is the frequency of the grand minima. The Malkus-Proctor model shows a number of lines close to the main cycle frequency. The shape of the spectrum indicates that the magnetic field appears rather irregularly. The spectrum of the model with all feedback terms and strong Λ-quenching shows a similar behavior. The average frequency of the grand minima is represented by the distance between the two highest peaks. Here, grand minima occur at a reasonable rate between 10 and 20 cycletimes. The cycle period varies by a factor of 3 or 4. The northern and southern

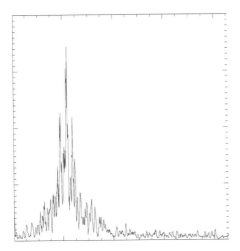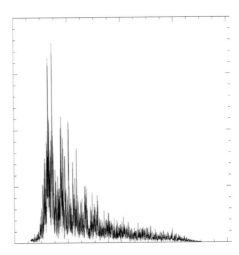

Figure 4.55: *Left*: Power spectrum of the magnetic-field amplitude variations for the Malkus-Proctor model of Fig. 4.53. The frequency is given in arbitrary units. *Right*: The same but for the model with strong Λ-quenching. The highest peaks are the basic cycle frequency.

hemispheres differ in their temporal behavior. This is a *general* characteristic of mixed-mode dynamo explanations of grand minima.

The magnetic Prandtl number of the model is $Pm = 0.1$, and it is noteworthy that grand minima do not appear for $Pm = 1$. The magnetic Prandtl number directs the intermittency of the activity cycle. Values smaller than unity are required for the existence of grand minima, but the occurrence of grand minima becomes more and more exceptional again for very small values of Pm (Knobloch, Tobias & Weiss 1998, Küker, Arlt & Rüdiger 1999, Pipin 1999).

5 The Magnetorotational Instability (MRI)

5.1 Star Formation

5.1.1 Molecular Clouds

Giant molecular clouds are the prime sites of star formation. Most stars in our galaxy were formed by gravitational instability in cold, self-gravitating, high-density clumps within such clouds. Their masses range from 10^5 to $10^{6.5}$ $M_\odot$, with a median of $3.3 \cdot 10^5$ $M_\odot$, and characteristic radius of 20 pc.

A molecular cloud is far from being homogeneous. Maps of molecular species like NH_3 or CS or H_2CO, which trace high-density regions, reveal a nested filamentary structure connected with magnetic fields. As part of this filamentary network dense clumps of up to 3000 $M_\odot$ are observed, reminiscent of the structures considered by Chandrasekhar & Fermi (1953), who studied the stability of an infinite cylinder of an incompressible, self-gravitating fluid supported by an aligned magnetic field. Modern computing facilities allow one to follow the emergence and (hydrodynamic) evolution of a whole network of such filaments subject to self-gravity from a prescribed initial turbulent velocity field (see Klessen, Heitsch & MacLow 2000).

The clouds are supported against self-gravity and the external pressure of the intercloud medium by both turbulent motions and magnetic fields. The contribution of thermal pressure and the rotational support are mostly negligible. These tiny cores, comprising a few solar masses only, give rise to the formation of single stars, binaries or multiple stellar systems (Fig. 5.1). The lifetime of such a clump therefore depends on the rate of leakage of magnetic fields and/or the decay of turbulence.

Stars are born in groups and clusters. The overall star formation efficiency seems to be low. Only a few per cent of the mass of a molecular cloud ends up in stars. In order to generate stars the collapsing portions of a molecular cloud must be able to fragment (even in the presence of a magnetic field). From low-mass star forming regions like the Taurus cloud it is known that star formation happens only in clumps exceeding some critical surface density. This observation makes the low efficiency plausible for dark clouds. In regions like the Orion Nebula Cluster, with many high-mass and low-mass stars being formed, a feedback mechanism may be at work. Previously formed stars (especially OB) energize the remaining cloud by radiation, winds, jets, and even supernova (SN) explosions, and may prevent the surrounding molecular gas from collapsing.

Unlike turbulent motions, the large-scale magnetic fields are rather long-lived. They may play a crucial role in supporting molecular clouds as a whole, and also their high-density

The Magnetic Universe: Geophysical and Astrophysical Dynamo Theory.
Günther Rüdiger, Rainer Hollerbach
Copyright © 2004 Wiley-VCH Verlag GmbH & Co. KGaA
ISBN: 3-527-40409-0

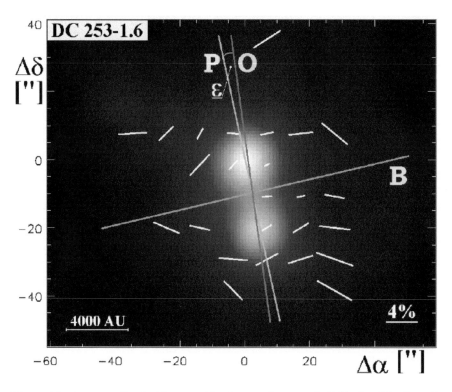

Figure 5.1: A protostellar double core with a large-scale magnetic field of $\sim 16\ \mu G$ amplitude and a direction perpendicular to the 'equatorial' plane. From Henning et al. (2001).

Table 5.1: Turbulence and magnetic fields in molecular clouds (Crutcher 1999). The last column gives the mass-to-flow ratio normalized by its critical value (5.1).

Cloud	$B\ [\mu G]$	$u_T\ [km/s]$	$T\ [K]$	$R\ [pc]$	μ/μ_{crit}
W3 OH	3100	1.5	100	0.02	0.8
DR 21 OH1	710	2.3	50	0.05	2.8
Sgr B2	480	15	70	22	2.6
M17 SW	450	4.0	50	1.0	1.4
W3 (main)	400	3	60	0.12	2.0
S106	400	1.6	30	0.07	0.8
DR 21 OH2	360	2.3	50	0.05	2.8

substructures. Measurements of the Zeeman splitting of the 18-cm OH lines, tracing the dense parts of the molecular gas, reveal large-scale ordered magnetic fields of about $30\ \mu G$ in starless clouds. Such fields, on scales of 0.1–10 pc, are dynamically important, as the magnetic energy is at least comparable with the potential energy of the clump (Crutcher 1999).

The virial theorem yields a critical mass above which a clump of gas cannot be supported against gravitational collapse by magnetic forces alone (Strittmatter 1966, Tscharnuter 1985). The critical mass-to-flux ratio $\mu = M/\Phi$ (where $\Phi = \pi R^2 B$ is the magnetic flux) is given by

$$\mu_{\text{crit}} = \frac{c_\Phi}{\sqrt{G}}. \tag{5.1}$$

The correction factor c_Φ measures the deviation from the virial-theorem analysis for a uniform sphere. The marginal value for gravitational collapse is $c_\Phi = 0.13$ (Mouschovias & Spitzer 1976). This value emerged from a sequence of numerical equilibrium models with evolution being driven by ambipolar diffusion of an initially dominant magnetic field (Mouschovias 1976a,b).

Ambipolar Diffusion

There are two possible outcomes with respect to the mass-to-magnetic-flux ratio. If this ratio exceeds (5.1), magnetic support alone can neither prevent the onset of collapse nor stop it at a later stage. For a frozen-in field the ratio does not change during the collapse. Dorfi (1982) demonstrated, with numerical simulations, how a magnetic field of 3 μG solves the angular momentum problem for 'supercritical' molecular clouds ($M = 10^4\ M_\odot$, $T = 100$ K, $\rho = 10\ \text{cm}^{-3}$, $\Omega = 10^{-15}\ \text{s}^{-1}$). The gas flows along the field lines, forming a disk-like structure (see Galli & Shu 1993). Most of the material has left after the free-fall time (explaining the low star formation rate), and the remaining magnetic field is small, but can constitute a fossil field for the newly formed star (Moss 2003a).

On the other hand, a magnetically subcritical cloud core is also able to become unstable, namely by ambipolar diffusion (Mestel & Spitzer 1956, Mouschovias 1976a,b, see Hujeirat et al. 2000). In a low-temperature, high-density core the ionization is extremely low, and the collisional coupling between ions and neutrals is only weak. The ions are attached to the magnetic field, but the neutrals can slip through it. The resulting induction equation

$$\frac{\partial \boldsymbol{B}}{\partial t} = \nabla \times \left(\boldsymbol{u} \times \boldsymbol{B} + \beta_{\text{ad}} (\boldsymbol{B} \cdot \nabla \times \boldsymbol{B}) \boldsymbol{B} - (\eta + \beta_{\text{ad}} \boldsymbol{B}^2) \nabla \times \boldsymbol{B} \right) \tag{5.2}$$

is highly nonlinear. One finds that (i) the magnetic dissipation is increased, and (ii) some kind of α-effect occurs, with α proportional to the current helicity.

Results of the temporal evolution of a collapsing cylindrical filament are given in Fig. 5.2. The magnetic field $B_\parallel$ is in the z-direction (along the filament), and depends only on s, so that Eq. (5.2) becomes

$$\frac{\partial B_\parallel}{\partial t} + \frac{1}{s} \frac{\partial}{\partial s} (s u B_\parallel) = \beta_{\text{ad}} B_\parallel^2 \left(\left(\frac{1}{s} - 2 \frac{d \log \rho}{ds} \right) \frac{\partial B_\parallel}{\partial s} + \frac{2}{B_\parallel} \left(\frac{\partial B_\parallel}{\partial s} \right)^2 + \frac{\partial^2 B_\parallel}{\partial s^2} \right) \tag{5.3}$$

where $\beta_{\text{ad}} \propto 1/(\rho_n \rho_i)$ is the diffusion coefficient with the densities of neutrals and ions (Shu 1992). The density and the radial inflow follow simply from the conservation of mass and momentum. The lines in the figure represent consecutive timesteps. Due to the collapse the density and the magnetic field grow everywhere, but the ambipolar-diffusion transports the magnetic field radially outward, so that its growth in the core is strongly reduced (Fig. 5.2).

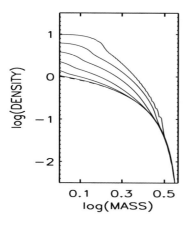 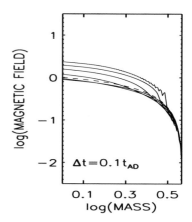

Figure 5.2: Density and magnetic field ($B_{\parallel}$) in an infinite cylindrical filament during the collapse. Initial configurations are given as dashed. By the collapse both the density and magnetic field grow, but due to ambipolar diffusion the magnetic field only grows slowly. Courtesy H.-E. Fröhlich.

It has been shown by Kim & Diamond (2002) that for the simultaneous existence of turbulence and ambipolar diffusion the turbulent decay of the field dominates; there is no enhancement of the decay rate of a large-scale magnetic field by the ambipolar diffusion.

The main effect of ambipolar diffusion is a prolongation of the collapse timescale by a factor of 10 compared with the free-fall time (Basu & Mouschovias 1994, Mouschovias 1996, Basu 1997). The ambipolar diffusion scenario, however, lacks observational support. Most clouds with known magnetic field amplitude are magnetically supercritical (Crutcher 1999). Very strong magnetic fields are rare (see Table 5.1).

According to Boss (1998) there is also an important extra effect due to ambipolar diffusion in rotating magnetized molecular clouds. The local magnetic field amplitude under the influence of ambipolar diffusion is simply modeled by

$$V_{\rm A} \propto 1 - \frac{t}{t_{\rm ad}}, \tag{5.4}$$

with $t_{\rm ad}$ as ~ 10 times the free-fall time. The resulting collapse differs strongly for slow and fast cloud rotation. Rapidly rotating clouds fragmented into binary protostars, while slowly rotating ones only formed bars or single protostars (Boss 2001). The importance of the gas's thermal physics for the resulting instability pattern was demonstrated by Durisen (2001), by means of nonlinear simulations adopting various versions of the equation of state and the important cooling processes.

MHD Turbulence

Besides being supported by magnetic fields, molecular clouds are partially supported by supersonic motions, as indicated by the observed line widths, but also by the relatively modest flattening of molecular cores (Klessen 2003). The thermal line broadening is negligible compared with the turbulent one. A typical Mach number is 5. High speeds are observed even in starless clouds without any internal source for turbulence. The turbulence could also be maintained by galactic shear motion or the concerted action of SN explosions, but it is not easy to see how the mechanical energy could then cascade down into small-scale high-density clumps.

Magnetic field measurements show the velocities to be roughly Alfvénic. The suggestion that MHD waves are responsible for the observed nonthermal line broadening opens up the possibility that the accompanying wave pressure could stabilize the cloud. Numerical simulations of MHD turbulence hint at a fast decay, with a timescale comparable with the free-fall time (Mac Low 1999, Ostriker, Stone & Gammie 2001). This strong damping suggests consideration of the loss of turbulent support rather than the loss of magnetic field by ambipolar diffusion as the primary cause of the collapse (Nakano 1998).

5.1.2   The Angular Momentum Problem

A very clear formulation of the stellar angular momentum problem is by Spitzer (1978). If a molecular cloud core of 1 $M_\odot$ and 1 lyr radius were to collapse to a solar-type star of 1 $R_\odot$ while conserving its angular momentum, the rotation rate would increase by 14 orders of magnitude. If the molecular cloud were to rotate with the (low) angular velocity of a galaxy (10^{-15} s^{-1}) the stellar rotation period would be 1 s. Stars, therefore, cannot be formed while conserving angular momentum. Table 5.2 gives values for the specific angular momentum $R^2 \Omega$ for the structures related to the star formation process. During the star formation the specific angular momentum is reduced by more than 4 orders of magnitude. The numbers also demonstrate the significant role that Jupiter appears to play in our own solar system. The difference in specific angular momentum of the molecular cloud core and that of Jupiter's orbit is only one order of magnitude.

The simplest way to get rid of the angular momentum is by viscous dissipation (v. Weizsäcker 1943). The general diffusion equation that governs the evolution of the distribution of the column density

$$\Sigma(s) = \int_{-\infty}^{\infty} \rho \, dz \tag{5.5}$$

in a thin Keplerian disk is

$$\frac{\partial \Sigma}{\partial t} = \frac{3}{s} \frac{\partial}{\partial s} \left(\sqrt{s} \frac{\partial}{\partial s} (\sqrt{s} \nu \Sigma) \right) \tag{5.6}$$

(v. Weizsäcker 1948, Lüst 1952, see Pringle 1981), which results from conservation of mass and angular momentum after elimination of the radial (accretion) flow. Disks are considered here because of the observational fact that many young stars are surrounded by circumstellar disks (Fig. 5.3). The characteristic diffusion time in this equation is $\tau_{\text{diff}} \simeq R^2/\nu$, so that a viscosity of more than 10^{20} cm^2/s is necessary to dissipate a ring of radius 1 lyr in (say) 10^6 years. Only turbulence can provide such enormous viscosities.

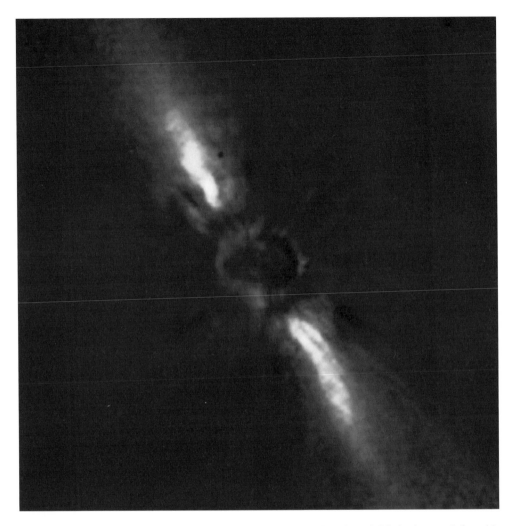

Figure 5.3: The A star β Pic with its circumstellar debris disk. The dust visible in the near infrared is due to the collisions of remaining planetesimals. The distance of the inner edge of the disk to the center is 24 AU. Credit J.-L. Beuzit, ESO.

Following Shakura & Sunyaev (1973) the eddy viscosity is often written with the sound speed c_{ac} as a characteristic velocity and the half-thickness H of the disk as a characteristic lengthscale, so that

$$\nu_{T} = \alpha_{SS}\, c_{ac}\, H, \tag{5.7}$$

from which the positive cross-correlation

$$Q_{s\phi} = 1.5\, \alpha_{SS}\, c_{ac}^{2} \tag{5.8}$$

Table 5.2: Characteristic values of specific angular momentum (Bodenheimer 1995). The Jupiter value concerns its orbit.

	MC [1 pc]	MC core [0.1 pc]	disk	TTau star	Sun	Jupiter
$R^2\Omega$ [cm^2/s]	10^{23}	10^{21}	$5\cdot 10^{20}$	$5\cdot 10^{17}$	10^{15}	10^{20}

results for thin Kepler disks[1]. Obviously, for subsonic turbulence the viscosity-α must not exceed unity.

Note also that $\nu_{\rm T} = \beta s^2\Omega$ has been used instead of Eq. (5.7). It is interesting to see how different the disk structure is in the two cases (Duschl, Strittmatter & Biermann 2000, Huré, Richard & Zahn 2001). With the β-viscosity the Reynolds stress becomes $Q_{s\phi} \simeq 1.5\beta u_\phi^2 \propto s^{-1}$, which is completely decoupled from the resulting disk structure.

Estimates of the required viscosity-α lead to values of order 10^{-3} to 1, which must be explained by the theory. The key problem of star formation is thus the explanation of the source of the turbulence. Some of the disks are convectively unstable. Simulations of this effect, however, often lead to negative $\alpha_{\rm SS}$. Ruden, Papaloizou & Lin (1988) started to study axisymmetric perturbations of a thin, convectively unstable but inviscid Keplerian disk. The nonaxisymmetric case is considered by Ryu & Goodman (1992), where negative correlations $Q_{s\phi}$ are obtained for convection in Keplerian disks. Cabot & Pollack (1992) also provide detailed computations for $Q_{s\phi}$. For large rotation rates it becomes negative near the walls. In this model one finds anisotropic turbulence with $\langle u_s'^2\rangle > \langle u_\phi'^2\rangle$.

Kley, Papaloizou & Lin (1993), with their 2D simulations of convection in a medium with a rather large eddy viscosity, also obtain a flow system appearing to yield an inward transport of angular momentum (see Goldman & Wandel 1995). If the 'correlation' tensor of the convection is computed with a time-averaging procedure using Kley's 2D hydrocode, a distinct anisotropy between the turbulence intensities in radial and azimuthal direction is found (Rüdiger, Tschäpe & Kitchatinov 2002). The radial rms velocity dominates the azimuthal one. As a consequence a radial Λ-effect appears and is negative ('inward transport'). It even seems to dominate the positive contribution of the eddy viscosity representing outward transport of angular momentum. The negative angular momentum transport can be explained by the dominance of the radial turbulence intensity, see Eq. (3.56).

Negative values for $Q_{s\phi}$ also appear in the 3D simulations of an inviscid fluid by Stone & Balbus (1996), probing the role of vertical convective motions in providing angular momentum transport in a Keplerian disk. Igumenshchev, Abramowicz & Narayan (2000) present 3D hydrodynamic simulations of convective advection-dominated accretion flows. They also report a strong tendency of the eddies toward axisymmetry, and an inward transport of angular momentum (see their Fig. 4).

One can ask whether an example is known that leads to the opposite type of anisotropy, i.e. for turbulence with dominant azimuthal intensity, and hence positive $Q_{s\phi}$. This is indeed the case for a Keplerian flow modulated with finite disturbances of a given wave number k. Ac-

---

[1] $H\Omega \simeq c_{\rm ac}$ is used resulting from the vertical hydrostatic disk equilibrium

cording to Dubrulle (1993) this hydrodynamic flow is unstable if the normalized amplitude of the flow disturbance exceeds ks. The resulting fluctuations are clearly azimuthally dominated, and the resulting angular momentum flux is outward (Rüdiger & Drecker 2001). However, if the imposed disturbance is switched off then after a few rotations the finite-amplitude disturbance also decays (see Fig. 5.4).

In the linear regime Keplerian flows are stable. The theory of this basic stability is given in the next section, where modifications are also discussed resulting from additional (vertical) stratifications of entropy and/or density.

5.1.3   Turbulence and Planet Formation

Turbulence rapidly dissipates large-scale concentration differences like those of chemicals and/or solid dust. On the other hand, its vortices may temporarily amplify the density, which is of interest for the dust growth in protoplanetary disks (Klahr & Henning 1997, Hodgson & Brandenburg 1998). A rapid growth of dust particles is the basis of the planetesimal theory of the planet formation (Lissauer 1993). Without turbulence the process of dust growth is far too slow (Kempf, Pfalzner & Henning 1999, $\gtrsim 10^6$ yr). Due to the electrical properties of the solids the electromagnetic forces in the disk are also of importance (Poppe, Blum & Henning 2000).

On the other hand, if the turbulence is too intense it would disturb the agglomeration of the dust particles (Völk et al. 1980, Dubrulle, Morfill & Sterzik 1995). The gravitational instability is also inhibited by turbulent mixing. However, Youdin & Shu (2002) demonstrate that the turbulence due to vertical shear between the gas and the thinner dust subdisk (pressure differences!) can be suppressed if the column density ratio of the dust and the gas is increased by a factor $\lesssim 10$ compared with the standard value. If correct then the dust disk may indeed become supercritical with respect to the Jeans instability, and form planetesimals in the very short orbital timescale ($\lesssim 10^3$ yr), without any need for the (slow) dust growth (Boss 1998, see Sect. 6.7). The timescale is clearly also short enough to overcome the rapid inward migration of the solids due to the natural friction with the gas particles. With a column density of 7.5 g/cm^2 for the dust disk, according to Eq. (6.21) the critical wavelength is 1000–10,000 km, which can also serve as an upper bound for the size of the planetesimals (Goldreich & Ward 1973).

5.2   Stability of Differential Rotation in Hydrodynamics

According to the Rayleigh criterion

$$\frac{\mathrm{d}j^2}{\mathrm{d}s} > 0, \tag{5.9}$$

with $j = s^2 \Omega$ being the angular momentum per unit mass, Keplerian disks are hydrodynamically stable configurations. However, if the disk plasma has a sufficiently high conductivity they are unstable under the influence of a (weak) magnetic field (Balbus & Hawley 1991, Hawley & Balbus 1991, Balbus 1995, Brandenburg et al. 1995). Linear studies of *global* configurations of disks threaded by magnetic fields were also carried out. Curry, Pudritz &

Sutherland (1994), Curry & Pudritz (1995) and Terquem & Papaloizou (1996) investigated the stability for vertical and azimuthal fields threading an inviscid and perfectly conducting global disk, basically considered in its midplane. They found that the actual initial field geometry does not depend strongly on field topology, as was also suggested by the numerical simulations of Hawley & Balbus (1991). The growth rates of the unstable modes are of the same order (see below) as the local growth rates, and the same is true for the largest allowed magnetic field strength. Kitchatinov & Mazur (1997) and Rüdiger et al. (1999) particularly addressed the angular momentum transport in their linear studies for thin accretion disks of real plasma. The first nonlinear global approach in 3D numerical simulations was by Armitage (1998), who omitted the density stratification in favor of a large radial extent of the disk. A global approach with stratification was then presented by Hawley (2000) and Arlt & Rüdiger (2001), who followed the evolution of a thick torus under the influence of an external magnetic field threading parts of the computational domain.

All local shearing-box numerical simulations to date have shown that Keplerian disks[2] are hydrodynamically stable to infinitesimal and finite-amplitude disturbances (Balbus, Hawley & Stone 1996). This raises the question whether other nonmagnetic influences exist, which in combination with the basic negative shear allow instabilities in the linear regime. We recall three recent suggestions. Urpin & Brandenburg (1998) and Arlt & Urpin (2004) consider the destabilizing action of a dependence of Ω on z, which may more than compensate the stabilization due to Eq. (5.9).

There is also the suggestion of Klahr & Bodenheimer (2003) that the *negative* radial entropy stratification in thin Kepler disks may act in a destabilizing manner. With the standard α-description of accretion disks, however, one finds the *positive* value $\partial S/\partial s \simeq C_{\mathrm{v}}/2s$, indicating stability.

Molemaker, McWilliams & Yavneh (2001) point out that a vertical density stratification in the Taylor–Couette flow may have a similar destabilizing effect as a global vertical magnetic field for a plasma. The sufficient condition that a (nonaxisymmetric) disturbance will be unstable is the same as for an unstratified ideal MHD Taylor–Couette flow in the presence of a vertical magnetic field, i.e.

$$\frac{\mathrm{d}\Omega}{\mathrm{d}s} < 0. \tag{5.10}$$

If this were also be true for nonmagnetic disks with vertical density stratifications, they would be hydrodynamically unstable, so that turbulence can develop, transporting the angular momentum inward or outward. In the following a stability analysis is presented, within the short-wave approximation and for *inviscid* media, which in particular considers the constellation of radial shear plus vertical stratification. How restrictive these approximations are is still an open question. Note that the Solberg–Høiland conditions for dynamic *stability* of constellations with vertical stratification take the form

$$\frac{1}{s^3}\frac{\partial j^2}{\partial s} - \frac{1}{C_{\mathrm{p}}\rho}\nabla P \cdot \nabla S > 0, \qquad \frac{\partial P}{\partial z}\left(\frac{\partial j^2}{\partial s}\frac{\partial S}{\partial z} - \frac{\partial j^2}{\partial z}\frac{\partial S}{\partial s}\right) < 0 \tag{5.11}$$

---

[2] without vertical gradients of the angular velocity

(Tassoul 1978). Here S is the specific entropy. For $\partial P / \partial z < 0$ (as is usual for accretion disks) the latter relation turns into

$$\frac{\partial j^2}{\partial s}\frac{\partial S}{\partial z} > \frac{\partial j^2}{\partial z}\frac{\partial S}{\partial s}, \tag{5.12}$$

which for Keplerian disks with $\Omega \propto s^{-1.5}$ leads to the usual Schwarzschild criterion,

$$\partial S / \partial z > 0 \tag{5.13}$$

for stability. Equation $(5.11)_1$ becomes

$$\frac{1}{s^3}\frac{\partial j^2}{\partial s} + \frac{g}{C_p}\frac{\partial S}{\partial z} - \frac{1}{\rho C_p}\frac{\partial P}{\partial s}\frac{\partial S}{\partial s} > 0, \tag{5.14}$$

which for $\partial S / \partial z > 0$ and $\partial S / \partial s > 0$ provides stability. It was adopted that $-g_z = g = \Omega^2 z > 0$. If, on the other hand, the standard stratifications of accretion disks are $j = j(s)$ and $S = S(z)$ with $\mathrm{d}j^2 / \mathrm{d}s > 0$ then the Solberg–Høiland criterion for stability reduces to

$$1 + \frac{z}{C_p}\frac{\mathrm{d}S}{\mathrm{d}z} > 0, \qquad\qquad \frac{\mathrm{d}S}{\mathrm{d}z} > 0 \tag{5.15}$$

(in the upper hemisphere). According to Eq. (5.11) $\mathrm{d}S / \mathrm{d}z > 0$ is evidently a *sufficient condition for stability* of Keplerian disks even for the combined action of vertical density stratification and differential (Kepler) rotation. This same result was previously derived by Livio & Shaviv (1977), Abramowicz et al. (1984) and Elstner, Rüdiger & Tschäpe (1989). Obviously, in the framework of *ideal* hydrodynamics the traditional Schwarzschild criterion $\mathrm{d}S / \mathrm{d}z > 0$ ensures stability for all Keplerian disks. If the disk can be considered as isothermal in the vertical direction then it follows that $\mathrm{d}S / \mathrm{d}z > 0$ for typical density stratifications. With $\mathrm{d}S / \mathrm{d}z > 0$ the condition (5.12) for a rotation law $\Omega = \Omega(z)$ yields stability for negative $\mathrm{d}\Omega / \mathrm{d}z$, so that only strong positive $\mathrm{d}\Omega / \mathrm{d}z$ might provide instability.

5.2.1 Combined Stability Conditions

In order to probe the central Eqs. (5.11) a local stability analysis of the equations of ideal hydrodynamics in cylindrical coordinates (s, ϕ, z) is necessary. Self-gravitation phenomena are excluded, but sound waves are allowed to exist. The three components of the momentum equation for axisymmetry are

$$\frac{\partial u_s}{\partial t} + u_s\frac{\partial u_s}{\partial s} + u_z\frac{\partial u_s}{\partial z} - \frac{u_\phi^2}{s} = -\frac{1}{\rho}\frac{\partial P}{\partial s} + g_s,$$

$$\frac{\partial u_\phi}{\partial t} + u_s\frac{\partial u_\phi}{\partial s} + u_z\frac{\partial u_\phi}{\partial z} + \frac{u_\phi u_s}{s} = 0,$$

$$\frac{\partial u_z}{\partial t} + u_s\frac{\partial u_z}{\partial s} + u_z\frac{\partial u_z}{\partial z} = -\frac{1}{\rho}\frac{\partial P}{\partial z} + g_z. \tag{5.16}$$

Mass conservation and the energy equation yield

$$\frac{\partial \rho}{\partial t} + \frac{\partial}{\partial s}(\rho u_s) + \frac{\rho u_s}{s} + \frac{\partial}{\partial z}(\rho u_z) = 0, \qquad \frac{\partial S}{\partial t} + u_s\frac{\partial S}{\partial s} + u_z\frac{\partial S}{\partial z} = 0. \tag{5.17}$$

The unperturbed state is $\bar{\boldsymbol{u}} = (0, s\Omega, 0)$, $\bar\rho = \bar\rho(s,z)$, $\bar{P} = \bar{P}(s,z)$, $\bar{S} = \bar{S}(s,z)$, $g_s = g_s(s,z)$ and $g_z = g_z(s,z)$ with

$$g_s = \frac{1}{\bar\rho}\frac{\partial \bar{P}}{\partial s} - s\Omega^2, \qquad\qquad g_z = \frac{1}{\bar\rho}\frac{\partial \bar{P}}{\partial z}. \tag{5.18}$$

Finally, $\mathrm{d}S = C_\mathrm{v}\mathrm{d}\log(P/\rho^\Gamma)$ with $\Gamma = C_\mathrm{p}/C_\mathrm{v} = 5/3$. The perturbations $\boldsymbol{u}'$, ρ', P' and S' to the basic state are assumed to be small, so that the linearized system

$$\frac{\partial u'_s}{\partial t} - 2\Omega u'_\phi + \frac{1}{\bar\rho}\frac{\partial P'}{\partial s} - \frac{\rho'}{\bar\rho^2}\frac{\partial \bar{P}}{\partial s} = 0, \qquad \frac{\partial u'_z}{\partial t} + \frac{1}{\bar\rho}\frac{\partial P'}{\partial z} - \frac{\rho'}{\bar\rho^2}\frac{\partial \bar{P}}{\partial z} = 0,$$

$$\frac{\partial u'_\phi}{\partial t} + \frac{1}{s}\frac{\partial(s^2\Omega)}{\partial s}u'_s + s\frac{\partial\Omega}{\partial z}u'_z = 0,$$

$$\frac{\partial S'}{\partial t} + u'_s\frac{\partial \bar{S}}{\partial s} + u'_z\frac{\partial \bar{S}}{\partial z} = 0,$$

$$\frac{1}{\bar\rho}\frac{\partial\rho'}{\partial t} + \frac{\partial u'_s}{\partial s} + \frac{u'_s}{s} + \frac{\partial u'_z}{\partial z} + \frac{\partial \log\bar\rho}{\partial s}u'_s + \frac{\partial\log\bar\rho}{\partial z}u'_z = 0 \tag{5.19}$$

results. The coefficients in the equations are assumed constant. The short-wave approximation $|k_s s| > 1$ is now applied (see Meinel 1983), so all perturbations can be expressed by the Fourier modes $\exp(\gamma t + \mathrm{i}\boldsymbol{kx})$, yielding

$$\gamma u_s - 2\Omega u_\phi + \mathrm{i}k_s\frac{P}{\bar\rho} - \frac{\rho}{\bar\rho^2}\frac{\partial \bar{P}}{\partial s} = 0,$$

$$\gamma u_\phi + \frac{1}{s}\frac{\partial(s^2\Omega)}{\partial s}u_s + s\frac{\partial\Omega}{\partial z}u_z = 0,$$

$$\gamma u_z + \mathrm{i}k_z\frac{P}{\bar\rho} - \frac{\rho}{\bar\rho^2}\frac{\partial \bar{P}}{\partial z} = 0,$$

$$\gamma\frac{\rho}{\bar\rho} + \frac{\partial\log\bar\rho}{\partial s}u_s + \frac{\partial\log\bar\rho}{\partial z}u_z + \mathrm{i}k_s u_s + \mathrm{i}k_z u_z = 0,$$

$$\gamma\left(\frac{P}{\bar{P}} - \Gamma\frac{\rho}{\bar\rho}\right) + \frac{1}{C_\mathrm{v}}\left(\frac{\partial\bar{S}}{\partial s}u_s + \frac{\partial\bar{S}}{\partial z}u_z\right) = 0, \tag{5.20}$$

after dropping the dashes. The determinant of the homogeneous system must vanish, yielding the dispersion relation

$$\gamma^4 + 2E\gamma^2 + F = 0, \tag{5.21}$$

where

$$E = \frac{1}{2}\left((k_s^2 + k_z^2)c_\mathrm{ac}^2 + \frac{1}{\bar\rho^2}\frac{\partial \bar{P}}{\partial s}\frac{\partial\bar\rho}{\partial s} + \frac{1}{\bar\rho^2}\frac{\partial \bar{P}}{\partial z}\frac{\partial\bar\rho}{\partial z} + \kappa^2\right), \tag{5.22}$$

and

$$F = \left(\frac{k_s}{\bar\rho}\frac{\partial \bar{P}}{\partial z} - \frac{k_z}{\bar\rho}\frac{\partial\bar{P}}{\partial s}\right)\cdot\left(\frac{k_s c_\mathrm{ac}^2}{C_\mathrm{p}}\frac{\partial S}{\partial z} - \frac{k_z c_\mathrm{ac}^2}{C_\mathrm{p}}\frac{\partial S}{\partial s} + \right.$$

$$+ \frac{\mathrm{i}}{\bar\rho^2}\left(\frac{\partial\bar{P}}{\partial z}\frac{\partial\bar\rho}{\partial s} - \frac{\partial\bar{P}}{\partial s}\frac{\partial\bar\rho}{\partial z}\right)\right) + \kappa^2\left(k_z^2 c_\mathrm{ac}^2 + \frac{1}{\bar\rho^2}\frac{\partial\bar{P}}{\partial z}\frac{\partial\bar\rho}{\partial z}\right) -$$

$$- s\frac{\partial\Omega^2}{\partial z}\left(k_s k_z c_\mathrm{ac}^2 + \frac{1}{\bar\rho^2}\frac{\partial\bar{P}}{\partial z}\frac{\partial\bar\rho}{\partial s} + \mathrm{i}\left(\frac{k_s}{\bar\rho}\frac{\partial\bar{P}}{\partial z} - \frac{k_z}{\bar\rho}\frac{\partial\bar{P}}{\partial s}\right)\right). \tag{5.23}$$

The epicyclic frequency κ and the adiabatic sound speed c_{ac} are

$$\kappa^2 = \frac{1}{s^3}\frac{\partial j^2}{\partial s}, \qquad\qquad c_{ac}^2 = \Gamma\frac{\bar{P}}{\bar{\rho}}. \tag{5.24}$$

The roots of the dispersion relation are simply $\gamma^2 = -E \pm \sqrt{E^2 - F}$. According to Eq. (5.22) the coefficient E is real. The flow is thus always unstable for negative E and complex F. Positive E and real F are, therefore, the *necessary* conditions for stability. According to Eq. (5.23), F is real if and only if

$$\frac{1}{\bar{\rho}^2}\left(\frac{\partial\bar{P}}{\partial z}\frac{\partial\bar{\rho}}{\partial s} - \frac{\partial\bar{P}}{\partial s}\frac{\partial\bar{\rho}}{\partial z}\right) - s\frac{\partial\Omega^2}{\partial z} = 0. \tag{5.25}$$

Inserting Eq. (5.25) into Eq. (5.18) we find

$$\frac{\partial g_s}{\partial z} = \frac{\partial g_z}{\partial s} \tag{5.26}$$

as the immediate consequence (Rüdiger, Arlt & Shalybkov 2002). Any *conservative force* is a solution of Eq. (5.26) and indicates stability. If – as it is in accretion disks – gravity balances the pressure and centrifugal forces, then Eq. (5.25) is automatically fulfilled. Note that from the Poincaré theorem for rotating media with potential force and $\Omega = \Omega(s)$ both the density and the pressure can be written as functions of a generalized potential, so that Eq. (5.25) is always fulfilled.

The magnetic field, however, is *not* conservative and can never fulfill Eq. (5.26). This is the basic explanation for the existence of the magnetorotational instability (MRI), which is driven by (weak) magnetic fields. With conservative forces fulfilling (5.26) the only possibility for any instability like that of Molemaker, McWilliams & Yavneh (2001) is that the short-wave approximation cannot be applied.

5.2.2   Sufficient Condition for Stability

Now the necessary condition (5.25) is assumed to be fulfilled. The flow is stable if both roots of the dispersion relation for γ^2 are real and negative. The sufficient conditions for stability are therefore $E > 0$ and $E^2 > F > 0$. Inserting Eq. (5.25) into Eq. (5.23) we find F to be positive[3] if

$$\frac{k_s^2}{k_z^2}N_z^2 + \frac{k_s}{k_z}\frac{2}{C_p\bar{\rho}}\frac{\partial\bar{P}}{\partial z}\frac{\partial S}{\partial s} + N_s^2 + \kappa^2 > 0, \tag{5.27}$$

where

$$N_s^2 = -\frac{1}{C_p\bar{\rho}}\frac{\partial\bar{P}}{\partial s}\frac{\partial S}{\partial s}, \qquad\qquad N_z^2 = -\frac{1}{C_p\bar{\rho}}\frac{\partial\bar{P}}{\partial z}\frac{\partial S}{\partial z} \tag{5.28}$$

are components of the Brunt–Väisälä frequency. Equation (5.27) is a quadratic expression in k_s/k_z. It is positive if (i) the expression is positive for some value k_s/k_z and (ii) the expression has no real roots, i.e. the discriminant is negative. The first condition is fulfilled if

$$N_s^2 + N_z^2 + \kappa^2 > 0. \tag{5.29}$$

---

[3] in the short-wave approximation

This is the Schwarzschild criterion for stability in the presence of rotation. The second condition leads to

$$\frac{\partial \bar{P}}{\partial z} \left(\kappa^2 \frac{\partial S}{\partial z} - s \frac{\partial \Omega^2}{\partial z} \frac{\partial S}{\partial s} \right) < 0. \tag{5.30}$$

These equations are exactly equivalent to the Solberg–Høiland conditions (5.11). The Schwarzschild criterion ensures that $E > 0$, and the short-wave approximation ensures that $F < E^2$, so these relations do not yield any additional conditions[4].

Without density stratification the sufficient stability condition $(5.11)_1$ for a rotating flow is

$$\kappa^2 > \frac{1}{\bar{\rho}^2 c_{ac}^2} \left(\left(\frac{\partial \bar{P}}{\partial s} \right)^2 + \left(\frac{\partial \bar{P}}{\partial z} \right)^2 \right). \tag{5.31}$$

This is the classical Rayleigh criterion (5.9) for stability generalized to the compressible case. Note that it is much more restrictive than the standard formulation (5.9).

5.2.3   Numerical Simulations

The main shortcoming of the linear stability analysis is the short-wave assumption. Without this assumption, Fourier modes are no longer solutions of the linearized equations. Numerical simulations were therefore undertaken to probe the stability of density-stratified Keplerian disks subject to finite but adiabatic disturbances.

The setup for 3D simulations applies a global, cylindrical computational domain with dimensionless radii 4 to 6, and a vertical extent of 2 density-scale heights on average. The ZEUS-3D code was used for the integration of the hydrodynamics with the adiabatic energy equation $(5.17)_2$. The source of gravitation is that of a point mass in the center. The initial configuration is isothermal with a sound speed c_{ac} of 7% of the Kepler velocity.

The conditions for the vertical boundaries are stress-free, i.e. the vertical derivatives of u_s and u_ϕ vanish. No matter can exit the computational domain in the vertical direction. The radial boundaries are also closed for the flow, and the radial derivative of u_z vanishes.

The initial conditions also involve nonaxisymmetric velocity perturbations of the form $u_z = A \sin m\phi$. Figure 5.4 (left) shows the evolution of the kinetic energies. A decay of the fluid pattern is observed for both $m = 1$ and $m = 200$ in the initial perturbation. Figure 5.4 (right) shows the evolution of α_{SS} according to Eq. (5.8) during 50 rotation periods (at the inner disk radius). A clear relaxation of the flow is seen. The fluctuations in the vertical direction reach $\pm 14\%$ of the Kepler velocity. Those in the radial direction even reach $\pm 27\%$. They are not small, and a nonlinear instability should appear, if one exists at all.

An average decay time in Fig. 5.4 is about 10 orbital periods. In terms of the Reynolds number $\mathrm{Re} \simeq 1500$ is achieved near the inner edge of the annulus.

5.2.4   Vertical Shear

Our dispersion relation is of fourth order, unlike the dispersion relation of second order one obtains in the Boussinesq approximation (Goldreich & Schubert 1967), and unlike the dispersion relation of fifth order found in nonideal hydrodynamics (Abramowicz et al. 1990). In this

---

[4] the conditions (5.11) can also be obtained in the Boussinesq approximation (Goldreich & Schubert 1967)

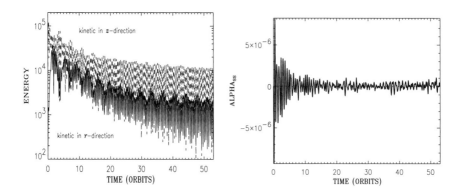

Figure 5.4: *Left*: Kinetic energies in the vertical and radial components of the flow after an initial perturbation with $m = 1$. The solid line is the energy of the radial direction, the dashed line that of the z-direction. *Right*: The α_{SS} in its temporal evolution after the disturbance. Times are given as orbit numbers of the inner boundary of the computational domain.

way we obtain a new necessary condition for stability, which requires the external force to be conservative, i.e. to possess a potential (see Eq. (5.26)).

When the angular velocity only depends on the vertical coordinate the conditions (5.11) transform to

$$\frac{\partial P}{\partial z} \frac{\partial S}{\partial s} \frac{\partial \Omega^2}{\partial z} > 0 \tag{5.32}$$

as the sufficient condition for stability, which can be fulfilled with $d\Omega/dz < 0$ for positive $\partial S/\partial s$. This exactly is also the constellation of the solar tachocline!

In accretion disks there is always a positive z-gradient of angular velocity. Arlt & Urpin (2004) have thus numerically simulated the stability of such accretion disks with the ZEUS-3D code. A vertical dependence of the angular velocity destabilizes the disk and leads to the generation of velocity fluctuations enhancing the angular momentum transport. The transport, however, appears to be negative. It remains an open question whether the closed radial boundary conditions may affect the direction of transport. The instability emerges for rather large radial wave numbers. The growth time is a few tens of orbital revolutions.

Another ingredient is the nonadiabaticity. Earlier simulations with nonaxisymmetric initial perturbations in an adiabatic disk did not show the onset of an instability, even though the disk possessed a small vertical shear. It is interesting to note that in the Boussinesq approximation for nonideal fluids, the Solberg–Høiland criterion changes to the Goldreich–Schubert–Fricke criterion (see Fricke & Smith 1971, Urpin & Brandenburg 1998), which no longer allows the existence of a vertical shear for stability, i.e. only $d\Omega/dz = 0$ is stable.

5.3   Stability of Differential Rotation in Hydromagnetics

The system is now assumed to be in a hydrostatic equilibrium in accordance with

$$\frac{\nabla P}{\rho} = \boldsymbol{g}^* + \frac{1}{\mu_0 \rho}(\nabla \times \bar{\boldsymbol{B}}) \times \bar{\boldsymbol{B}}, \tag{5.33}$$

with $\boldsymbol{g}^* = \boldsymbol{g} + \Omega^2 \boldsymbol{s}$ and $\boldsymbol{g}$ the gravity. The magnetic field $\bar{\boldsymbol{B}}$ is assumed to be weak in the sense that its Alfvén speed V_A is small compared to the sound speed. The Lorentz force is small compared with the pressure force, so that the structure is mainly determined by the balance between $\boldsymbol{g}$, centrifugal force and pressure, i.e. $\nabla P/\rho = \boldsymbol{g}^*$. The magnetic pressure does *not* contribute to the zero-order stratification. The large-scale magnetic field is assumed to be uniform. The instability theory of stratified media under the influence of *toroidal* magnetic fields has been presented by Terquem & Papaloizou (1996) and Papaloizou & Terquem (1997). It is not considered here. For the strong field amplitude the Parker (magnetic buoyancy) instability for a local toroidal flux tube appears, while for weak nonaxisymmetric fields the shear of the differential rotation becomes unstable (see below).

With $\boldsymbol{g} = -\nabla \psi$ one finds the condition (5.26) fulfilled. The induction equation with Hall effect included reads

$$\frac{\partial \boldsymbol{B}}{\partial t} = \nabla \times \left(\boldsymbol{u} \times \boldsymbol{B} - \eta \nabla \times \boldsymbol{B} - \beta(\nabla \times \boldsymbol{B}) \times \boldsymbol{B} \right), \tag{5.34}$$

where β represents the Hall effect[5].

Again axisymmetric short-wave perturbations are considered with the modal expansion $\exp(\gamma t + \mathrm{i}\boldsymbol{k}\cdot\boldsymbol{x})$, where $\boldsymbol{k} = (k_s, 0, k_z)$, and with viscosity neglected. The linearized momentum equation in the Boussinesq approximation then becomes

$$\gamma \boldsymbol{u}' + 2\boldsymbol{\Omega} \times \boldsymbol{u}' + \hat{\boldsymbol{e}}_\phi s(\boldsymbol{u}' \cdot \nabla)\Omega =$$
$$-\frac{\mathrm{i}\boldsymbol{k}P'}{\bar{\rho}} - \frac{T'}{\bar{T}}\boldsymbol{g}^* - \frac{\mathrm{i}}{\mu_0 \bar{\rho}}\left((\bar{\boldsymbol{B}} \cdot \boldsymbol{B}')\boldsymbol{k} - (\boldsymbol{k} \cdot \bar{\boldsymbol{B}})\boldsymbol{B}' \right), \tag{5.35}$$

with $\boldsymbol{k} \cdot \boldsymbol{u}' = 0$. Here $\hat{\boldsymbol{e}}_\phi$ is the unit vector in the azimuthal direction. It is assumed that the density perturbation in the buoyancy force is determined by the temperature perturbation, and the linearized adiabatic equation (3.67) turns into

$$\gamma T' + \boldsymbol{u}' \cdot \frac{\bar{T}}{C_\mathrm{p}}\nabla S = 0. \tag{5.36}$$

The linearized induction equation (5.34) reads

$$(\gamma + \eta k^2)\boldsymbol{B}' = \mathrm{i}(\boldsymbol{k} \cdot \bar{\boldsymbol{B}})\boldsymbol{u}' + s(\boldsymbol{B}' \cdot \nabla\Omega)\hat{\boldsymbol{e}}_\phi + \beta(\boldsymbol{k} \cdot \bar{\boldsymbol{B}})(\boldsymbol{k} \times \boldsymbol{B}'), \tag{5.37}$$

with $\boldsymbol{k} \cdot \boldsymbol{B}' = 0$. From these equations with the modal expansion one obtains the dispersion relation

$$\gamma^5 + a_4\gamma^4 + a_3\gamma^3 + a_2\gamma^2 + a_1\gamma + a_0 = 0, \tag{5.38}$$

---

[5]   here and in chapters 7 and 8 β means the Hall coefficient (not confuse with $\beta = B/B_\mathrm{eq}$ after (4.43))

with $a_0 = \eta k^2 \omega_A^2 \omega_g^2$ and with

$$a_1 = \left(\omega_H(\omega_H + \omega_{sh}) + \eta^2 k^4\right)(\omega_g^2 + \omega_{rot}^2) +$$
$$+ \omega_A^2\left(\omega_g^2 + \omega_A^2 + \omega_H\omega_{sh} + \omega_C(2\omega_H + \omega_{sh})\right),$$
$$a_2 = 2\eta k^2(\omega_g^2 + \omega_A^2 + \omega_{rot}^2),$$
$$a_3 = \omega_H(\omega_H + \omega_{sh}) + 2\omega_A^2 + \omega_g^2 + \omega_{rot}^2 + \eta^2 k^4, \tag{5.39}$$

and $a_4 = 2\eta k^2$, where

$$\omega_{rot}^2 = \omega_C^2 + \omega_C\omega_{sh}, \qquad \omega_g^2 = -\frac{\nabla S}{C_p} \cdot \left(\mathbf{g}^* - (\mathbf{k} \cdot \mathbf{g}^*)\frac{\mathbf{k}}{k^2}\right), \tag{5.40}$$

the definitions (2.55) and

$$\omega_{sh} = \frac{s}{k}\left(k_z\frac{\partial\Omega}{\partial s} - k_s\frac{\partial\Omega}{\partial z}\right), \qquad \omega_H = \beta k(\mathbf{k} \cdot \bar{\mathbf{B}}) \tag{5.41}$$

have been used (see Fricke 1969). For ω_C only the positive sign is considered. ω_g is the frequency of buoyancy waves[6], ω_A is the Alfvén frequency, ω_{rot}^2 represents the effects of the differential rotation, and ω_H and ω_{sh} are the characteristic frequencies of the Hall and the shear-driven processes. For pure Kepler flow we have $\omega_{rot} = k_z\Omega/k$.

5.3.1   Ideal MHD

For very high electrical conductivity and for $\omega_H = 0$ we have $a_0 = a_2 = a_4 = 0, a_1 = \omega_A^2(\omega_g^2 + \omega_A^2 + \omega_C\omega_{sh})$, and $a_3 = \omega_g^2 + \omega_{rot}^2 + 2\omega_A^2$. Then Eq. (5.38) can be rewritten as

$$(\gamma^2 + \omega_A^2)^2 + A_1(\gamma^2 + \omega_A^2) + A_2 = 0 \tag{5.42}$$

with

$$A_1 = \omega_g^2 + \omega_{rot}^2, \qquad A_2 = -\omega_C^2\omega_A^2. \tag{5.43}$$

The solutions of this equation have the form

$$\gamma^2 = -\frac{1}{2}(\omega_g^2 + \omega_{rot}^2 + 2\omega_A^2) \pm \frac{1}{2}\sqrt{(\omega_g^2 + \omega_{rot}^2)^2 + 4\omega_C^2\omega_A^2}. \tag{5.44}$$

Negative (positive) γ^2 indicates stability (instability) of the flow. The flow is therefore unstable if $\omega_g^2 + \omega_{rot}^2 + 2\omega_A^2 < 0$. Since ω_A^2 is always positive the necessary condition for instability is $\omega_g^2 + \omega_{rot}^2 < 0$. This is again the same condition as given above for the hydrodynamic case.

The flow is also unstable if $\omega_g^2 + \omega_{rot}^2 + 2\omega_A^2 > 0$ but if

$$(\omega_g^2 + \omega_{rot}^2)^2 + 4\omega_C^2\omega_A^2 > (\omega_g^2 + \omega_{rot}^2 + 2\omega_A^2)^2. \tag{5.45}$$

Hence,

$$\omega_A^2 < \omega_C^2 - \omega_g^2 - \omega_{rot}^2 \tag{5.46}$$

---

[6] it is $\omega_g^2 > 0$ for convectively stable layers, see Eq. (5.13)

holds for instability so that

$$\omega_g^2 + \omega_{\text{rot}}^2 < \omega_C^2 \tag{5.47}$$

is *necessary* for instability. This condition differs from the same condition without magnetic field only by its RHS. This analysis leads to a new 'MHD Høiland criteria' instead of (5.11) where the gradients of the angular momentum $j = s^2\Omega$ are replaced by the gradients of the angular velocity Ω, i.e.

$$s\frac{\partial\Omega^2}{\partial s} - \frac{1}{C_p\rho}\nabla P \cdot \nabla S > 0, \qquad \frac{\partial P}{\partial z}\left(\frac{\partial\Omega^2}{\partial s}\frac{\partial S}{\partial z} - \frac{\partial\Omega^2}{\partial z}\frac{\partial S}{\partial s}\right) < 0. \tag{5.48}$$

The *sufficient* condition for instability, Eq. (5.46), for $\Omega = \Omega(s)$ then reads

$$\frac{k_z^2}{k^2}\frac{\mathrm{d}\Omega^2}{\mathrm{d}\log s} < -\omega_A^2 - \omega_g^2. \tag{5.49}$$

The instability only exists for sufficiently steep subrotation or, in other words, for weak fields and not too stably stratified layers. In its simplest form, for a Kepler flow without buoyancy, the condition is reduced to the relation

$$k < \sqrt{3}\frac{\Omega}{V_A}, \tag{5.50}$$

with the Alfvén velocity $V_A = \bar{B}/\sqrt{\mu_0\rho}$ (Velikhov 1959, Fricke 1969, Balbus 1991). The critical wave number thus depends on the magnetic field, being inversely proportional to the field strength. The maximal possible magnetic field amplitude follows from $V_A \simeq \Omega D$ with D as the characteristic size of the considered domain.

5.3.2   Baroclinic Instability

Equation (5.38) describes five low-frequency modes. The condition for instability (at least one of its roots has a positive real part) is equivalent to one of the four conditions[7]

$$a_0 < 0, \qquad A_1 < 0, \qquad A_2 < 0, \qquad A_3 < 0, \tag{5.51}$$

with

$$A_1 = a_3a_4 - a_2, \qquad\qquad A_2 = a_2(a_3a_4 - a_2) - a_4(a_1a_4 - a_0),$$
$$A_3 = (a_1a_4 - a_0)\big(a_2(a_3a_4 - a_2) - a_4(a_1a_4 - a_0)\big) - a_0\big(a_3a_4 - a_2\big)^2. \tag{5.52}$$

Since $\omega_A^2 > 0$, the condition $(5.51)_1$ is equivalent to $\omega_g^2 < 0$, or with Eq. $(5.40)_2$

$$g_z^*\frac{\partial S}{\partial z}\xi^2 - \left(g_z^*\frac{\partial S}{\partial s} + g_s^*\frac{\partial S}{\partial z}\right)\xi + g_s^*\frac{\partial S}{\partial s} > 0, \tag{5.53}$$

with $\xi = k_s/k_z$. It is easy to see that Eq. (5.53) has two real roots for ξ, so that negative parts of the curve always exist for special wave numbers, independent of the signs of g^* and ∇S (Urpin & Brandenburg 1998). This (unavoidable) instability is called the barocline instability; interestingly enough only 1D configurations can be stable in this regime.

---

[7] since ηk^2 is positive definite the fifth condition ($a_4 < 0$) will never apply

5.4 Stability of Differential Rotation with Strong Hall Effect

A detailed analysis of MHD modes in accretion disks demonstrates a wider variety of instabilities than MRI (see Keppens, Casse & Goedbloed 2002). The situation is particularly uncertain in cold and dense protostellar disks, where the electrical conductivity is small because of the low ionization degree. The magnetic field therefore cannot be considered as frozen into the gas (Gammie 1996). The behavior of the magnetic shear instability with nonvanishing Ohmic dissipation has been considered in both the linear (Jin 1996) and nonlinear (Sano, Inutsuka & Miyama 1998) regimes. However, as was pointed out by Wardle (1999), poorly conducting protostellar disks can be strongly magnetized if electrons are the main charge carriers. If the field is sufficiently strong then the main contribution to the magnetic-dissipation tensor is provided by the Hall effect. This component drastically changes the geometry of the magnetic field. The linear stability analysis by Wardle (1999) shows that the Hall effect can provide an additional influence upon a shear flow depending on the *direction* of the magnetic field. A more general consideration of the magnetic shear instability in the presence of Hall currents is by Balbus & Terquem (2001), with the result that the Hall effect qualitatively changes the stability properties of rotating plasma.

The magnetization of the electron gas is characterized by the product of the electron gyrofrequency $\omega_B = eB/m_{\mathrm{e}}c$ and the relaxation time τ of electrons (see Frank-Kamenezki 1967, Spitzer 1978). In protostellar disks, τ is determined by the scattering of electrons on neutrals, so $\tau = 1/n\langle\sigma v\rangle$, where $\langle\sigma v\rangle$ is the average product of the cross-section and velocity, and n is the number density of neutrals. Using the estimate of $\langle\sigma v\rangle$ by Draine, Roberge & Dalgarno (1983) for electron-neutral collisions, the magnetization parameter a_{e} of electrons can be represented by

$$a_{\mathrm{e}} \equiv \omega_B\tau = 2.1 \cdot 10^{16}\frac{B}{n\sqrt{T}}, \tag{5.54}$$

with B in Gauss, all others in c.g.s. (Shalybkov & Urpin 1995). If this parameter exceeds unity (see Fig. 5.5) then the Hall effect dominates. The Hall-originated magnetic diffusivity is given by $a_{\mathrm{e}}\eta = \beta B$ with $\beta = c/4\pi e n_{\mathrm{e}}$, where η is the microscopic magnetic diffusivity and n_{e} is the number density of electrons. The magnetic diffusivity is $\eta = 234\sqrt{T}/f$ in cm^2/s, where $f = n_{\mathrm{e}}/n$ is the ionization fraction, which is only of order 10^{-12} in cold protoplanetary disks.

5.4.1 Criteria of Instability of Protostellar Disks

Now following Urpin & Rüdiger (2004) we discuss Eq. (5.38) in more detail with the instability conditions (5.51). Note that the condition $A_2 < 0$ does not provide new results.

The Condition $A_1 < 0$

This becomes $\omega_{\mathrm{A}}^2 + \omega_{\mathrm{H}}^2 + \omega_{\mathrm{H}}\omega_{\mathrm{sh}} + \eta^2 k^4 < 0$, or, substituting frequencies,

$$\beta s(\boldsymbol{k}\cdot\bar{\boldsymbol{B}})\left(k_z\frac{\partial\Omega}{\partial s} - k_s\frac{\partial\Omega}{\partial z}\right) < -\omega_{\mathrm{A}}^2 - \beta^2 k^2(\boldsymbol{k}\cdot\bar{\boldsymbol{B}})^2 - \eta^2 k^4. \tag{5.55}$$

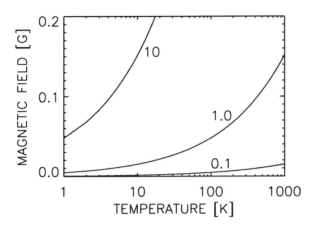

Figure 5.5: The condition $a_e = \omega_B \tau = 1$, where the lines are marked with the values of the number density $n_{14} = 10^{-14} n$. For magnetic fields above the lines the Hall effect dominates the Ohmic dissipation. From Urpin & Rüdiger (2004).

There is no instability without shear. The condition (5.55) describes an instability due to the combined influence of shear and Hall effect. This instability differs from the MRI because the only term that can provide a destabilizing influence is on the LHS of Eq. (5.55), and this term vanishes for $\beta \to 0$. For any dependence of Ω on s and z, and for any direction of $\bar{\boldsymbol{B}}$, wave vectors exist satisfying

$$(\boldsymbol{k} \cdot \bar{\boldsymbol{B}}) \left(k_z \frac{\partial \Omega}{\partial s} - k_s \frac{\partial \Omega}{\partial z} \right) < 0, \tag{5.56}$$

so that the LHS of Eq. (5.55) becomes negative. Therefore, any form of differential rotation[8] can be unstable for sufficiently strong Hall effect (Balbus & Terquem 2001).

If the Hall parameter is large, $a_e \gg 1$, then ηk^2 is negligible in Eq. (5.55). Assuming that the gas is weakly ionized, then for very small f in a large fraction of the disk volume, the Hall term can also dominate the Alfvén term, $|\omega_H| > |\omega_A|$. The necessary condition for instability is then simply

$$|\omega_{\rm sh}| > |\omega_H|. \tag{5.57}$$

The shear must (slightly) dominate the Hall term. The sign of $\omega_{\rm sh}$ plays no crucial role in the condition, so that the instability can appear for any $\nabla \Omega$ if Eq. (5.57) is fulfilled. Assuming $|\partial \Omega / \partial s| \sim \Omega / s$ this condition can be represented by

$$\bar{B} < \frac{\Omega}{\beta k^2} \simeq 10^{-3} \left(\frac{\tau_{\rm rot}}{\rm yr} \right)^{-1} \frac{n_e}{100} \left(\frac{\lambda}{10^{11}\ {\rm cm}} \right)^2 \ {\rm G}, \tag{5.58}$$

where $\tau_{\rm rot}$ is the rotation period. However, the condition for dominating Hall effect ($a_e > 1$) requires rather strong magnetic fields, so that it is not easy to find a consistent parameter range for a 'shear-Hall instability'. One finds for $T = 100$ K and $f = 10^{-12}$ the condition

$$\left(\frac{\lambda}{\rm AU} \right)^2 > 0.003 \left(\frac{\tau_{\rm rot}}{\rm yr} \right), \tag{5.59}$$

---

[8] but not any amplitude

so that the disturbances must have long wavelengths in protostellar disks. Global models must therefore be considered to assess the importance of the Hall effect in protostellar disks.

The Condition $A_3 < 0$

The criterion $A_3 < 0$ yields for $\omega_H > \eta k^2$ (or, which is the same, $\omega_B \tau > 1$) and $\Omega > \omega_H$ the relation

$$\omega_H(\omega_H + \omega_{sh})\left((1+q)(\omega_g^2 + \omega_{rot}^2) - q\omega_A^2\right) + \omega_A^4 +$$

$$+ \omega_A^2\left(\omega_{sh}(\omega_C + \omega_H) + 2\omega_C\omega_H\right) + q\omega_A^2\left(\frac{1}{2}\omega_g^2 + \omega_{rot}^2\right) < 0, \quad (5.60)$$

with

$$q = \frac{1}{2}\frac{\omega_g^2}{(\omega_C - \omega_H)^2 + \eta^2 k^4}. \quad (5.61)$$

With the (realistic) condition $\Omega > \omega_A > \sqrt{\omega_H \Omega}$ then from Eq. (5.60)

$$\omega_{sh}\omega_C + q\left(\frac{\omega_g^2}{2} + \omega_C^2 + \omega_{sh}\omega_C\right) < 0 \quad (5.62)$$

arises. In a convectively stable disk with $\omega_g^2 > 0$ an instability thus only appears if

$$2\omega_C\omega_{sh} < -\omega_g^2. \quad (5.63)$$

For $\omega_g \sim \Omega$ this condition is more restrictive than the criterion (5.50) for MRI. The 'negative' buoyancy of the stable vertical stratification requires stronger shear than without gravity. The Hall frequency no longer appears in this relation.

5.4.2   Growth Rates

A general analysis of the roots of Eq. (5.38) is very complicated. Simple expressions can be obtained, however, for some cases of astrophysical interest. For small ηk^2 only the four roots of

$$\gamma^4 + c_2\gamma^2 + c_0 = 0 \quad (5.64)$$

exist, with $c_0 = \omega_A^2\left(\omega_g^2 + \omega_C\omega_{sh}\right) + \omega_H(\omega_H + \omega_{sh})(\omega_g^2 + \omega_{rot}^2)$ and $c_2 \simeq \omega_g^2 + \omega_{rot}^2$. Without a magnetic field c_0 vanishes, so that $\gamma^2 = -c_2$ solves (5.64), leading to $\gamma^2 \simeq -\omega_C^2/4$ for nonmagnetic Kepler flow, i.e. stability.

The general solution of Eq. (5.64) is given by $\gamma^2 = -(c_2 \pm \sqrt{c_2^2 - 4c_0})/2$. If $\Omega > \max(\omega_H, \omega_A)$ then $c_2^2 > 4c_0$ and

$$\gamma_{1,2}^2 \approx -\frac{c_0}{c_2}. \quad (5.65)$$

One of the modes $\gamma_{1,2}$ is unstable if $c_2 > 0$ and $c_0 < 0$. The remaining modes describe convection and are stable in convectively stable disks. There are two possibilities:

i) For $\sqrt{\omega_H \Omega} > \omega_A$ we have for the magnetically driven modes $\gamma_{1,2}^2 \simeq -\omega_H(\omega_H + \omega_{sh})$. If the condition (5.57) is fulfilled then one of these modes is unstable. Since $\omega_{sh} \sim \Omega$ it is

$$\frac{\gamma}{\Omega} \sim \sqrt{\frac{\omega_H}{\Omega}} \tag{5.66}$$

for the Hall-driven modes.

ii) If $\sqrt{\omega_H \Omega} < \omega_A$ then $\gamma_{1,2}^2 \simeq -\omega_A^2(\omega_g^2 + \omega_C \omega_{sh})/(\omega_g^2 + \omega_{rot}^2)$. This is the dispersion equation for magnetic shear-driven modes (Balbus 1995, Urpin 1996) in the limit $\omega_g > \omega_A$. One of these modes is unstable for small ω_g^2. For $\omega_g = 0$ one finds

$$\gamma \simeq \sqrt{3}\,\omega_A. \tag{5.67}$$

The growth rate is of order ω_A, which with Eq. (5.50) leads to $\gamma \simeq \Omega$, which is a very fast growth rate.

The critical wavelength can easily be calculated with the instability conditions. For $B = 1$ G, $n = 10^{14}$ and $\tau_{rot} = 1$ yr they are ~ 0.1 AU (Urpin & Rüdiger 2004).

5.5   Global Models

The existence of the MRI in global models is now considered. Spherical geometry allows a completely global formulation also of the magnetic part of the system[9]. The instability needs a negative shear $\partial\Omega/\partial s$. Such a decrease may well be present in stellar radiative cores with their extremely small microscopic viscosities, because of the loss of angular momentum through the surface by stellar winds or by the outward transport of angular momentum by radiation (Kippenhahn 1958). We do not assume axial symmetry and include finite diffusivities, but neglect the effects of compressibility and stratification (see Stone et al. 1996, Balbus 1995, Spruit 1999).

We shall find that the boundaries impose an upper limit B_{max} on the magnetic field amplitude. The instability is present only for fields smaller than B_{max}, which grows with the rotation rate. The finite diffusivities also impose a lower limit B_{min}, which approaches a constant value for sufficiently rapid rotation. Close to B_{min} the axisymmetric modes are preferred, while close to B_{max} the nonaxisymmetric modes are more unstable, at least for stress-free boundaries. The symmetries relative to the equatorial plane can also be distinguished. The modes with equatorially antisymmetric flow and symmetric magnetic field ('quadrupoles') always prove to be preferred.

5.5.1   A Spherical Model with Shear

A rotating sphere of conducting fluid with radius R is considered by Kitchatinov & Rüdiger (1997). Its angular velocity only varies with s. Outside the sphere is vacuum. A uniform magnetic field $\boldsymbol{B}_0$ is imposed parallel to the rotation axis. Such a field satisfies the steady induction equation inside the sphere, Maxwell's equations outside, and the continuity condition

---

[9] see Priklonsky et al. (2000) for the problems with magnetized disks

on the boundary. Let the rotation law be prescribed by

$$\Omega(s) = \Omega_0 \left(1 + \left(\frac{s}{s_0} \right)^{2q} \right)^{-1/2}, \tag{5.68}$$

with $s_0 = R/2$ and q as a free parameter (Donner & Brandenburg 1990). According to the Rayleigh criterion the rotation law for $q = 1$ is clearly subcritical. The case $q = 2$ is close to the critical state but is still subcritical. The majority of our results belongs to $q = 2$. With $q = 3$ there is an (outer) region in the sphere where the Rayleigh criterion is violated. Hence, for sufficiently low viscosity the rotational state with $q = 3$ is expected to be unstable even in the nonmagnetic case.

We start with the MHD equations for incompressible fluids, linearized around the reference state with $\bar{u}$ and B_0, i.e.

$$\frac{\partial u'}{\partial t} + (u' \cdot \nabla) \bar{u} + (\bar{u} \cdot \nabla) u' = \frac{1}{\rho} J' \times B_0 - \frac{1}{\rho} \nabla P' + \nu \Delta u',$$

$$\frac{\partial B'}{\partial t} = \nabla \times (u' \times B_0 + \bar{u} \times B') + \eta \Delta B' \tag{5.69}$$

with $\nabla \cdot u' = \nabla \cdot B' = 0$. The pressure can be eliminated by *curling* Eq. (5.69)$_1$, hence

$$\frac{\partial \omega}{\partial t} = \nabla \times (\bar{u} \times \omega + u' \times \bar{\omega} + J' \times B_0/\rho) + \nu \Delta \omega, \tag{5.70}$$

where $\omega = \nabla \times u'$ and $J' = \nabla \times B'/\mu_0$ are the vorticity and current density, and $\bar{\omega} = \hat{e}_z s^{-1} \mathrm{d} \left(s^2 \Omega \right)/\mathrm{d}s$ is the vorticity due to the basic rotation. A natural normalization is used so that normalized equations result. The dimensionless parameters that finally define the model are

$$C_\Omega = \frac{\Omega_0 R^2}{\eta}, \qquad \mathrm{Ha} = \frac{B_0 R}{\sqrt{\mu_0 \rho \nu \eta}}, \qquad \mathrm{Pm} = \frac{\nu}{\eta}, \tag{5.71}$$

i.e. the magnetic Reynolds number of the rotation, the Hartmann number and the magnetic Prandtl number.

Both the vector fields in the present problem are divergence-free. They can thus be expressed in terms of the scalar potentials

$$u' = x \times \nabla \left(\frac{w}{x} \right) + \nabla \times \left(x \times \nabla \left(\frac{\Psi}{x} \right) \right),$$

$$B' = x \times \nabla \left(\frac{B}{x} \right) + \nabla \times \left(x \times \nabla \left(\frac{A}{x} \right) \right), \tag{5.72}$$

defining the toroidal and poloidal parts of the fields. The resulting relations include the operator that has the spherical Legendre functions as eigenfunctions. This makes expansions in spherical harmonics convenient, e.g. $A = \sum A_{nm}(x, t)e^{im\phi} P_n^{|m|}(\cos \theta)$. The vacuum boundary conditions for the magnetic field are now easy to formulate in terms of the amplitudes of the expansions, i.e. $\partial A_{nm}/\partial x + n A_{nm}/x = B_{nm} = 0$. Krause & Rädler (1980, their Chapt. 13) presented all the necessary details of these formulations.

For stress-free surfaces it is required that the zonal cross-components of the Reynolds stress tensor (Maxwell tensor not included!) vanish, i.e.

$$\frac{\partial}{\partial x}\left(\frac{w_{nm}}{x^2}\right) = 0, \qquad\qquad w_{nm} + \frac{2}{x}\frac{\partial \Psi_{nm}}{\partial x} = 0. \qquad\qquad (5.73)$$

The radial velocity must also be zero at the surface, hence $\Psi_{nm} = 0$.

In the linear theory the axisymmetric modes and nonaxisymmetric modes can be considered separately. For each value of m the equation system splits into two independent subsystems governing the modes with different types of equatorial symmetry. Note that the symmetry notation relates to the magnetic field. The symmetry of the flow is opposite. The S-modes combine symmetric magnetic fields with antisymmetric flow, and vice versa for the A-modes.

Figure 5.6 presents the geometry of the model and the resulting stability map for the axisymmetric magnetic A-modes. The same for the S-modes is given in Fig. 5.7 (left).

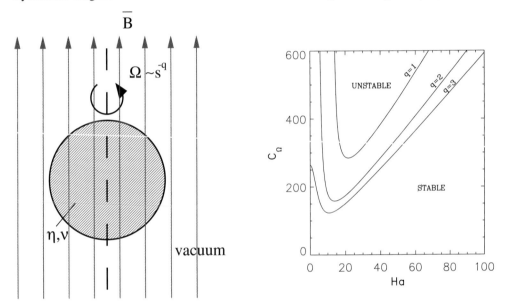

Figure 5.6: *Left*: The model of Kitchatinov & Rüdiger (1997) for a sphere in differential rotation embedded in vacuum and subject to a homogeneous vertical magnetic field. *Right*: The resulting neutral stability lines for the modes with (magnetic) dipolar symmetry and for $\mathrm{Pm} = 1$.

The curves for symmetric and antisymmetric modes are similar, but the A-modes are excited at (slightly) higher rotation rates. The equatorially *symmetric* (magnetic) modes are thus preferred. Only the lines for $q = 3$ cross the C_Ω-axis. The neutral stability lines here represent the magnetic modification of the hydrodynamic instability. A weak magnetic field leads to a subcritical excitation. The critical rotation rate becomes smaller. The system becomes stable, however, for sufficiently large Hartmann number.

The curves for $q = 1$ and $q = 2$ represent the MRI. The two lines are similar; with less shear ($q = 1$) the instability requires higher C_Ω. For sufficiently high but fixed rotation rate a finite interval exists, $B_{\min} < B_0 < B_{\max}$, where the rotation law is unstable. The instability

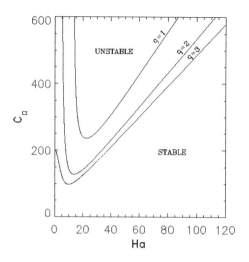 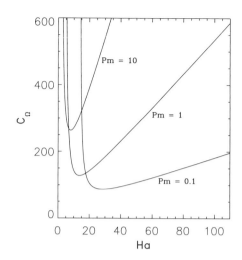

Figure 5.7: The influence of rotation laws with $q = 1, 2, 3$ (*left*, for $\mathrm{Pm} = 1$) and magnetic Prandtl number (*right*, for $q = 2$) on the excitation of the (magnetic) quadrupolar modes.

only exists for fields that are neither too strong nor too weak. If the magnetic Reynolds number C_Ω is too small then the system is stable for all magnetic fields.

The local stability analysis of ideal fluids, Eq. (5.50), predicts that only excitations with wavelengths larger than the critical value $\lambda_{\min}$ are unstable, i.e.

$$\lambda > \lambda_{\min} \simeq 2\pi \frac{V_A}{\Omega_0}. \tag{5.74}$$

In the global formulation, however, the critical wavelengths, Eq. (5.74), are limited by the size of the system, hence $V_{A,\max} \simeq R\Omega_0/2\pi$ defines the maximum field amplitude. This relation agrees well with the estimate of $B_{\max}$ by Papaloizou & Szuszkiewicz (1992) and Curry & Pudritz (1995) for thick disks.

Figure 5.7 (right) illustrates the dependence on the magnetic Prandtl number. The reason for the change of the slope of the right-hand branches in the plot is obvious. One easily finds the relation $C_\Omega/\mathrm{Ha} \simeq 2\pi\sqrt{\mathrm{Pm}}$, resulting in

$$\frac{B_0}{\sqrt{\mu_0 \rho}} = \frac{\Omega_0 R}{2\pi} \tag{5.75}$$

for the maximum field. The behavior of the left-hand branches in Fig. 5.7 (right) in dependence on the magnetic Prandtl number is also interesting. One finds $\mathrm{Ha}_{\min} \propto \mathrm{Pm}^{-0.5}$, so that

$$\frac{B_0}{\sqrt{\mu_0 \rho}} \simeq \frac{4\eta}{R} \tag{5.76}$$

for the minimum field amplitude, which reflects the influence of the magnetic dissipation.

Nonaxisymmetric Modes

Figure 5.8 presents the results for the nonaxisymmetric modes with $m = 1$ and $m = 2$. The slopes of the right-hand branches of the neutral stability lines for $m \neq 0$ are smaller than for the axisymmetric solution. For sufficiently fast rotation and sufficiently strong magnetic fields the excitation of nonaxisymmetric modes is therefore preferred. In Fig. 5.8 the crossover of the $m = 1$ modes can be observed. The smaller slopes of the right-hand branches of the lines for $m = 2$ suggest that these modes may eventually dominate for sufficiently large C_Ω and Ha[10].

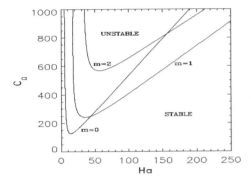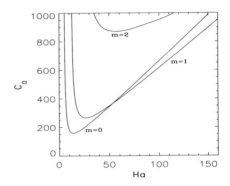

Figure 5.8: Neutral stability for S-modes (*left*) and A-modes (*right*). Nonaxisymmetric modes are dominating only for higher magnetic fields. $\mathrm{Pm} = 1$, $q = 2$.

The existence of a region in parameter space where nonaxisymmetric excitations are preferred is very promising for a dynamo effect by the shear instability. After Cowling's theorem self-excitation of magnetic fields only exists for nonaxisymmetric magnetic fields. Note that with an azimuthal background field *only* nonaxisymmetric modes of the magnetic shear instability are excited (Ogilvie & Pringle 1996).

For a nonlinear analysis the parameters $\mathrm{Ha} = 50$ and $C_\Omega = 714$ are used (Drecker, Hollerbach & Rüdiger 1998). With such simulations we proceed to the turbulent behavior of the model. The averaging scheme for the correlation tensor, consisting of both the Maxwell and the Reynolds parts, uses an integration over the horizontal coordinates and keeps the radial dependence. The eddy viscosity becomes positive throughout the whole shell. The eddy viscosity is mainly due to the influence of the Maxwell stress tensor.

In many turbulence models it is assumed that α_{SS} is constant. For rigid boundary conditions[11] we find in Fig. 5.9 (left) that there exists a region $0.3 < s < 0.7$ where this parameter has the constant value $\alpha_{SS} \simeq 3 \cdot 10^{-3}$. For stress-free boundary conditions the values are one order of magnitude smaller. It changes its sign above $s = 0.8$.

---

[10]   for rigid boundary conditions the axisymmetric modes always have the lowest C_Ω

[11]   rigid boundaries have been considered by Drecker, Hollerbach & Rüdiger (1998), with the striking result that in this case the axisymmetric modes are always preferred

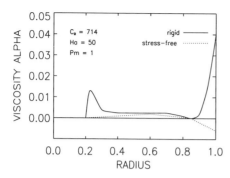 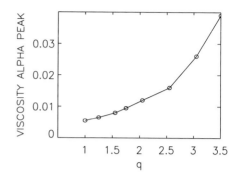

Figure 5.9: Nonlinear results. *Left:* α_{SS} for stress-free (solid) and rigid (dotted) boundary conditions. For stress-free boundary conditions the values are much smaller. *Right:* α_{SS} vs. the differential parameter q for rigid boundaries. From Drecker, Hollerbach & Rüdiger (1998).

Next the influence of the parameter q on the eddy viscosity is considered. Abramowicz, Brandenburg & Lasota (1996) calculated α_{SS} in accretion disks and found that it becomes infinite for $q = 2$. We have varied the value of q in the range from 1 to 3.5. The corresponding nonlinear solutions appeared to be axisymmetric and also antisymmetric with respect to the equator. Figure 5.9 (right) seems to reveal a linear relation between the maximum of α_{SS} and q until a critical point $q \approx 2$ where hydrodynamical instability sets in. Beyond that point the values for α_{SS} grow rapidly, but remain finite.

5.5.2   A Global Disk Model

A similar computation exists for disk geometry with the boundary conditions for thin disks

$$B = \frac{\partial A}{\partial z} = 0 \qquad\qquad (5.77)$$

by Kitchatinov & Mazur (1997). Note that $B_s \propto \partial A / \partial z$. This is a simplification, but to match a magnetic field to a vacuum field in cylindrical geometry is not easy. Priklonsky et al. (2000) have shown that Eq. (5.77) holds exactly for extremely thin disks, while for thick disks it must be replaced by a nonlocal relation resulting from the form $A \sim \exp(-\lambda k|z|) J_1(kr)$ with k as an arbitrary wave number (see their Eqs. (10) and (11)). Rädler & Wiedemann (1990) have already used similar equations.

The rotation law is Eq. (5.68) with $s_0 = 5H$. The eigensolutions for B are formed from the functions $\sin n\pi z$ and $\cos(n - 0.5)\pi z$, and with similar expressions for the potential A. The neutral stability lines for $Pm = 1$ are similar to those for the sphere, but it appears easier to excite the MRI in disks rather than in spheres. Note that the definition $C_\Omega = \Omega_0 H^2 / \eta$ has been used for disks.

The stability lines for quadrupolar symmetry are given in Fig. 5.10 (left). In opposition to the spherical case the modes with the lowest magnetic Reynolds number are always axisymmetric. Figure 5.10 (right) shows an example of a magnetically induced subcritical excitation of the instability in the case where a hydrodynamic instability also exists. The rotation law with $q = 2.1$ (slightly) violates the Rayleigh stability criterion Eq. (5.9). The marginal line crosses the vertical axis at $C_\Omega = 66$. The subcritical excitation works as long as the magnetic field is still weak. For larger Ha (not shown) the magnetic field starts to suppress the instability[12] and $C_\Omega \to \infty$ for Ha $\to \infty$.

The two branches in Fig. 5.10 (left), for given C_Ω, differ strongly in the geometry of the resulting magnetic pattern. The pitch angle varies from $0°$ (left branch) to $90°$ (right). In both the extreme cases the magnetic angular transport $B_s B_\phi$ therefore vanishes, and we expect its maximum somewhere *between* the two branches. It indeed exists and provides an outward transport of angular momentum by the Maxwell stress.

A similar model was considered by Rüdiger et al. (1999), but with variations of the magnetic Prandtl number. The simulations yielded a relation

$$\frac{C_\Omega}{\text{Ha}} \propto \sqrt{\text{Pm}} \tag{5.78}$$

for the strong-field branch of the stability lines, which along this line leads to the relation (5.75) with $\Omega H \simeq c_{\text{ac}}$ for the strongest vertical field for which the MRI is possible. For protostellar disks with $T \simeq 10^3$ K and $\rho \simeq 10^{-10}$ g/cm^3 this maximal magnetic field strength $\simeq 10$ G. The remnant magnetization of meteorites suggests the existence of magnetic fields of 1–10 G in the disk (Levy & Sonett 1978, see Birk, Wiechen & Lesch 2002).

Concerning the angular momentum transport also the Reynolds stress and the Maxwell stress proved to be positive and of the same order. The same holds for both the kinetic and magnetic energy.

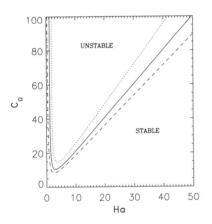
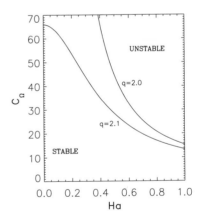

Figure 5.10: *Left*: The stability map for the slab model for axisymmetric magnetic quadrupoles excited for Kepler rotation (solid), $q = 1$ (dotted) and $q = 2$ (dashed). *Right*: The model shows hydrodynamical instability, i.e. for Ha = 0, already for $q > 2$. Pm = 1. From Kitchatinov & Mazur (1997).

---

[12] see Chapt. 8 for quite a similar behavior in magnetic Taylor–Couette flow experiments

5.6   MRI of Differential Stellar Rotation

5.6.1   T Tauri Stars (TTS)

Doppler imaging of the weak-line T Tauri star V 410 Tau shows cool spots at low latitudes as well as in the polar regions (Joncour, Bertout & Ménard 1994, Hatzes 1995, Rice & Strassmeier 1996 [13]). The spots are very large compared to sunspots ($\simeq 0.1R$) and their distribution on the stellar surface is nonaxisymmetric. The rotation law derived from the evolution of the spot distribution is almost rigid. While the average rotation period is 1.87 days, the rotation rate at the equator exceeds that of the polar caps by maximally 0.2°/day (Rice & Strassmeier 1996). Johns-Krull (1996) found from high-resolution spectra of a sample of TTS that the rotation law of these stars is either rigid-body or antisolar. As the same spectral feature that indicates antisolar rotation is also caused by the presence of cool spots in the polar region of the star it is concluded that the rotation is most likely rigid. The findings of Rice & Strassmeier and Johns-Krull indicate rigid or a weak differential rotation for TTS as predicted by Küker & Rüdiger (1997). After Küker & Stix (2001) for a TTS with 1 $M_\odot$ the *total* horizontal shear, $\delta\Omega^{\text{lat}} = \Omega_{\text{eq}} - \Omega_{\text{pole}}$, is similar to the solar value.

Guenther et al. (1999) determined the photospheric field strengths for a sample of TTS, containing both classical and weak-line TTS, by means of the Zeeman effect. Field strengths up to 4.3 kG and filling factors of about 0.5 were found. Johns-Krull et al. (1999) derived a value of 2.6 ± 0.3 kG for the magnetic field of BP Tau. The large filling factors indicate the field is not dipolar. Also at the solar surface the magnetic activity is the result of the dynamics of closed field structures. As, on the other hand, the field is dominated by the dipole component at large distances from the star, coupling forces between the star and the surrounding gas are probably weaker than measurements of the surface field strength suggest.

The finding of a period of 5.4 yr in the light curve of V410 Tau by Stelzer et al. (2003) seems to rule out a purely nonaxisymmetric field for this star. As the differential rotation is very weak and the spot pattern stable, the field-generating dynamo is probably a more complex mechanism than a simple $\alpha\Omega$-dynamo or α^2-dynamo.

A comprehensive study of X-ray emitting pre-MS stars in the Orion Nebula Cluster by Feigelson et al. (2003) provided an anticorrelation between Coriolis number Ω^* and X-ray luminosity for these stars, i.e. the fastest rotators show the lowest activity ('supersaturated'). This behavior is also known from the most rapidly rotating MS stars, while for slowly rotating MS stars the X-ray luminosity and Ω^* are correlated (Hempelmann, Schmitt & Stępień 1996). One could believe that the dynamo process in the TTS is completely different from that in MS stars.

The X-ray activity of WTTS is significantly higher than that of CTTS. This has been shown for Taurus (which is one of the best studied star formation regions) with the ROSAT All Sky Service by Neuhäuser et al. (1995) and König, Neuhäuser & Stelzer (2001). The WTTS are rotating faster than the CTTS which are magnetically coupled to (the outer and slower parts) of their accretion disks (Fig. 5.13).

---

[13] Rice & Strassmeier (1996) also found hot spots

5.6.2   The Ap-Star Magnetism

Schüssler (1975) considered the dynamo regime during the transition from the fully convective Hayashi phase of a pre-MS star to its mostly radiative MS state for (rigidly rotating) A stars. The ratio of the magnetic eddy diffusivity to the microscopic one was 10^3. The dynamo-built magnetic field survives the fast (1 Myr) development but it changes its geometry. No new computations are known for this interesting problem in order to approach the more realistic value of (say) 10^{7-8} for the diffusivity ratio and including nonaxisymmetric field modes.

The magnetic variations seen in Ap stars are indeed incompatible with a simple dipolar magnetic geometry (Fig. 5.11). In many Ap stars a significant quadrupolar component is indicated (Oetken 1979, see Rüdiger & Scholz 1988). Spectral variability indicates a nonuniform distribution of certain chemical elements over the stellar surface.

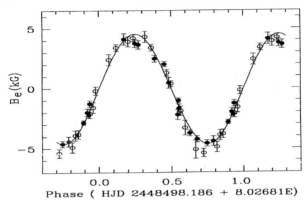

Figure 5.11:   The rather regular magnetic field of 53 Cam. Rotation period is 8 d. From Hill et al. (1998).

Compared to the Sun, A stars are rapidly rotating. If stratified turbulence is present in the interior its α-tensor becomes highly anisotropic. As a consequence the corresponding α^2-dynamo yields a highly nonaxisymmetric solution with a slowly drifting S1-mode. This solution survives a rather strong differential rotation surprisingly well, as well as the nonlinear magnetic quenching of the α-effect (Rüdiger & Elstner 1994, Moss & Brandenburg 1995). Recent empirical results by Landstreet & Mathys (2000) confirm the concept of the oblique rotator by Krause & Oetken (1976). Almost all Ap stars with longer rotation periods have a small obliquity between the magnetic axis and the rotation axis, while for the more rapid rotators the two axes form a much bigger angle (Fig. 5.12, left). The slower rotators also have the stronger magnetic fields (Hubrig, North & Mathys 2000). The stars with the axes parallel lost much more angular momentum than the stars with the axes perpendicular, which is explained by Stępień & Landstreet (2002) by the strong effect of magnetic wind-induced angular momentum transport in the pre-MS phase of these stars. For the TTS Lamm (2003) found the opposite behavior (Fig. 5.12, right). The faster the rotation the smaller is the rotation-induced photometric variation. Obviously, two very different mechanisms are at work here.

On the other hand, the observational results show a high degree of randomness. Field geometry and amplitude differ from star to star. There is only one rather strict rule: the Ap stars are slow rotators and slow rotators are Ap stars (Abt & Willmarth 2000). Stępień (2000) suggested to compare the situation in the A star group with that of the T Tauri stars. Among them there is also a subgroup (CTTS) with slow rotation that are systems with accretion disks

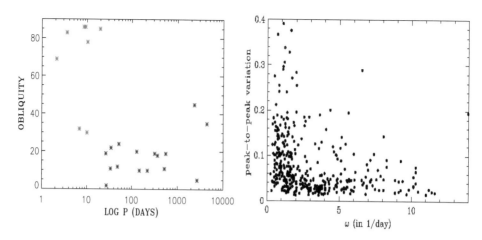

Figure 5.12: Two different kinds of Ω-dependencies for rotational modulations. *Left*: The magnetic obliquity of Ap star magnetic fields is maximal for large Ω (Landstreet & Mathys 2000). *Right*: The amplitude of the rotation-induced photometric variations of TTS is maximal for small Ω. Courtesy M. Lamm.

(Fig. 5.13). Magnetic star-disk interaction might easily spin-down the central object which finally becomes the Ap star.

There is, however, the puzzling observation that magnetic fields only appear in A stars that long ago left the ZAMS, more precisely after about 30% of their MS-lifetime (Hubrig, North & Mathys 2000). If it is true that the slow rotators become magnetic after some time then the rotation distribution of young and old A stars (without Ap!) should differ strongly. If, on the other hand, the rotation is suddenly reduced with the occurrence of the magnetic fields, then the rotational distribution of young and old A stars (with Ap!) should be different. However, Hubrig, North & Medici (2000) do not find significant differences for either of these statistics,

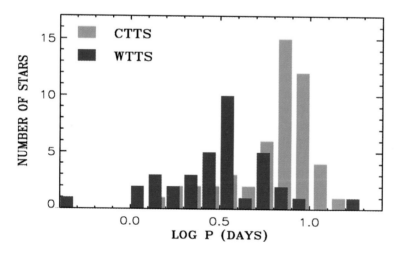

Figure 5.13: Rotation period differences for TTS with (CTTS) and without (WTTS) accretion disks. (Bouvier et al. 1993).

possibly because the number of candidates of known age is too small. The question remains why slowly rotating A stars should become magnetic. The possibility that is favored here is the formation of a differential rotation within the extended radiative stellar envelope. Such a rotation law becomes unstable under the influence of a (weak) magnetic field (see Spruit 1999, for a detailed discussion).

Figure 5.6 (left) presents the model for the global MHD instability. The model is unstratified, but its outer part is differentially rotating with $\Omega \propto s^{-q}$. The innermost sphere is considered as vacuum in order to simulate the very high magnetic diffusivity of convective cores. The entire sphere is embedded in vacuum, and is initially threaded by a large-scale magnetic field in the axial direction. The resulting stability diagram is similar to those of Fig. 5.7. For a given Reynolds number of the rotation there are two limiting magnetic field amplitudes between which the rotation law is not stable. Excitation of fluctuating flows and fields are the immediate consequence. Only nonlinear simulations can provide the spectrum of both quantities. An initial field is necessary to start the instability, but if after a while the external field is no longer necessary then one can speak of an MHD shear-flow dynamo.

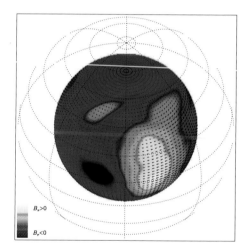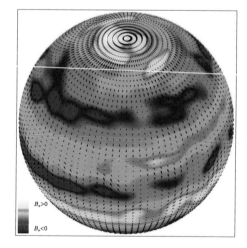

Figure 5.14: Magnetic field structure from a direct spherical dynamo simulation shown at $x = 0.65$ (*left*) and at the surface (*right*). Arrows are the horizontal projections of the magnetic field; gray levels depict the strength of the radial component. Large-scale structures in the interior do not penetrate the surface. Courtesy R. Arlt.

Figure 5.14 shows a snapshot of the nonlinear simulation of a model like that shown in Fig. 5.6 but with a convective ($\sim$ vacuum) core. The resulting Fourier spectrum of the azimuthal magnetic components runs as $m^{-5/2}$ and it also includes global modes up to $m = 0$ (see Drecker, Rüdiger & Hollerbach 2000). The dominance of the kinetic energy is only due to the axisymmetric components but at small scales the magnetic modes dominate.

5.6.3   Decay of Differential Rotation

Considered as an initial value problem the rotation law within the radiative stellar envelope decays after some time. A similar model as in Fig. 5.6 (left) is considered to find the decay time. The simulations of Arlt, Hollerbach & Rüdiger (2003) are fully global and nonlinear but the (negative) buoyancy is not included. This makes the configuration different from the linear analysis by Balbus & Hawley (1994) which was linear but included buoyancy.

The inner radius of the spherical shell is at $x_{\rm in} = 0.2$. The time-dependent, incompressible equations in the inertial system are solved using the potentials as in Eq. (5.72). The initial condition for the velocity represents a rotation profile in which the angular velocity decreases outward according to Eq. (5.68), with the magnetic Reynolds number $C_\Omega = R^2\Omega_0/\eta$ as its amplitude. Initially $q = 2$, which is hydrodynamically stable.

First, the evolution of the rotation flow without magnetic fields is probed. The stress-free boundary conditions are not compatible with the initial azimuthal velocity profile $\Omega(s)$. The rotation profile will thus lead to meridional circulations, which equalize the differential rotation on the viscous timescale. A model with $\rm Rm = 50,000$ gradually decays and reaches $q = 1$ after 200 rotation periods.

The values of C_Ω, Pm and B_0 are the free parameters of the model. Stellar gases possess magnetic Prandtl numbers of $\rm Pm \simeq 0.01$. The timescales for the diffusive processes in velocity and magnetic fields differ greatly and are thus hard to cover by one simulation, so that it makes sense to find the scaling with Pm.

The Fourier spectra for the magnetic fields (Fig. 5.15, left) show satisfying power contrast all through the simulation. The angular momentum transport is a result of stresses from both velocity and magnetic field fluctuations. The averages $\langle u'_s u'_\phi \rangle$ and $\langle B'_r B'_\phi \rangle$ taken in the equatorial plane show a clear dominance of magnetic stresses over kinetic stresses, occasionally by a factor 10.

The decay time of the differential rotation is measured by the time that the system needs to approach a $q = 1$ profile (say). The decay of q with time is given by the solid line in Fig. 5.15 (right). A short transition period can be seen. At the end of the computation the system oscillates around an equilibrium state with $q = 0$ where magnetic and kinetic energies decay exponentially.

The rotation profile seems to decay on the *rotational* timescale. This is far too fast for stars with radiative envelopes. The reason is that there are physical quantities that cannot be matched in the simulation. The diffusive timescale is many orders of magnitudes longer than the rotational timescale. The magnetic Reynolds number in stellar radiative zones is about 10^{13}. Our highest value is 50,000. The dependence of the decay time $\tau_{\rm dec}$ (given in rotation times) on the magnetic Reynolds number and magnetic Prandtl number is shown in Fig. 5.16. No significant dependence of the decay times on Pm has been found.

With a series of computations with various C_Ω the decay time may be rescaled with

$$\frac{\tau_{\rm dec}}{\tau_{\rm rot}} \simeq \frac{C_\Omega}{2000} \simeq \frac{\rm Pm}{2000}\,\rm Re \tag{5.79}$$

to stellar conditions. With $\rm Pm = 10^{-2}$ we see that the differential-rotation decay as a consequence of MRI is 10^5 times faster than the viscous decay scaling with Re. Equation (5.79)

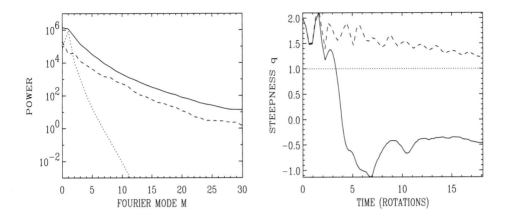

Figure 5.15: *Left*: Longitudinal Fourier spectra of the azimuthal magnetic field energy for the model with $C_\Omega = 10{,}000$ and $\mathrm{Pm} = 1$ at $t = 0.001$ (dotted), 0.003 (solid), and 0.010 (dashed). *Right*: Development of the steepness q of the rotation profile for a model with density stratification. The dashed line is for $B = 0$.

leads to a decay time of $3 \cdot 10^8$ yr. This is of the order of the lifetime of an A star. The extrapolation, of course, involves considerable uncertainty.

5.7 Circulation-Driven Stellar Dynamos

According to Cowling's theorem a dynamo mechanism cannot generate stationary axisymmetric magnetic fields. It does not exclude, however, that nonaxisymmetric fields are generated by axisymmetric velocity fields. The existence of dynamo solutions from axisymmetric flows has often been studied before. Dudley & James (1989) presented a detailed compilation of such dynamo solutions. All their flows are a combination of a meridional circulation with

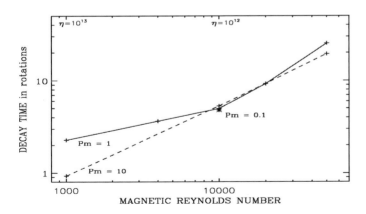

Figure 5.16: Decay time of differential rotation versus magnetic Reynolds number. The decay time is measured in rotation times. Solid: $\mathrm{Pm} = 1$, dashed: $\mathrm{Pm} = 10$. An asterisk indicates the decay time for the only computation with $\mathrm{Pm} = 0.1$. From Arlt, Hollerbach & Rüdiger (2003).

differential rotation. Dynamo action was confirmed for most of these flow patterns but only for certain parameters.

A key question is whether a meridional circulation alone is able to operate a dynamo without differential rotation. Gailitis (1970) announced the existence of such solutions. This configuration was adopted to a stellar dynamo model by Moss (1990). He worked with the meridional circulation in the radiative envelopes of stars of early spectral types predicted by Eddington and Sweet (see Tassoul 1978). Moss reported dynamo action for magnetic Reynolds numbers (of the meridional flow) larger than about 30, translating into a circulation velocity of only 10^{-6} cm/s.

We also start with the problem to excite a dynamo from axisymmetric flows without differential rotation. In a second step, differential rotation is added to extend the models of Dudley & James (1989), and in a third step, a consistent flow pattern for a radiative star is constructed.

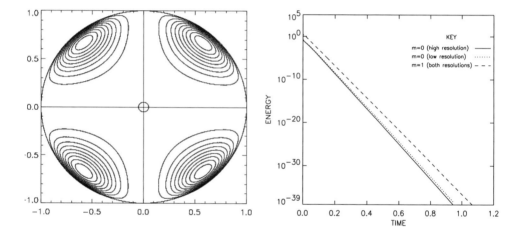

Figure 5.17: *Left*: The meridional flow pattern used by Gailitis (1994). Note that u_θ does not vanish at the surface. *Right*: The evolution of two initial magnetic fields at $\mathrm{Rm} = 10{,}000$. The decay of the $m = 0$ mode is shown as a test for possible numerical artifacts. The runs on $m = 1$ with different spatial resolutions are here indistinguishable (dashed line). From Arlt, Hollerbach & Rüdiger (2003).

5.7.1 The Gailitis Dynamo

The computational domain is a spherical shell embedded in vacuum with an inner boundary at $x_{\mathrm{in}} = 0.05$. A velocity field is prescribed adopted from Gailitis (1994) that does not penetrate the boundary at $x_{\mathrm{in}} = 0.05$ (Fig. 5.17, left). The basic difference from the Gailitis model is the vacuum exterior. The feeling is that any finite diffusivity in the exterior would imply infinite shear at $x = 1$, leading to unpredictable results.

The nondimensional induction equation is integrated in time, where the magnetic Reynolds number $\mathrm{Rm} = u_0 R / \eta$. Figure 5.17 (right) shows two time series of a measure for the magnetic energy. We first integrated the evolution of an axisymmetric ($m = 0$) mode for which it is clear that no dynamo can exist. Solid and dotted lines show the $m = 0$ series for two different spatial resolutions. No numerical artifacts exist.

The dashed line represents the runs for $m = 1$ with the same resolutions. The models shown there employed a configuration with odd l; we have also made the same experiment with the opposite symmetry with identical results. The orientation of the circulation was also reversed, and again led to decay of the $m = 1$ mode. We did not find any dynamo action up to magnetic Reynolds numbers of 10,000 for this and the reversed circulation.

5.7.2 Meridional Circulation plus Shear

With shear the chance is strongly enhanced to find dynamo action in axisymmetric flows. Our models combine a differential rotation and a meridional circulation that are symmetric with respect to the equator, i.e.

$$\boldsymbol{u} = \nabla \times (E\hat{\boldsymbol{e}}_r) + \epsilon \nabla \times (\nabla \times F\hat{\boldsymbol{e}}_r), \tag{5.80}$$

with a circulation amplitude ϵ and the potentials $E = x \sin(\pi x) P_1(\cos \theta)$ for the differential rotation and $F = x^2 \sin(\pi x) P_2(\cos \theta)$ for the meridional flow. For $\epsilon = 0.13$ Dudley & James (1989) find a critical magnetic Reynolds number of about 95. A second flow uses $E = F = x^2 \sin(\pi x) P_2(\cos \theta)$ for which Dudley & James obtain $\mathrm{Rm_{crit}} \simeq 54$ when $\epsilon = 0.14$. If the meridional circulation is reversed, no dynamo action is found for the first flow. Such a *reversed* circulation, however, exists in the radiative envelopes of hot MS stars. The reversed circulation of the second example does show dynamo action with a critical magnetic Reynolds number of $\mathrm{Rm_{crit}} = 474$. Such strong flows were not considered by Dudley & James.

However, we do not find any dynamo for $\epsilon > 0.20$. Dynamos only exist for sufficiently strong differential rotation. These findings thus provide further evidence that dynamo excitation from axisymmetric circulation without differential rotation can hardly be expected (Arlt, Hollerbach & Rüdiger 2003).

Following Kippenhahn (1954), we assume that the *radiation* of Ap stars transports angular momentum, forming a radial gradient of Ω. Such a gradient must lead to a meridional circulation. We ask whether this axisymmetric flow is able to work as a dynamo. For simplicity we assume a homogeneous sphere with no density gradients. The hydrodynamic solutions for the meridional circulation are obtained with a few thousand timesteps. No high spectral truncations are required since smooth large-scale flows emerge. The velocity field is plotted in Fig. 5.18. The flow resembles the first flow taken from Dudley & James (1989), but with reversed circulation, $\epsilon < 0$. No dynamo action of the circulation pattern could be found. Recalling the onset of dynamo action in the flow of Dudley & James, we (artificially) reverse the meridional flow excited by the differential rotation. Now the model has turned into a configuration that is very similar to Eq. (25) of Dudley & James, but still no dynamo action is found for the $m = 1$ mode.

5.8 MRI in Kepler Disks

The early models of rotating turbulence of Gough (1978), Hathaway & Somerville (1983), Durney & Spruit (1979) and Gailitis & Rüdiger (1982) all led to negative Λ-effect in the

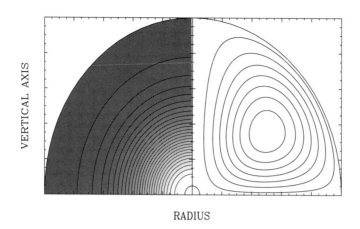

RADIUS

Figure 5.18: Velocity field (*right*) generated from a background differential rotation (*left*) with $\mathrm{Re} = 10{,}000$. The pattern is used as the flow for a possible kinematic dynamo. Subrotation ($\mathrm{d}\Omega/\mathrm{d}r < 0$) is connected with counterclockwise flow (see Fig. 3.13).

Reynolds stress relation, i.e. $\alpha_{\mathrm{SS}} < 0$ in the general expression (5.8) for the radial transport of angular momentum. With their linear normal mode analysis for a thin, differentially rotating disk Ryu & Goodman (1992) found the angular momentum flux to be nonzero only for non-axisymmetric modes and to be predominantly *inward*. The nonlinear numerical simulations by Ruden, Papaloizou & Lin (1988), Cabot & Pollack (1992), Kley, Papaloizou & Lin (1993) and Stone & Balbus (1996) also yielded negative values for the cross-correlation $Q_{s\phi}$. The quasilinear model with Lorentz force-driven turbulence in Sect. 4.2.4 with magnetic buoyancy included also leads to negative Reynolds stress under the influence of a global but rigid rotation (see Balbus, Hawley & Stone 1996).

On the other hand, Brandenburg (1998) argues that in magnetic-dominated shear flows the two α's should have opposite signs[14], i.e. $\alpha_{\mathrm{SS}} \cdot \alpha_{\mathrm{dyn}} < 0$. Positive α_{SS} thus require negative α_{dyn}, with consequences for the dynamo-excited large-scale magnetic fields. Positive α_{dyn} leads to magnetic fields with quadrupolar symmetry with respect to the equator, while negative values lead to dipolar symmetry. Only in the latter case does the field geometry favor the generation of jets according to the Blandford & Payne (1982) mechanism. As jets are commonly associated with accretion disks (see Livio 1997 for a detailed discussion) dynamos with negative α_{dyn} will play a particular role in the MHD theory of accretion disks (see also Brandenburg & Donner 1997). We shall demonstrate that the idea of the simultaneous existence of positive α_{SS} and negative α_{dyn} may indeed work for the case of Keplerian disks.

5.8.1 The Shearing Box Model

To study the nonlinear evolution of the MRI in a differentially rotating disk we make use of the shearing box formalism. The MHD equations are solved with NIRVANA in a corotating Cartesian frame of reference (Ziegler & Rüdiger 2000). The governing ideal fluid equations for this local ansatz are

$$\rho\left(\frac{\partial \boldsymbol{u}}{\partial t} + (\boldsymbol{u}\nabla)\boldsymbol{u}\right) = -\nabla P + \boldsymbol{J} \times \boldsymbol{B} - 2\rho\Omega\hat{\boldsymbol{e}}_z \times \boldsymbol{u} + 2\rho\Omega^2 q\boldsymbol{x} - \rho\Omega^2 \boldsymbol{z} \qquad (5.81)$$

---

[14] in order to avoid confusion, the α-effect of the dynamo theory – which is antisymmetric with respect to the equator – is represented in the following by its characteristic value from the upper disk plane, $\alpha_{\mathrm{dyn}} = \alpha_{\phi\phi}[\mathrm{north}]$

and the remaining equations like in Eq. (3.89) with $\rho = \nu = \chi = 0$. Here $q = -\mathrm{d}\log \Omega/\mathrm{d}\log s$ is a measure of the local shear rate; $q = 1.5$ for a Keplerian disk. The term $-\rho\Omega^2 z$ represents the vertical gravitational force of the central object in the thin disk approximation. The $2\rho q\Omega^2 x$ force term results from the radial expansion of the effective (gravitational+centrifugal) potential in the corotating reference frame. The remaining equations are given in Eq. (3.89) with $\nu = \eta = \chi = 0$. In the adiabatic models the gas pressure is $P = (\Gamma - 1)e$ with $\Gamma = 5/3$.

All simulations start with a configuration that is an exact stationary solution of the hydrodynamic equations. The initial magnetic field is purely vertical, but varies sinusoidally in the x-direction. The plasma β^* parameter decreases with z ranging from $\beta^* = 100$ at the disk midplane to a value of $\beta^* = 1.9$ at $z = \pm 2$.

The same set of parameters is used by Brandenburg et al. (1995) and Stone et al. (1996), but the codes are completely different. The standard resolution is $32 \times 64 \times 64$. Figure 5.19 shows the time evolution of the volume-averaged cross-components of the Reynolds and Maxwell stress tensors. The stresses are scaled to the horizontally averaged midplane pressure $\bar{P}(t)$, which is a function of time in the adiabatic model. The instability grows rapidly at first. Turbulence starts to develop then around orbit 3, and persists up to the latest simulated time. The flow shows a highly irregular behavior.

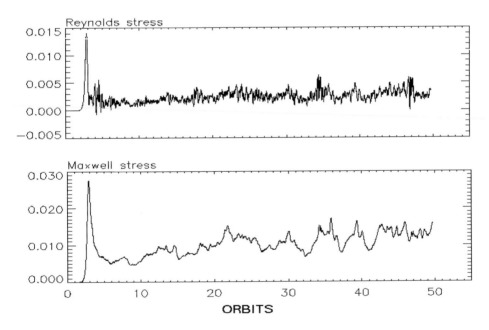

Figure 5.19: Time evolution of the volume-averaged Reynolds stress (*top*) and Maxwell stress (*bottom*). The stresses are scaled to the midplane pressure $\bar{P}(t)$. The magnetic stress generally dominates.

At the end, the magnetic energy has been amplified by a factor of roughly 16 relative to its initial value. Most of the energy is stored in the azimuthal component ($\langle B_y^2\rangle/\langle B_z^2\rangle = 46$).

Buoyancy effects due to the density stratification seem to play no essential role. This is in agreement with the findings of Brandenburg et al. (1995) and Stone et al. (1996).

The Reynolds stress and Maxwell stress

$$\left\langle \rho u'_x u'_y - \frac{B'_x B'_y}{\mu_0} \right\rangle = \alpha_{SS} \bar{P}(t) \tag{5.82}$$

are of considerable interest because of their relation to the viscosity-α. For the time-averaged values (denoted by an overbar) of the volume-averaged stresses one finds

$$\frac{\overline{\langle \rho u'_x u'_y \rangle}}{\bar{P}(t)} = 0.003, \qquad -\frac{\overline{\langle B_x B_y \rangle}}{\mu_0 \bar{P}(t)} = 0.012. \tag{5.83}$$

The total value is 0.015. Hawley, Gammie & Balbus (1995), Brandenburg et al. (1995) and also Stone et al. (1996) quote similar values for the mean stresses (see also Abramowicz, Brandenburg & Lasota 1996). Matsumoto & Tajima (1995) and Torkelsson et al. (1996) even report such rather high values as $\alpha_{SS} \simeq 0.1$ for special cases discussed in the papers. These results clearly confirm former statements that angular momentum transport is dominated by correlations in the fluctuating magnetic field rather than the velocity field.

To explore the time evolution of the vertical disk structure, Fig. 5.20 presents (t, z)-images of α_{SS}. The stress varies drastically in the vertical direction. At later times, the vertical disk structure can be represented by a weakly magnetic core surrounded by a strongly magnetic corona. Most of the magnetic energy is confined to the outer region. The stress peaks in the corona, and the outward angular momentum transport occurs primarily away from the disk midplane.

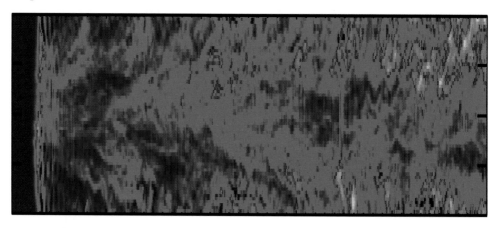

Figure 5.20: The horizontally averaged total stress in the t-z-plane. Red (blue) stands for positive (negative) angular momentum flow. In the equatorial plane sometimes the transport is inward.

One can even observe the appearance of a magnetic field of the same sign over a rather long time. We take averages over the entire box, and also in the upper and lower disk planes separately. Concerning the azimuthal field we have $\mathcal{E}_y = \alpha_{yy} \bar{B}_y$ neglecting higher derivatives. The main issue is the equatorial antisymmetry, which is realized exactly (Fig. 5.21). The α-effect proves to be small and highly noisy, but it exists. It is negative in the upper disk plane and positive in the lower (see Brandenburg et al. 1995). This is opposite to the expected situation in a stellar convection zone. One can explain this fundamental difference by the action of the shear in Kepler disks (Rüdiger & Pipin 2000). The same difference occurs in the simulations by Brandenburg et al. (1995), where the Maxwell stress also dominates the Reynolds stress (by about a factor 5). The dynamo-α is negative in the upper disk plane and positive in the lower. Local simulations including diffusivity by Fleming, Stone & Hawley (2000) proved the onset of instability even for low conductivity. The angular momentum transport is strongly reduced for $\mathrm{Rm} \lesssim 10^4$ though. As shown by Ziegler & Rüdiger (2001) the magnetic Reynolds number has a significant influence on the magnetic energy level of the resulting fluctuations. It is reduced for small Rm. The α-effect remains negative (positive) in the upper (lower) disk hemisphere. It vanishes completely for weak shears, i.e. for $q \lesssim 0.6$.

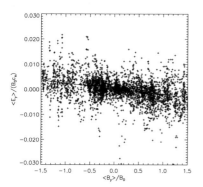 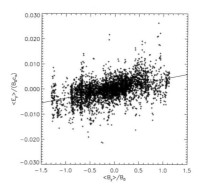

Figure 5.21: Snapshots of the correlation between the turbulent azimuthal EMF and the mean toroidal magnetic field. *Left*: Upper disk plane. *Right*: Lower disk plane. From Ziegler & Rüdiger (2000).

5.8.2   A Global Disk Dynamo?

Hawley (2000) followed the evolution of a thick torus under the influence of an external vertical magnetic field. The global MRI indeed developed eddy viscosity. Arlt & Rüdiger (2001) applied the ZEUS-3D code to a disk with uniform sound speed and finite eddy diffusivity. The initial field has vanishing total flux through the upper and lower (stress-free) surfaces, and the magnetic boundaries are pseudovacuum. The mass loss at the inner boundary is compensated for a slow infall at the outer boundary.

The resulting flow and density patterns are highly nonaxisymmetric (Fig. 5.22). After saturation of the turbulence the temporal behavior of α_{SS} changes into an oscillatory behavior. The Reynolds stress alone is then negative throughout the whole simulation. The time series of the angular momentum transport in the high-η model are shown in Fig. 5.23 (left). For high conductivity the resulting α_{SS} is much smaller[15]. We find the total angular momentum transport is outward, dominated by the Maxwell stresses.

---

[15]   for an ideal fluid Steinacker & Papaloizou (2002) obtain $\alpha_{SS} \lesssim 0.04$

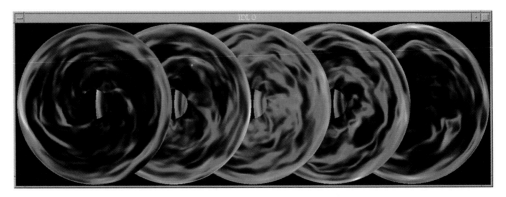

Figure 5.22: Density fluctuations after 19 orbits in horizontal slices parallel to the midplane (middle panel). The magnetic field applied is vertical but of vanishing flux. The disk is isotherm and density stratified. (Arlt & Rüdiger 2001).

Since the stress is dominated by the Maxwell stresses we also ask about the ratio of the flow and the field energies. The temporal evolutions of these two quantities clearly differ between the low- and high-η models. In Fig. 5.23 (right) magnetic dominance is found for the case of low electrical conductivity ('high-η'), i.e. for cold disks.

The amplification of the total magnetic energy by a factor of 10^3 from the initial magnetic field perturbation may indicate dynamo action in the disk. The time-averages of $\mathcal{E}_\phi$ and $\bar{B}_\phi$ for both hemispheres are thus plotted in Fig. 5.24, but only with their resulting signs. The quantities are averaged in the azimuthal direction and plotted in the meridional plane. For

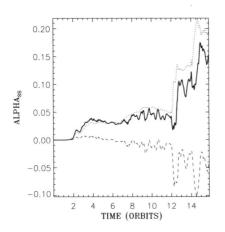
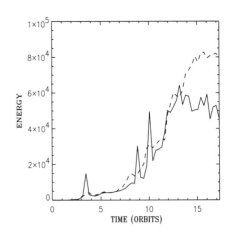

Figure 5.23: High-η model. *Left*: Angular momentum transport by Reynolds (dashed) and Maxwell (dotted) stresses. The total stress is the solid line, the Reynolds stress as negative. *Right*: Temporal evolution of kinetic (solid) and magnetic (dashed) energies. The magnetic energy dominates. From Arlt & Rüdiger (2001).

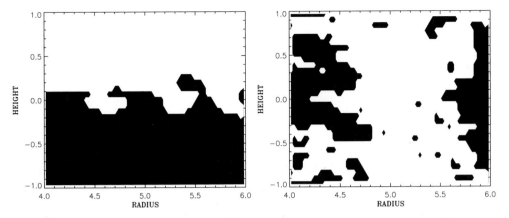

Figure 5.24: Time-average toroidal magnetic field (*left*) and the azimuthal EMF (*right*). White areas denote negative signs.

the upper hemisphere both quantities have the same sign, which is not true for the lower hemisphere. The dominance of negative EMF indicates a positive (negative) α^{dyn} in the upper (lower) hemisphere.

This result is opposite to the findings of the box simulations. The temporal development of the parity of the toroidal magnetic fields for the high-η model indeed shows a tendency to leave the initial dipolar perturbation. However, with only 160 revolutions the simulation was not long enough for a definitive statement.

5.9   Accretion-Disk Dynamo and Jet-Launching Theory

5.9.1   Accretion-Disk Dynamo Models

In several papers, starting with Tout & Pringle (1992), magnetic disk dynamos have been constructed that produce magnetic fields leading to a Maxwell stress $\alpha_{\mathrm{SS}} = -\bar{B}_s\bar{B}_\phi/\mu_0\rho c_{\mathrm{ac}}^2$ of the desired (positive) sign and order. While Tout & Pringle combined the Parker instability with MRI, Ma & Biermann (1998) also included the α-effect, following Stepinski & Levy (1991) that $\tau_{\mathrm{corr}} \simeq \tau_{\mathrm{rot}}$. Though it would be tempting to follow this concept for dynamo-driven accretion disks, we turn to another question: How important is the result that the α-effect due to MRI seems to be negative? Is it true that there are strong consequences for the jet-launching theory?

A mean-field disk-dynamo with positive α-effect (in the upper disk plane) always generates a stationary magnetic field with quadrupolar equatorial symmetry (Fig. 5.25, right). The field is confined to the disk with the maximum value at the midplane of the disk, close to the inner edge. After their first formulation by Pudritz (1981) such findings were later confirmed several times. The dynamo with *negative* α-effect, as due to the MRI in accretion disks, always generates a stationary magnetic field with *dipolar* equatorial symmetry (Deinzer, Grosser &

Schmitt 1993, v. Rekowski, Rüdiger & Elstner 2000, Bardou et al. 2001)[16]. The field is predominantly toroidal, and its maximum value is now outside of the disk, depending on the ratio
$\epsilon = \sigma_{\mathrm{disk}}/\sigma_{\mathrm{halo}}$. The greater ϵ the smaller is the toroidal field in the halo. Increasing ϵ to
infinity leads to the solution for vacuum where the toroidal component in the halo vanishes.
Simulating the vacuum case by setting the pseudovacuum condition directly at the surface of
the disk, one gets the chaotic oscillating mode of quadrupolar type predicted by Campbell
(1997).

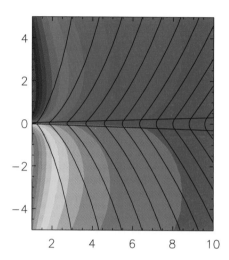 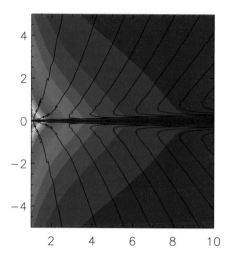

Figure 5.25: The poloidal field lines for negative $\alpha_{\phi\phi}$[north] as realized in accretion disks (*left*, dipolar
symmetry) and for positive $\alpha_{\phi\phi}$[north] (*right*, quadrupolar symmetry). The colors represent sign and
amplitude of the toroidal field that dominates in the central part. From v. Rekowski, Rüdiger & Elstner
(2000).

The conductivity of the halo is still unknown and certainly not zero. Note that the geometry of the disk is also important for the symmetry of the dynamo-generated field. An
overestimation of the disk height leads to solutions with even parity, as found by Torkelsson
& Brandenburg (1994).

The numerical calculation shows that for the model with negative dynamo-α the resulting
poloidal component outside the disk is of the same order as within the disk. Figure 5.25 (left)
shows the poloidal field structure with prescribed Keplerian velocity in the disk as well as in
the halo. Due to the differential rotation, the toroidal field does not vanish in the disk halo, but
it does gradually decrease to zero in the vertical direction.

The influence of vertical outflows on disk dynamos has been investigated by Bardou et
al. (2001). The outflow can be imagined as due to a wind or due to turbulent pumping (see

---

[16] the dipoles found by Stepinski & Levy (1988) for thick torus geometry and positive α-effect, however, are oscillatory

Sect. 4.2.1). The influence of the vertical outflow differs for negative $\alpha_{\phi\phi}$ (dipoles) and positive $\alpha_{\phi\phi}$ (quadrupoles). While the growth rates of the quadrupoles are amplified, the growth rates of the dipoles are reduced. The dipole excitation is even stopped for magnetic Reynolds numbers of the vertical wind of order unity. As the next step in this direction v. Rekowski et al. (2003) presented an axisymmetric model of an accretion disk dynamo that for sufficiently strong magnetic field production drives an outflow system with a fast axial flow in the resulting wind system.

5.9.2 Jet-Launching

A radio emission along the double jet structure of the T Tau system was observed by Ray et al. (1997). The emission shows strong circular polarization, indicating the existence of a large-scale ordered magnetic field (~ 1 G). The idea is suggested that such systems are typically formed as a triplet of a star, a disk and a jet – linked together by a large-scale magnetic field (Fig. 5.26). The magnetic field can arise from an accretion disk dynamo or from the central object. The latter often have their own magnetic fields, ranging from 10^3 G for protostars to 10^{12} G for neutron stars.

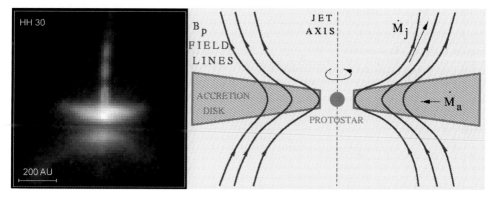

Figure 5.26: *Left*: HH 30 as a prototype of a young star with a dark disk and a gaseous jet. The central star is hidden but its light is reflected by the disk. Note the jet perpendicular to the disk (Burrows et al. 1996). Credit: NASA. *Right*: The star-disk-jet connection of a magnetized accretion disk system. Courtesy M. Čemeljić.

Ménard & Duchêne (2003) demonstrated that the geometric constellation in the HH 30 system (Fig. 5.26) is not exceptional. Comparing the magnetic field orientation in Taurus with the orientation of those TTS with disks (CTTS) they found that the orientation of the CTTS with jets and without jets strongly differs. The disk plane of CTTS with known jet is often perpendicular to the magnetic field direction. This is not true, however, for CTTS without jet. If an accretion disk has a poloidal field and a hot corona it can launch a magnetically driven wind or jet. Once driven by thermal pressure gradients, mass is accelerated centrifugally along the magnetic field lines, assuming that the matter along the field line rotates with the angular velocity of the footpoint. As long as the kinetic energy density of the outflowing material is small compared to the poloidal magnetic energy density the field line will be only slightly distorted by the flow. This situation changes at the Alfvén points. If the flow reaches the Alfvén velocity the field line will be dragged outward with the flow collimating the jet.

For a Keplerian disk the effective potential ψ at a point on a field line, corotating with the footpoint s_0, is

$$\psi = -\frac{GM_*}{s_0}\left[\frac{s_0}{(s^2+z^2)^{1/2}} + \frac{1}{2}\left(\frac{s}{s_0}\right)^2\right]. \tag{5.84}$$

Blandford & Payne (1982) and Campbell (1997) showed that there is a characteristic angle of launching cold winds or jets. If the inclination angle between the vertical and the field line at the disk surface is greater than $30°$, the equilibrium at $s = s_0$ is unstable. If the angle is smaller than $30°$ a potential barrier exists so that the equilibrium is stable.

Ogilvie & Livio (1998) showed that a certain potential difference must be overcome even for an angle greater than $30°$ if the magnetic field is strong enough to influence the rotation law. The potential difference is minimized for an angle of $38°$. They proposed that there has to be an additional source of energy to launch an outflow from a magnetized disk.

In principle, an inclination of the poloidal field can be the result of the action of an accretion flow to a field originally aligned with the rotation axis. This 'dragging' effect has been considered by Lubow, Papaloizou & Pringle (1994) and Reyes-Ruiz & Stepinski (1996). A considerable inclination results only for low dissipation, i.e. for unrealistically high magnetic Prandtl number. This problem has been attacked by Shalybkov & Rüdiger (2000) by reconsidering the vertical structure of the accretion disk under the influence of a $\bar{B}_z$-field, which by the shearing at the disk surface always acquires an azimuthal field component as well. Such a field configuration automatically leads to the generation of a radial inclination. The resulting inclination angles depend only weakly on the magnetic Prandtl number if the latter is not too small.

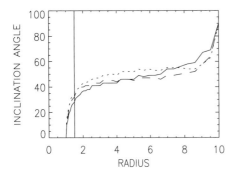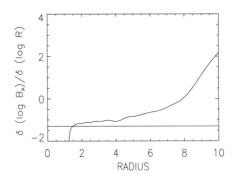

Figure 5.27: *Left*: The inclination of the poloidal field to the vertical for $\mathrm{Pm} = 1$ and various ϵ from 1–100 (only small differences). *Right*: In the bulk of the disk the dynamo yields very small μ. The horizontal line marks the limit $\mu = 1.3$. $\mathrm{Pm} = 1, \alpha_{\mathrm{SS}} = 0.01$. (v. Rekowski, Rüdiger & Elstner 2000).

On the other hand, the accretion-disk dynamo with negative α-effect shows no dependence of the inclination angle on the ratio ϵ of the magnetic diffusivities between halo and disk.

There exists a broad region where the inclination angle is large enough ($\approx 45°$) such that a magnetically driven wind or jet is possible (see Campbell, Papaloizou & Agapitou 1998, v. Rekowski, Rüdiger & Elstner 2000, Fig. 5.27, left).

Spruit, Foglizzo & Stehle (1997), however, argue that the toroidal fields are too unstable to contribute much to collimation. If the toroidal field is the only collimation mechanism the jet would be decollimated. They propose that the collimation mechanism of a jet is due to the magnetic pressure of a *poloidal* field anchored in the disk, where the radial profile is of great importance. The criterion that $B_{\mathrm{p}} \propto s^{-\mu}$ with $\mu \leq 1.3$ is sufficient for collimation. It is indeed possible to generate with accretion disk dynamos such large-scale fields of dipolar polarity with a radial profile of the poloidal field with $\mu \leq 1.3$ (Fig. 5.27, right).

Such fields cannot be produced by a dipolar magnetic field of the central object; only an accretion disk dynamo can be the source of such fields. In Fig. 5.28 examples are given of how strong the influence of the exponent μ is. The models are started with a force-free magnetosphere. Initially the magnetosphere is at rest, and the disk rotates Keplerian. There is a density jump of factor 100 between the disk and the halo. The steep profile ($\mu = 1.2$) is not collimating, but the rather flat one ($\mu = 0.5$) is.

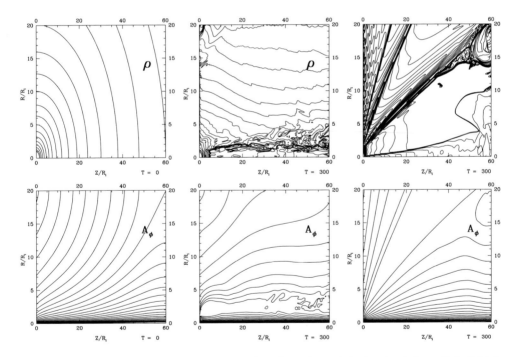

Figure 5.28: Isodensity lines (*top*) and magnetic field lines (*bottom*) for jet simulations. *Left*: Initial configuration. *Middle*: $\mu = 0.5$ in confirmation of Spruit, Foglizzo & Stehle (1997). *Right*: $\mu = 1.2$. Collimation only occurs for $\mu = 0.5$. Courtesy C. Fendt.

5.9.3 Accretion-Disk Outflows

A *stellar* dipole surrounded by an accretion disk could also be a model for a variety of astrophysical sources (CTTS, cataclysmic variables or high-mass X-ray binaries). The evolution of such a stellar dipole in interaction with a diffusive accretion disk has been numerically investigated by Hayashi, Shibata & Matsumoto (1996), Miller & Stone (1997) and Goodson, Böhm & Wingley (1999). The crucial point in these models is to consider a proper disk model taking into account viscous effects for the dynamics. Otherwise, a breakdown of the simulation after only a couple of rotational periods happens. On the other hand, simulations by Romanova et al. (2002) modeling the time-dependent accretion process along the magnetic field lines between star and disk have found a stationary flow within the accretion funnel.

Time-dependent simulations lasting only a short time period depend strongly on the initial conditions, so that simulations of stellar magnetospheres over many rotational periods are necessary. Therefore another approach might be reasonable, i.e. to consider the accretion disk as a boundary condition for the mass injection. As the disk structure itself is not treated, such simulations may last over thousands of Keplerian periods. These simulations have been performed using the ZEUS-3D code[17] with the axisymmetry option for solving the time-dependent ideal MHD equations (Fendt & Elstner 2000). The model also takes into account the effect of an entropy jump between disk and corona, i.e the corona is warmer than the disk. The basic model setup represents a central star and a Keplerian disk separated by a gap. The initial field distribution is chosen as a current-free stellar magnetic dipole, deformed by the effect of field-dragging in the disk. The poloidal field is therefore inclined toward the disk surface. The initial disk toroidal magnetic field is also force-free, i.e. $B_\phi \propto s^{-1}$, but the stationary solution of the induction equation leads to toroidal fields that are not force-free[18] so that the outflow can produce significant gradients of the toroidal field collimating the centrifugal wind (Sakurai 1985, Pudritz & Norman 1986, Pelletier & Pudritz 1992).

As a general behavior, the initial dipolar magnetic field structure disappears on spatial scales larger than the inner disk radius and a two component wind structure (a disk wind and a stellar wind) evolves. For certain parameters one finds a *quasistationary* final state of a spherically radial mass outflow. This two-component outflow obtained in the simulations shows almost no evidence of collimation. In the example presented by Fendt & Elstner (2000) of an initially dipolar magnetosphere, the final spherically radial outflow encloses a neutral line with poloidal and toroidal magnetic field reversal. The toroidal field reversal implies a reversal of the electric current, and thus only a weak *net* poloidal current. This is in agreement with the analysis of Heyvaerts & Norman (1989), who showed that only outflows carrying a net poloidal current will collimate to jets of cylindrical shape.

The TTS phase is the final stage in the star formation process. During this phase the star is fully convective (see Sect. 5.6.2). Magnetic activity has been observed with fields of ~ 1 kG (Guenther & Emerson 1996, Guenther et al. 1999, Johns-Krull et al. 1999). The accretion rates found for CTTS range from 10^{-9} to 10^{-7} $M_\odot/\mathrm{yr}$ (Valenti, Basri & Johns 1993, Gullbring, Barwig & Schmitt 1997). Figure 5.30 shows the evolution of the magnetic

---

[17] Krause & Camenzind (2001) have presented the systematic study of the convergence behavior of both hydrodynamic and magnetohydrodynamic jet simulations with various codes

[18] the corona is nonrotating initially

field and the gas flow in the inner region of a CTTS. The initial setup consists of a Keplerian accretion disk threaded by a purely dipolar stellar magnetic field. The outer parts of the disk rotate slower than the star, the inner parts faster. The rotational shear along the field line will generate a toroidal field; the Lorentz force is thus directed outward for radii greater than the corotation radius. When the toroidal field has become strong enough the gas in the halo is pushed outward and the magnetic field is stretched in the radial direction. The field lines connecting the polar caps of the star with the outer parts of the disk in the initial state break up, and only the inner parts of the disk remain connected to the star. Far away from the star the field is essentially radial, pointing away from the star at high latitudes and toward the star at low latitudes. Finally an outflow is launched from the disk. The mass loss rate is typically of the order of 10% of the accretion rate but varies strongly with time.

Figure 5.29 shows the magnetic torque for field amplitudes of 100 G and 1 kG as functions of time. The poloidal field energy increases as the field is stretched in the radial direction and saturates when the final field configuration is reached. The magnetic torque is negative in the initial phase, which means that the angular momentum flux is directed toward the star, i.e. the star is spun up. The torque is strongly reduced once the field energy saturates, and varies strongly on a shorter timescale because of reconnection. It sometimes even becomes negative but the sign remains positive on average. By the magnetic field, obviously, the stellar rotation is decelerated.

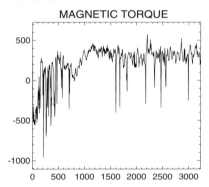
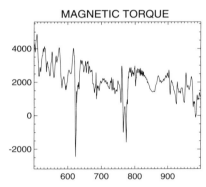

Figure 5.29: The magnetic torque for 100 G (*left*) and 1 kG (*right*) within the corotation radius vs. time. For 100 G the magnetic field appears to be saturated, which is not true for the strong-field case.

5.9.4   Disk-Dynamo Interaction

As we can see from Fig. 5.29, the magnetic torque appears to be highly fluctuating. This fact could be important for the behavior of the rotation of the central object. Nelson et al. (1997) report that many (but not all) X-ray pulsars oscillate dramatically between periods of spin-up and spin-down. The periods between torque reversals reach from 10 days to 10 yr. Vela X-1 is one of the sources and showed phases of 6 yr with acceleration and 10 yr with deceleration. In the UV line spectrum of the companion of Vela X-1 one can find wind characteristics, so that the phenomenon is discussed in the context of accretion of a stellar wind supplied by the

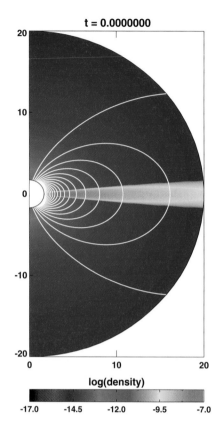 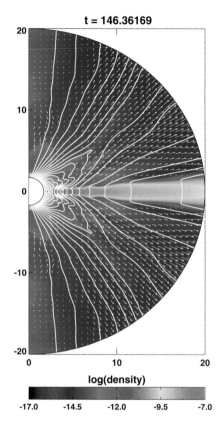

Figure 5.30: A solar-mass CTTS with a disk with $\dot{M} = 10^{-7}\ M_\odot/\mathrm{yr}$. The colors indicate different density. The white lines denote (poloidal) field lines. *Left*: The initial configuration with the disk threaded by a stellar magnetic field. The field strength on the stellar surface is 1 kG at the poles. *Right*: The final field configuration. The arrows indicate the gas motion. From Küker, Henning & Rüdiger (2003).

companion. As, however, the resulting torque of the accreted mass remains very small (Anzer, Börner & Monaghan 1987) there must be some extra mechanism to explain its long-lasting fluctuations (Anzer & Börner 1992).

Torkelsson (1998) reports the X-ray pulsars are the best laboratories to study the angular momentum exchange between a magnetic central star and a magnetized accretion disk. He considers the magnetic torque $B_s \cdot B_\phi$ as a combination of the time-independent stellar field and the dynamo-induced accretion disk-field that may fluctuate or even oscillate with a timescale of about 0.5 d (also Ma & Biermann 1998).

6 The Galactic Dynamo

Five per cent of the galactic baryonic mass is interstellar gas and dust, organized in a clumpy cloud structure dominated by cold molecular clouds of only 10 K. In the neighborhood of stars warm clouds of 10^4 K also exist. The kinetic energy without rotation of the turbulent ISM is about 10^{54-55} erg, corresponding to the energy of 5000 SN explosions. The characteristic turbulent velocity u_T approaches 10 km/s with correlation lengths of 100 pc. The corresponding eddy viscosity[1] of about 10^{26} cm^2/s would destroy any coherent structure of a size of order kpc after only 0.7 Gyr. To maintain a global magnetic structure for a longer time a special excitation mechanism must exist.

The large-scale magnetic fields of spiral galaxies, but also flocculent galaxies (Fig. 6.1), must be maintained by the inducing action of the partly ionized interstellar gas. Although the maximal field strength does not exceed (say) 10 μG, because of their huge dimensions galaxies are, *the* supermagnets in the Universe. The magnetic flux $\pi R^2 B$ of a galaxy exceeds the total flux of all (say) 10^{11} stars by many orders of magnitude. Apart from the impressive numbers characterizing galactic magnetism, galaxies are also a very interesting realization of a dynamo machine as they are rather transparent, providing detailed information about the internal flow structures.

6.1 Magnetic Fields of Galaxies

The interpretation of radio-polarization data of spiral galaxies reveals the existence of large-scale magnetic fields with very special properties. Their explanation is of considerable interest because galaxies are astrophysical configurations with observable internal flow systems. In particular, there is a problem of understanding the relation between the considered flow field and the associated α-effect. The observed butterfly diagram of the solar activity seems to indicate a very small α compared with the action of differential rotation in the solar convection zone. On the other hand, the observed large-scale structure of the known galactic magnetic fields requires only small differential rotation, or in other words, a rather large α-effect. The key properties of the galactic large-scale magnetic field pattern are

- field amplitude is $\lesssim 10$ μG,
- field lines have pitch angles up to 35°,
- some field geometries exhibit a bisymmetric spiral structure (M 81) or a distinct vertical orientation (NGC 4631, NGC 5775),
- the magnetic fields already exist in very young galaxies.

---

[1] the microscopic value of $\sim 10^{18}$ cm^2/s is extremely high

The Magnetic Universe: Geophysical and Astrophysical Dynamo Theory.
Günther Rüdiger, Rainer Hollerbach
Copyright © 2004 Wiley-VCH Verlag GmbH & Co. KGaA
ISBN: 3-527-40409-0

NGC 4414 8.44 GHz TP/PI B-vect

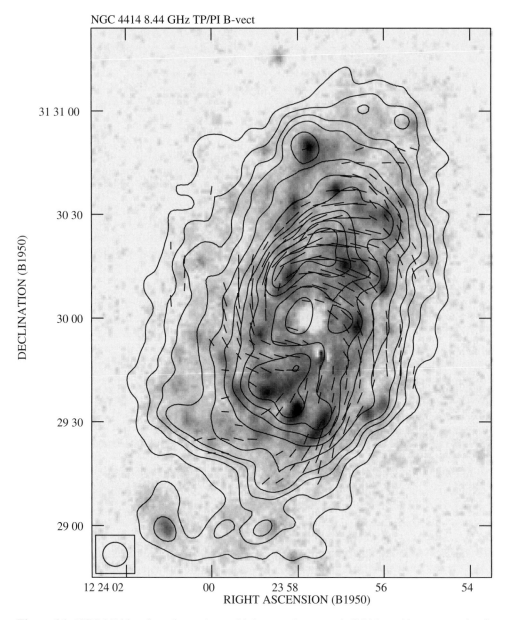

Figure 6.1: NGC 4414 is a flocculent galaxy with large-scale magnetic field but without strong density waves. Note the large pitch angles existing almost everywhere. From Soida et al. (2002).

These topics are now presented in more detail. See also the review papers by Wielebinski & Krause (1993), Beck et al. (1996) and Kulsrud (1999), particularly for the details of special galaxies. Krause & Beck (1998) also discuss the empirical background to explore the symme-

Figure 6.2: The magnetic geometry of the edge-on galaxy NGC 5775 indicates even (quadrupolar) symmetry with respect to the equator. Courtesy M. Urbanik & M. Soida.

try of the magnetic field with respect to the galactic midplane. For many cases a quadrupolar symmetry of the magnetic fields is derived as dominating together with symmetry with respect to the rotation axis (Fig. 6.2). In four of five cases the magnetic field lines are directed *toward the center* of the spiral galaxies. The orientation of the magnetic field does not seem to be random. In this connection Krause & Beck (1998) mention an interesting difference of dipolar and quadrupolar field symmetry. If the galactic magnetic fields had dipolar symmetry, the magnetic field amplitudes averaged over the whole sky must be zero. This is not true, however, for quadrupoles with a fixed magnetic orientation in the galactic plane but, of course, for random or oscillating orientations.

Table 6.1: Average energy densities in 10^{12} erg/cm^3 in the inner disk of NGC 6946 (Beck 2002).

Warm medium	Magnetic field	Cosmic ray	Turbulence	Gas rotation
1	13	13	10	5000

6.1.1 Field Strength

The observed magnetic field energy is of the order of the energy of the interstellar turbulence. The equipartition field strength

$$B_{\mathrm{eq}} = \sqrt{\mu_0 \rho}\, u_{\mathrm{T}}, \tag{6.1}$$

with density of order 10^{-24} g/cm^3 and the turbulence velocity of about 10 km/s is 3.5 μG. As the observed values are indeed of this order a turbulent origin of the induced fields has been suggested. The theory here meets the discussion about the η_{T}-quenching. This is based on the philosophy that the magnetic feedback on the turbulence completely quenches any diffusion even with field strengths much below its equipartition value. The coincidence between observed fields and their theoretical equipartition value, however, strongly favors the canonical eddy diffusivity concept with quenching in a way as described above for various turbulence models.

The numbers of the energy budgets are known for NGC 6946. The neutral-gas turbulence and the magnetic fields are roughly in energy equipartition (Table 6.1), but both dominate the thermal energy of the gas. In the outermost parts of the galaxy the magnetic field may even dominate the basic rotation, so that even the rotation profile may be influenced by the Lorentz force (Battaner & Florido 1995).

6.1.2 Pitch Angles

The pitch angles $p = \tan^{-1}(\bar{B}_s/\bar{B}_\phi)$ reflect the ratio of the radial and the toroidal magnetic field strengths. Dynamos of $\alpha\Omega$-type are characterized by very small pitch angles, e.g. $\sim$ 0.05° for the Sun. For galaxies pitch angles of 10–40° are reported (Fig. 6.3). The pitch angles mostly decrease outward (Beck 1993). Such observed values indicate that the differential rotation in galaxies does *not* play a dominant role in the induction equation. Frozen-in fields as a consequence of a small magnetic diffusivity are wound up by any differential rotation up to very small pitch angles. Only for relatively large diffusivities can one thus expect to explain the observed large pitch angles. In NGC 4414 pitch angles up to 45° have been found.

Of special interest is the case of NGC 6946, possessing pitch angles between 20° and 30°. NGC 6946 is a standard example insofar as there is no companion, no strong density wave and no active nucleus. The large-scale magnetic fields are *concentrated between* the optical spiral arms (Fig. 6.4) but the azimuthal field orientation is of the axisymmetric ASS type (Beck & Hoernes 1996). The turbulent component of the magnetic field in the spiral arm reaches 15 μG, while a regular field of 10 μG is located in the interarm region.

The large pitch angles are the most characteristic property of the galactic fields. Obviously, the galactic dynamo differs strongly from the solar/stellar $\alpha\Omega$-dynamo. They cannot

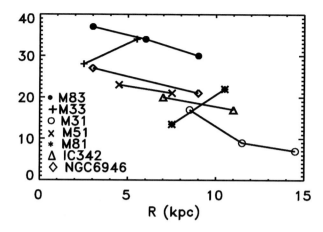

Figure 6.3:   Pitch angle statistics for various galaxies (Beck 1993).

be interpreted as *oscillating* stellar $\alpha\Omega$-dynamos that are made stationary by some extra effect like turbulent pumping (see Brandenburg, Moss & Tuominen 1992, Ferrière & Schmitt 2000).

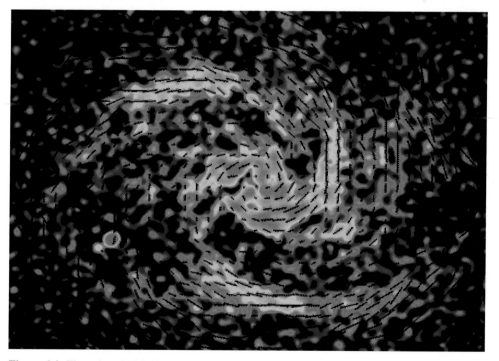

Figure 6.4: The galaxy NGC 6946 with its *magnetic spiral arms* between the gaseous spiral arms. The optical spiral arms are not shown. Courtesy R. Beck.

6.1.3   Axisymmetry

NGC 6946 has a nonaxisymmetric field geometry, but the magnetic polarity does not reverse along the azimuth ('ASS'). For at least one case (M 81) there is a clear bisymmetric azimuthal structure with polarity reversals, so that in one magnetic arm the field spirals into the center and in the other it spirals outward (Fig. 6.5). It is not easy to explain such an asymmetry (called BSS) with a mean-field dynamo theory. Rüdiger, Elstner & Schultz (1993) present models with anisotropic α-effect and large rigid-rotation core (compared to the vertical thickness of the galaxy) which indeed produces BSS magnetism. So far there is still no nonlinear confirmation of this result, due to the complexity of the 3D dynamo codes (see Moss, Tuominen & Brandenburg 1991, Panesar & Nelson 1992, Moss & Brandenburg 1995).

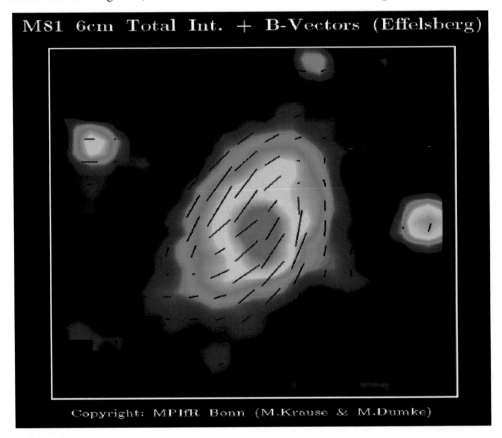

Figure 6.5: M 81, the magnetically bisymmetric spiral (BSS) standard example. Courtesy M. Krause & M. Dumke.

Almost no azimuthal structure exists for the flocculent galaxy NGC 4414 (see Fig. 6.1). It is hard to imagine its large-scale magnetism as a result of the inducing action of the galactic differential rotation starting with an external uniform magnetic field. An external uniform magnetic field not parallel with the rotation axis subject to a differential rotation will always produce a BSS-type field. An ASS-type field can only be produced from an initially even-m field with radial components in the equatorial plane, e.g. quadrupoles of type S0. The existence of such a rather artificial seed field is unlikely though, so that NGC 4414 seems to require the existence of a large-scale galactic dynamo (Beck 1996).

6.1.4 Two Exceptions: Magnetic Torus and Vertical Halo Fields

Radio observations of M 31 revealed a 20-kpc-sized torus of magnetic fields (Fig. 6.6) with fields aligned in one single direction (Beck 1982), which has been interpreted by Deinzer, Grosser & Schmitt (1993) as a clear indication for a torus-dynamo. M 31 has a low star-formation rate and no spiral structure. As cold gas and young stars are also restricted to the torus, one might speculate whether the magnetic field determines the star formation activity in M 31 (Beck 2000). No other case of a such toroidal field structure, however, has yet been found (see Moss et al. 1998 for details).

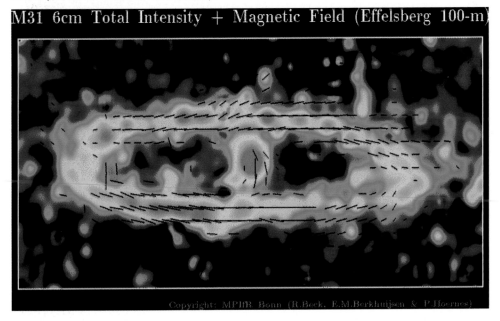

Figure 6.6: M 31, the Andromeda nebula, with its distinct torus structure. Courtesy R. Beck, E.M. Berkhuijsen & P. Hoernes.

Another exciting exception is also NGC 4631, which provides both evidence for an outflow from its disk into the halo *and* fields of considerable strength perpendicular to its midplane. In the halos of a few edge-on galaxies synchrotron emission has been detected. Among them is NGC 4631, a galaxy exhibiting a huge synchrotron halo extending to 10 kpc off the midplane. Based on multifrequency radio observations of the polarized emission with the VLA, Hummel, Beck & Dahlem (1991) and Golla & Hummel (1993) derived the magnetic field orientation in the halo above the central region of NGC 4631 as radially outgoing from the central region. In Fig. 6.7 a map is shown of the radio emission of NGC 4631 as obtained with the Effelsberg

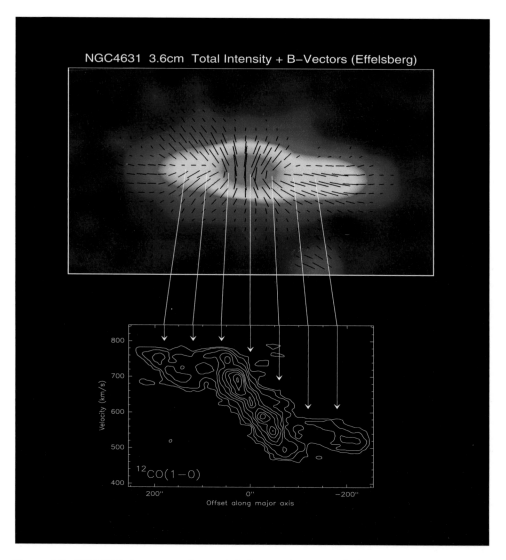

Figure 6.7: NGC 4631 shows very clear and distinct vertical magnetic halo field components. Also given is the rotation law of this galaxy with the rigidly rotating core and the outer $V \simeq$ const. domain. Courtesy M. Dumke.

100-m telescope, including the magnetic field orientation. It shows the complete large-scale emission. It thus also shows the field orientation in and above the outer disk, wherever the degrees of polarization were high enough. Meanwhile, the number of known galaxies with vertical magnetic field structure is growing (Tüllmann et al. 2000).

Figure 6.7 also shows the velocity diagram derived along the major axis of NGC 4631 by Golla & Wielebinski (1994). It shows rigid rotation in the central region and differential

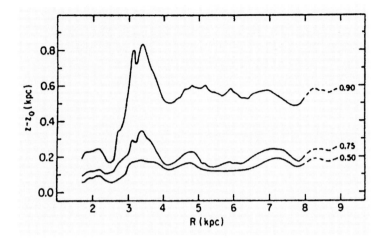

Figure 6.8: Contours enclosing 90%, 75%, and 50% of the galactic H I layer (Lockman 1984, his Fig. 3).

rotation with a constant rotation velocity of ~ 140 km/s in the outer regions. The comparison suggests a connection between the rotation of the galaxy and its magnetic field orientation in the halo. The magnetic fields are perpendicular to the galactic plane only above the central rigidly rotating region.

Evidence for outflow from the central region of NGC 4631 has been found from the studies of Hummel & Dettmar (1990), Rand, Kulkarni & Hester (1992) and Golla & Hummel (1994). This outflow must be related to the enhanced star forming activity of the central region. Star-formation induced outflow has its origin in individual star forming regions in the disk.

Brandenburg et al. (1993) mentioned that the simplest possibility to produce vertical field components in a galaxy might be the inclusion of a galactic wind as a component of the mean flow. They find some modification of the field configuration. The strength of the off-disk fields, however, remains low. The reason is the following: a vertical wind induces a vertical magnetic field proportional to $(\boldsymbol{B}\nabla)w$, with w as the wind velocity. Only the s-dependence of the wind, $w = w(s)$, can thus create (axisymmetric) B_z-components of any significant strength. The induction of wind with an extra azimuthal profile (i.e. real spikes) would be much more effective (Elstner et al. 1995).

6.1.5   The Disk Geometry

There is strong evidence for substantial amounts of highly turbulent gas at heights above 1 kpc (Münch & Zirin 1961, Lockman 1984, Reynolds 1989). Inclusion of this halo gas has changed the traditional view of a very thin gas layer. Despite its low density this high-velocity component (Anantharamaiah, Radhakrishnan & Shaver 1984) is energetically important because it contains a significant fraction of the whole kinetic energy in the interstellar medium (Kulkarni & Fich 1985). The question arises as to how this corotating halo gas will influence a disk dynamo.

The gas as a whole has a nearly constant scale height of 230 pc. In a strict sense this holds for H I only. In the inner part of the galaxy, the interstellar medium consists mostly of molecular gas concentrated in a very thin layer (but see Dame & Thaddeus 1994, Malhotra 1994). If the gas is considered the thickness decreases toward the center. More than 13% of HI emissions are from gas that lies mostly above 500 pc from the plane. In Fig. 6.8 the heights are marked below which 50%, 75%, and 90% of the gas are enclosed. Note that the Lockman layer is absent to 3 kpc distance to the center. The thickness of the HI layer is low there, below only 100 pc.

6.2 Nonlinear Winding and the Seed Field Problem

We start with computations of the induction equation without α-effect. The fields are thus considered as decaying, but with an unknown decay rate. While the galactic differential rotation always amplifies the magnetic fields, the interstellar turbulence will destroy them. The latter, however, is nonlinearly quenched by the magnetic field itself, and its EMF possesses a complex tensorial structure.

6.2.1 Uniform Initial Field

We consider galaxies to be differentially rotating turbulent disks embedded in a plasma of given conductivity. In the simplest case, the 'plasma' is vacuum and the conductivity therefore vanishes. The half-thickness of the galaxy is H. The rotation law is known: beyond a rigidly rotating core with $s = s_0$ the angular velocity is inversely proportional to s, i.e. Eq. (5.68) is applied with $q = 1$. The velocity of the outer part is uniform, $V = s_0 \Omega_0$. We assume this velocity as $V = 100$ km/s in the models. The dimensionless ratio $\tilde{s}_0 = s_0/H$ determines the geometry of the problem. Thin disks have large values of $\tilde{s}_0$ and vice versa. In the computations $\tilde{s}_0 = 4$ is used (see Table 6.2). The real radial size of the galaxy plays very little role in the computations.

In order to simulate the spiral arms the profiles

$$Q = 1 + \frac{\varepsilon - 1}{2} \left(1 + \cos \left(2(\phi - \Omega_{\mathrm{p}}t) + 2 \log \left(\frac{s}{R_0} \cot p_{\mathrm{p}} \right) \right) \right) \tag{6.2}$$

by Otmianowska-Mazur & Chiba (1995) are adopted, varying with values between 1 and ε. This is used for the density and turbulence intensity. Ω_{p} is the angular pattern speed of the $m = 2$ spiral. Its pitch angle p_{p} is taken as $40°$. The turbulence velocity is 10 km/s, and the density contrast may be strong, e.g. $\varepsilon = 5$.

Only a vertical magnetic field is allowed to penetrate the boundary. This pseudovacuum boundary replaces the global vacuum boundary condition by a local one. The evolution of the mean magnetic field $\bar{\boldsymbol{B}}$ is governed by Eq. (4.5) with

$$\mathcal{E}_i = \epsilon_{ijk} \gamma_j \bar{B}_k + \eta_{ijk} \bar{B}_{j,k}, \tag{6.3}$$

and $\boldsymbol{\gamma} = \boldsymbol{U}^{\mathrm{dia}} + \boldsymbol{U}^{\mathrm{buo}}$. The microscopic diffusivity is neglected, and the α-effect is ignored. The advection velocity $\boldsymbol{\gamma}$ plays the role of a large-scale mean flow and the η-tensor represents the eddy diffusivity. Via magnetic feedback all of them are influenced ('quenched') by the induced magnetic field.

For slow rotation and weak magnetic field the eddy diffusivity tensor takes the simple and well-known form $\eta_{ijk} = \eta_{\mathrm{T}} \epsilon_{ijk}$. As described in Sect. 4.3 the η_{T}-tensor for strong fields

Table 6.2: The galaxy models of Elstner, Meinel & Beck (1992).

Galaxy	$V_{\max}$ (km/s)	s_0 (kpc)	Ω_0 (km/s/kpc)	s_0/H	C_Ω
M 31	230	8	30	8	30
M 33	120	4	30	4	30
M 51	220	1.2	180	1.2	180
M 81	240	4	60	4	60
IC 342	180	6	30	6	30
NGC 6946	280	8	30	8	30
NGC 891	230	3	80	3	80

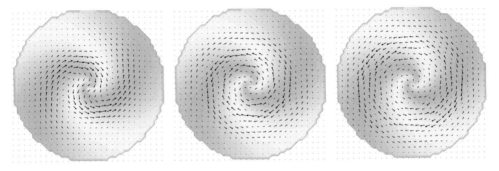

Figure 6.9: Time evolution of the magnetic field with $c_\eta = 0.03$. *Left*: After 0.07 Gyr. *Middle*: After 0.14 Gyr. *Right*: After 0.21 Gyr. Light gray means large density. From Rohde, Elstner & Rüdiger (1998).

becomes much more complex. The reference value η_0 of eddy diffusivity is Eq. (4.70) with the standard value $c_\eta \simeq 0.3$ (see Parker 1979, also Ruzmaikin, Shukurov & Sokoloff 1988). The true c_η is unknown. The standard value is used here, but we also discuss values one order of magnitude greater or smaller. The correlation time is always fixed as 30 Myr.

The initial field is considered as being nonaxisymmetric with equatorial symmetry (i.e. S1). It fulfills the pseudovacuum boundary condition. The result of Otmianowska-Mazur & Chiba (1995) is a lifetime of 200 Myr, similar to Parker's (1979) estimate. It is thus doubtful whether a nonaxisymmetric seed field could still give an observable field after 1 Gyr (see Parker 1992, Moss et al. 1993, Camenzind & Lesch 1994). It remains to check whether the nonlinear η_T-effect will change this situation. If the answer is no then we have to explain the existence of galactic magnetic fields with a dynamo mechanism.

In Fig. 6.9 the temporal evolution of the magnetic field geometry is given for various times and for $c_\eta = 0.03$. The field geometry in Fig. 6.9 is S1, i.e. of BSS-type. It fulfills the relation $\bar{B}_s\bar{B}_\phi < 0$, i.e. the Maxwell stress transports the angular momentum outward.

In early epochs we find the magnetic fields concentrated between the gaseous spirals, close to the observations. Rather quickly, however, the magnetic arms are wound up by the

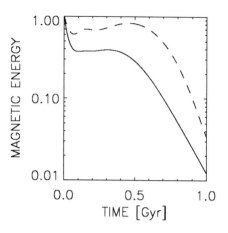 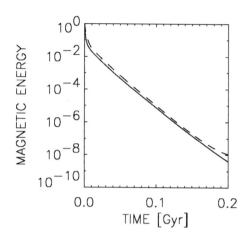

Figure 6.10: Time evolution of the magnetic energy for small (solid) and large (dashed) initial field amplitude with $c_\eta = 0.03$ (*left*) and $c_\eta = 3$ (*right*).

differential rotation. The magnetic pitch angles become smaller and smaller so that in the final states the (still existing) magnetic spirals change their form, remaining between the gaseous spirals only in the outer part of the galaxy.

The initially nonaxisymmetric configuration develops into ring-like structures (Moss & Brandenburg 1992). These are formed by magnetic fields of opposite polarity, hence the magnetic dissipation is continually reducing the field. At first the central part of the galaxy becomes field-free. Figure 6.10 demonstrates the decay of the magnetic energy. There is no amplification at all – not even at early times (see Weiss 1966, Moffatt 1978, Spencer 1994). The differences for weak and strong seed fields are surprisingly small. The maximal decay time (for very small diffusivity, $c_\eta = 0.03$) is about 300 Myr, for the standard diffusivity value ($c_\eta = 0.3$) it is 100 Myr and for $c_\eta = 3$ one obtains only 25 Myr. There are no decay times exceeding 1 Gyr. The decay of the pitch angles is even faster. No influence of ambipolar diffusion was observed in the simulations.

The magnetic decay is thus rather fast. Observations of μG fields in galaxies of high redshift show that there must be some dynamo mechanism, and the dynamo must be saturated after only 1–3 Gyr (Kim et al. 1990, Wolfe, Lanzetta & Oren 1992, Kronberg, Perry & Zukowski 1992, Oren & Wolfe 1995). Based on the observation of Faraday rotation of high redshift objects ($z > 2$) by Athreya et al. (1998) and Carilli & Taylor (2002), Lesch & Hanasz (2003) even discuss magnetic field generation (to a few μG) in intervals of a few 100 Myr. If correct, such a situation must be considered as a big challenge to the standard dynamo concept.

6.2.2   Seed Field Amplitude and Geometry

The exponential growth of magnetic fields in the dynamo theory is not too effective. Imagine that four or five rotation periods of about (say) 100 Myr form the minimum growth time. Then after 2.5 Gyr we have an amplification of only two orders of magnitude, i.e. from 10^{-8} G to

10^{-6} G. Obviously, the (kinematic) dynamo theory runs into trouble unless there are seed fields stronger than 10^{-8} G at the formation of the (young) galaxy (see Beck et al. 1996)[2].

There is another critical argument concerning the seed fields. In order to form a quadrupolar geometry of the dynamo-induced fields the seed field must already have a (small) quadrupolar component. However, a battery effect, caused by the early collapse of the material to the galactic midplane, cannot generate such an initial field. The battery effect forms electrical currents with *equatorial symmetry*, so that the corresponding magnetic fields are always antisymmetric with respect to the galactic midplane (Krause & Beck 1998). As even the smallest magnetic field that is basic for any dynamo theory must be explained (Widrow 2002), we have no explanation for a quadrupolar seed field – not even of the smallest amplitude. There is no clear route from the formation of a single galaxy to the present-day large-scale, dynamo-generated quadrupolar magnetic fields of a few μG amplitude. The typical seed field strength in the early Universe is given as 10^{-18} G (Hanasz & Lesch 1997).

Poezd, Shukurov & Sokoloff (1993) and Beck et al. (1994) suggest considering small-scale dynamo action to produce the necessary seed fields. Such dynamos too need seed fields, but they can be small. The turnover time of such fluctuations with $l \simeq 100$ pc and $u' \simeq 10$ km/s reaches only 10 Myr. The magnetic amplitude for balancing kinetic and magnetic energy is about 3 μG. Beck et al. (1996) finally favor a value of 10^{-8} G as the mean-field component of such a fluctuating field. Its dipolar part will decay and its quadrupolar part (if it exists) will be amplified. On the other hand, Deiss et al. (1997) and Thierbach, Klein & Wielebinski (2003) report a few μG magnetic field strength in the intracluster medium of Coma. Even higher field amplitudes are given for the intergalactic medium for two radio source samples by Kronberg et al. (2001) and Carilli & Taylor (2002).

Ruzmaikin, Shukurov & Sokoloff (1988) also discuss the possibility of magnetic field amplification by a collapsing protogalactic cloud. If the marginal field has an amplitude of 10^{-10} G then the compressed field might easily have an amplitude of 10^{-7} G. The geometry of such fields remains antisymmetric with respect to the equator though, so they cannot serve as a seed field for quadrupoles.

The solution of this problem might only result from the MRI. As shown in Sect. 6.10 the instability that results from the interaction of a vertical magnetic field (antisymmetric with respect to the galactic midplane) and the rotation law $\Omega \propto s^{-1}$ leads to the excitation of *quadrupolar* magnetic field geometry.

The minimum field for such a mechanism could be more important as the absolute minimum seed field necessary for dynamo action. The results for the smallest possible magnetic field in Fig. 6.33 (left) is given by Eq. (6.48). The Lundquist number Ha* is a magnetic Reynolds number with the Alfvén velocity V_A as the velocity. With the galactic microscopic diffusivity of $\eta \simeq 10^7$ cm²/s the resulting minimum field for the MRI is only $\sim 10^{-25}$ G – much smaller than the above-mentioned 10^{-18} G.

---

[2] after 10 Gyr the amplification is by (only) 8 orders of magnitude

6.3   Interstellar Turbulence

The galactic dynamo operates in a completely different geometry from the spherical stellar dynamo. The vertical stratification differs strongly from the radial one, and the observed shear is in accordance with $d\Omega/ds \simeq -q\Omega/s$ (with $q \simeq 1.03$, see Ziegler 1996). The vertical half-thickness H of the disk is used to normalize the α-effect and the shear, i.e.

$$C_\alpha = \frac{\alpha H}{\eta_T}, \qquad\qquad\qquad C_\Omega = -\frac{\Omega H^2}{\eta_T}. \qquad (6.4)$$

If the standard expression (4.8) is used together with the estimate $\eta_T \simeq c_\eta \ell_{corr}^2/\tau_{corr}$ one finds

$$C_\alpha \simeq \frac{1}{c_\eta}\tau_{corr}\Omega, \qquad\qquad |C_\Omega| \simeq \frac{1}{c_\eta}\tau_{corr}\Omega\left(\frac{H}{\ell_{corr}}\right)^2, \qquad (6.5)$$

so that

$$C_\alpha|C_\Omega| \simeq \frac{(\tau_{corr}\Omega)^2}{c_\eta^2}\left(\frac{H}{\ell_{corr}}\right)^2, \qquad\qquad \frac{C_\alpha}{|C_\Omega|} \simeq \left(\frac{\ell_{corr}}{H}\right)^2 \qquad (6.6)$$

results. The latter relation directs the pitch angle between the radial and the azimuthal component of the magnetic field (see Fig. 6.14, right). The observed large pitch angles can only be understood with $C_\alpha \simeq |C_\Omega|$, i.e. with large correlation length, $\ell_{corr} \simeq H$. This finding, however, immediately leads to $C_\alpha|C_\Omega| \propto (\tau_{corr}\Omega)^2/c_\eta^2$, which for $c_\eta \approx 1$ only exceeds any threshold value of order unity if $\tau_{corr} \gtrsim \tau_{rot}/2\pi \simeq 40$ Myr. Turbulent patterns with shorter correlation (turnover) times cannot work as a turbulent dynamo for galaxies. This argument is based on the assumption that $c_\eta \simeq 1$. For smaller eddy diffusivity, of course, it is easier to make the dynamo operate.

6.3.1   The Advection Problem

There is another problem for the galactic dynamo theory. As we shall demonstrate, disk dynamos are highly sensitive to the amplitude of the vertical advection effect γ, which in cylindrical coordinates is simply $\gamma_z = \alpha_{\phi s}$. It is *not* created by the basic rotation. If the rotation, therefore, is too slow, the α-effect becomes too low compared with the own diamagnetic pumping and cannot work. This happens already for $|\hat{\gamma}| \simeq 10$ with $\hat{\gamma}$ from $\hat{\gamma} = \alpha_{\phi s}/\alpha_{\phi\phi}$ (Schultz, Elstner & Rüdiger 1994). The longer the correlation time the smaller the value of $\hat{\gamma}$. Many of the α-effect computations on the basis of SN explosions lead to $\hat{\gamma}$-values exceeding 10, e.g. Ferrière (1992a, $\hat{\gamma} \simeq 50$) and Ziegler, Yorke & Kaisig (1996, $\hat{\gamma} \simeq 30$). If the phenomenon of networked SN ('superbubbles', see Fig. 6.11) is used as the source of interstellar turbulence (with an increase of the correlation time from 2 Myr to 16 Myr), Ferrière (1996) reaches α-amplitudes of 400 m/s and $\hat{\gamma} \simeq 15$. It is hard to imagine how such a dynamo may operate[3]. The cause of this complication for galactic dynamos is the slow galactic rotation compared with the correlation time of the interstellar turbulence. The typical values $\tau_{corr} \simeq 10$ Myr and $\Omega = 10^{-15}$ s^{-1} lead to $\Omega^* \simeq 0.5$ (Parker 1979). With the

---

[3] $\hat{\gamma}$ is reduced to about 8 if 3000 SN simultaneously explode instead of only 300 (Ferrière 1998)

SOCA-expressions for intensity-stratified turbulence

$$\alpha_{ss} = \alpha_{\phi\phi} = -\frac{4}{15}\tau_{corr}^2\Omega\frac{du_T^2}{dz}, \qquad \alpha_{zz} = \frac{8}{15}\tau_{corr}^2\Omega\frac{du_T^2}{dz},$$

$$\alpha_{s\phi} = -\alpha_{\phi s} = \frac{1}{6}\tau_{corr}\frac{du_T^2}{dz}, \tag{6.7}$$

with the density stratification neglected, one obtains quite small values for the advection effect, such as

$$\hat{\gamma} \simeq 1.25/\Omega^* \simeq 2.5. \tag{6.8}$$

With the given numbers for both correlation time and rotation rate, and for a half-thickness of the galactic disk of about 100 pc, the resulting α-effect is of order 1 km/s, in extreme contrast to the values of about 50 m/s for isolated SN explosions (see Sect. 6.9.2).

It is challenging to ask whether the $s\phi$-component of the α-tensor must be antisymmetric as is formulated in Eq. (6.7). The answer differs for stars and galaxies. For the latter the density gradient G and the rotation axis Ω are parallel, which is not true for stars. In general one can construct for sufficiently weak fields the pseudotensor

$$\alpha_{ij} = \gamma_0\epsilon_{ijk}G_k + \gamma_1\epsilon_{ipk}G_p\Omega_k\Omega_j + \gamma_2\epsilon_{jpk}G_p\Omega_k\Omega_i +$$
$$+\gamma_3\epsilon_{ipk}G_k\Omega_p\Omega_j + \gamma_4\epsilon_{jpk}G_k\Omega_p\Omega_i \tag{6.9}$$

for the advection terms. However, if $G \parallel \Omega$ all terms vanish except the first, which is antisymmetric. There is no symmetric part of $\alpha_{s\phi}$.

The argument does not hold for stars with G as the radial direction (Ossendrijver et al. 2002, their Fig. 6) and it also does not hold if shear is involved. In her evaluations of SN-explosions under the influence of the galactic differential rotation Ferrière (1998), however, did not find any symmetric part of the $s\phi$-component of the α-tensor even if shear is involved.

6.3.2 Hydrostatic Equilibrium and Interstellar Turbulence

Rotating turbulence does not produce any α-effect if the turbulence is not stratified. A homogeneous rotating field of isotropic SN explosions produces eddy diffusivity but does not produce any α-effect. Mean-field dynamo theory is thus the theory of density and/or intensity stratifications. There is strong observational evidence for substantial amounts of highly turbulent gas in heights above 1 kpc (see Sect. 6.1.5). This suggests using the equation of vertical hydrostatic equilibrium to derive the anisotropies in the turbulence field (Fröhlich & Schultz 1996). The main questions for this concept concern the sign of the α-effect and the resulting amplitude.

For the density stratification the empirical HI distribution from Dickey & Lockman (1990) has been taken. The extended ionised gas, which is the source of the diffuse Hα emission at high latitudes, has been described by an exponential with a scale height 1500 pc and a midplane density of 0.025 cm^{-3} (Reynolds 1989). This normalised gas density profile, which is appropriate to describe the gas in the solar neighborhood, is assumed to be valid also in other parts of the galaxy. For smaller radial distances the same shape of the density profile

Figure 6.11: Superbubble N70 in the Large Magellanic Cloud, with a diameter of 100 pc blown by stellar winds and SN explosions. Courtesy FORS Team, ESO.

has been taken, but the vertical scale height is scaled down linearly in such a manner that it diminishes at $s = 0$.

The turbulence pressure occurs in the equation for the vertical momentum, $\mathrm{d}P_\text{tot}/\mathrm{d}z = -\rho\, k_z$, with

$$P_\text{tot} = \rho\langle u_z'^2\rangle + \frac{\bar{B}_x^2 + \bar{B}_y^2 - \bar{B}_z^2}{2\mu_0} + P_\text{CR}. \tag{6.10}$$

Only the intensity of the vertical turbulence contributes. The pressure due to cosmic rays is P_CR (Parker 1992). Self-gravitation only plays a minor role. Equation (6.10) for prescribed

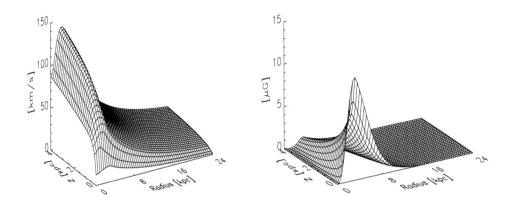

Figure 6.12: *Left:* Turbulent intensity in the vertical direction. *Right:* Magnetic equipartition field. $c_\tau = 1$.

turbulence and magnetic field gives the density profile $\rho = \rho(z)$. On the other hand, if the density can be considered as given, the vertical profile $\langle u_z'^2 \rangle$ results from Eq. (6.10).

The k_z-force is essentially due to a self-gravitating isothermal sheet of stars with constant half-thickness $z_0 = 600$ pc. For z-values exceeding the thickness of the stellar disk the contribution of the more spherical components (bulge, spheroid, and dark matter halo) of the galaxy becomes important, i.e.

$$k_z(s, z) = \frac{2u_T^2}{z_0} \tanh\left(\frac{z}{z_0}\right) + \varepsilon(s)\frac{z}{z_0}, \tag{6.11}$$

with u_T^2 being the vertical turbulence intensity of the old disk stars[4]. For its radial variation $u_T^2 = (20\,\text{km/s})^2 \exp((s_\odot - s)/0.44\,s_\odot)$ with $s_\odot = 8.5$ kpc. The separation of the disk potential into a vertical and a radial component is only reasonable in the thin-disk approximation.

In order to calculate unquenched α- and η-coefficients via Eqs. (6.7) and (6.10) one has to specify the correlation time τ_{corr} for the interstellar turbulence. The approach $\tau_{\text{corr}} \simeq c_\tau H(s)/u_T$ is used, where $H(s)$ is the scale height of the gas and u_T the midplane velocity. The mixing-time factor c_τ is now the only free parameter; its role is very similar to the role of the mixing-length-α in convection theory.

The results of Fröhlich & Schultz for $c_\tau = 1$ are given in Figs. 6.12 and 6.13. The main result is that the turbulence intensity increases with the height z. This is due to the exponential tail of the vertical density profile.

The resulting α-effect and eddy diffusivity are rather high. Characteristic are values of 10 km/s for $\alpha_{\phi\phi}$ and 10^{27} cm^2/s for η_T, so that for a vertical scale of 1 kpc the dynamo number, Eq. (6.4)$_1$, takes values of about unity. With $c_\eta = 1$ the eddy diffusivity might

---

[4] ε gives the contribution of the dark matter halo

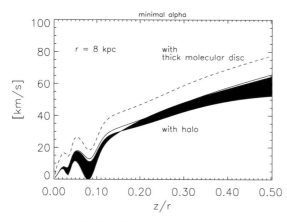

Figure 6.13: The α-effect at $s = 8$ kpc after Fröhlich & Schultz (1996). The standard case (full line) lies generally within the filled region.

be overestimated. Such values are also quite characteristic for several completely different models of rotating stratified turbulence in the interstellar medium (see Hanasz & Lesch 1997). Note that the α-effect should *not* exceed the value of the turbulence intensity, so that already with this idea one finds $C_\alpha \lesssim H/\ell_{\mathrm{corr}}$, which should always (slightly) exceed unity. All the computations, however, lead to *positive* values of $\alpha_{\phi\phi}$ in opposition to the negative values resulting from MRI. No bridges seem to exist between the presented results for interstellar turbulence and the alternative models of SN-driven interstellar turbulence (Sect. 6.9.2) and for MRI (Sect. 6.10).

6.4   From Spheres to Disks

A galaxy might be modeled by an axisymmetric disk-like structure with axisymmetric functions $\alpha, \eta_{\mathrm{T}}$ and $\bar{\boldsymbol{u}}$. While η_{T} and $\bar{\boldsymbol{u}}$ are taken as symmetric with respect to the galactic plane, the components of the α-tensor are antisymmetric. On the other hand, the eigensolutions of the linear induction equation in a domain with such symmetries are either symmetric (S) or antisymmetric (A) with respect to the galactic midplane, and depend on the azimuth ϕ according to $\mathrm{e}^{im\phi}$. The field modes are again denoted by Am or Sm.

The mathematics of thin but finite disks is complicated. The research has been started considering oblate spheroids embedded in vacuum in order to find clean models for the necessary match of the outer to the internal magnetic fields (Stix 1975, White 1978, Soward 1978). The magnetic modes considered were axisymmetric, and the resulting magnetic geometry proved to be symmetric with respect to the galactic midplane ('quadrupoles'). Stix (1975) in the spherical limit for his oblate spheroidal coordinates confirmed the result that oscillating dipoles then occur (Roberts 1972). If the spheroid, however, becomes more and more oblate the mode with the lowest dynamo number (only $\alpha\Omega$-dynamos were considered) is a *steady quadrupole*. Obviously, the geometry of the dynamo massively influences the geometry of the induced magnetic field. The oscillating dipole (which for spheres has the lowest dynamo number) for ellipsoids with (say) 1 kpc and 15 kpc for the two half-axes needs a dynamo number that exceeds the lowest one (for the steady quadrupole) by a factor of 50.

There is another surprise considering nonaxisymmetric magnetic modes. The flat geometry seems to suppress nonaxisymmetric modes. For artifical models with $\alpha = \mathrm{const.}$ (no

Table 6.3: Marginal values C_α for eigenmodes for α^2-disk dynamos with $H/R_{\max} = 0.1$ surrounded by a perfect conductor. The minimum values are given in bold.

$\alpha_{zz}/\alpha_{\phi\phi}$	S0	A0	S1	A1
0.1	6.0	**4.1**	5.9	**4.1**
1	**3.2**	4.1	**3.2**	4.1
10	**1.1**	1.4	**1.1**	1.4

difference between above and below the equator) for spheres the eigenvalue C_α naturally does not differ for $m = 0$ and $m = 1$. For oblate ellipsoids, however, the $m = 0$ solution is always preferred and this is more so the lower the conductivity of the outer space (Krause et al. 1990). It is only the symmetry breaking that leads to this result.

However, if the disk is embedded in a perfect conductor then the differences between $m = 0$ and $m = 1$ disappear. For a finite disk without differential rotation but with equatorially antisymmetric α-effect embedded in a perfect conductor Meinel, Elstner & Rüdiger (1990) find the results given in Table 6.3. The standard solution for isotropic α-effect is of S0-type. With differential rotation and for isotropic α-effect the lowest C_α for dipoles and quadrupoles become equal[5]. Only for α_{zz} small compared with $\alpha_{\phi\phi}$ are the nonaxisymmetric solutions preferred. The differential rotation, however, generally disturbs the excitation of nonaxisymmetric modes (Rädler 1986b) and this is also the case for the galactic dynamo. A characteristic result is given by Rüdiger, Elstner & Schultz (1993): Only if the rigidly rotating core extends to a radius of about $16\,H$ for $\alpha_{zz} = 0$ does the nonaxisymmetry start to survive the smoothing action of differential rotation (which is then restricted to the outer parts of the galaxy).

6.4.1 1D Dynamo Waves

The dynamo equations are studied in a 1D approximation (Parker 1971). The integration region extends only in the z-direction, where the remaining radial and azimuthal components of the magnetic field depend on z only, so that from the divergence condition one has $B_z = $ const. Normalizing time with the diffusion time H^2/η_T and vertical distances with H yields

$$
\frac{\partial A}{\partial t} = C_\alpha \hat\alpha(z)\,\Psi(B)B + \frac{\partial^2 A}{\partial z^2},
$$
$$
\frac{\partial B}{\partial t} = -C_\alpha \frac{\partial}{\partial z}\left(\hat\alpha(z)\,\Psi(B)\frac{\partial A}{\partial z}\right) - C_\Omega \frac{\partial A}{\partial z} + \frac{\partial^2 B}{\partial z^2}. \tag{6.12}
$$

A (pseudo)vacuum surrounds the disk. Priklonsky et al. (2000) also include the radial derivatives in Eqs. (6.12) and the corresponding nonlocal boundary conditions (see Sect. 5.5.2).

The α depends on the magnetic field as well as on the location in the object, expressed simply by $\hat\alpha(z) = -\sin 2\pi z$. The lower and upper boundaries are located at $z = 0$ and $z = 1$. The galactic rotation laws lead to negative C_Ω, it is $|C_\Omega| = \Omega_0 H^2/\eta_T$. The boundary conditions are Eq. (5.77). Positive C_α means positive (negative) α-effect above (below) the disk midplane.

---

[5] Proctor (1977) and Krause et al. (1990) find for the same model embedded in vacuum that the A0 mode is preferred

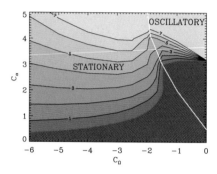
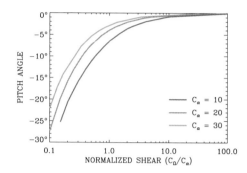

Figure 6.14: *Left*: The growth rates if the time is measured in rotation times for the 1D disk dynamo. In the blue area the magnetic field decays. *Right*: The pitch angles for the 1D disk dynamo depend mainly on the ratio C_Ω/C_α. The pitch angles are large only if C_α dominates C_Ω.

The α-quenching by magnetic fields has often been expressed by functions such as $1/(1+(B/B_{max})^2)$, with a cutoff field strength B_{max} related to the energy of velocity fluctuations or the gas pressure. Here the magnetic quenching of the α-effect is modeled by Ψ according to the quenching function (4.44).

The solutions of the system (6.12) are represented in Fig. 6.14. For negative shear in Fig. 6.14 (left) the growth rates are given. By definition the marginal stability has a zero growth rate. For strong C_Ω the well-known eigenvalue $\mathcal{D} = C_\alpha C_\Omega = -2.08$ for steady and marginal quadrupoles is found. For small C_Ω, however, the quadrupole with the lowest eigenvalue oscillates. We can speak in this (disk!) approximation about an α^2-dynamo producing oscillating quadrupoles. For negative α-effect (not shown) all the solutions with negative C_Ω lead to oscillating quadrupoles. In the $\alpha\Omega$-regime the dynamo number $\mathcal{D}$ is 44.1.

Also the growth times are interesting for discussion. They can be taken from Fig. 6.14 (left) given in rotation periods[6]. It is infinity at the marginal line but reaches values of a few rotation periods close to the line of neutral stability. For galaxies this time is not much smaller than 1 Gyr. As is also shown by Fig. 6.14 (left) for highly supercritical dynamos the growth times are formally smaller than the rotation period. Also the radial distribution of the initial magnetic fields, however, plays an important role and the diffusion time can become rather long. Here only nonlinear calculations provide the temporal development of the model.

For $C_\alpha < |C_\Omega|$ the pitch angles are much too low for galaxies. In order to produce large pitch angles one has to fulfill a relation such as $C_\Omega \lesssim 0.5 C_\alpha$ leaving the $\alpha\Omega$-regime so that the oscillating solutions appear. In consequence, the galactic dynamo can *not* be an $\alpha\Omega$-dynamo; it exists close to the limit where the α^2-regime prevails. Obviously, despite their slow rotation galaxies seem to develop a rather high α-effect. If they really exist in the (narrow?) transition region between the $\alpha\Omega$ and the α^2-regime then it should not be surprising that each galaxy seems to have rather individual magnetic properties. The condition $C_\Omega \simeq C_\alpha$ leads to $\alpha \simeq \Omega H$, which is of the order of 10 km/s.

---

[6] note that $C_\Omega = 2\pi \tau_{diff}/\tau_{rot}$

6.4.2   Oscillatory vs. Steady Solutions

The question arises from Fig. 6.14 how the dynamo regime changes from stationary to oscillatory solutions. For dominant C_Ω the steady S0-mode is expected to be realized. Decreasing C_Ω, however, it is not obvious which dynamo regime prevails. If, additionally, the α-tensor is anisotropic, nonaxisymmetric modes could dominate for small C_Ω.

Table 6.4: Time regimes for the 'toy model' with $\mathcal{D} = 5000$. The last line concerns dynamos with artificially isotropic α-effect. Stationary is 'stat', oscillatory is 'osc'. From Elstner, Rüdiger & Schultz (1996).

C_Ω	5000	625	500	50	5
C_α	1	8	10	100	1000
time-regime	stat	trans	osc	osc	osc
isotropic α-tensor	stat	stat	stat	stat	osc

This question can be attacked with a very simple galactic dynamo model. The only radial dependence may result from the rotation law. The applied vertical profiles are very simple, i.e. $\alpha_{\phi\phi} = \alpha_0 \Omega(s) \sin 2\pi z$, but $\alpha_{\phi\phi} = 0$ for $z > 0.5$, and the eddy diffusivity of the halo is 100 times the disk value. Additionally, the α-tensor is anisotropic, $\alpha_{zz} = -2\alpha_{\phi\phi}$, and the rotation law is given by Eq. (5.68). In Table 6.4 one finds the numerical results. For increasing C_α – or decreasing influence of differential rotation – the solution becomes indeed oscillatory. For isotropic α-effect this is only true for the smallest given C_Ω, which is the same result as presented in Fig. 6.14 for thin slab dynamos. It is opposite to the behavior of spherical dynamos, where α^2-dynamos are known to oscillate only for anisotropic α. It should thus be possible to find oscillatory solutions for galactic fields. Brandenburg et al. (1993) have found the same kind of oscillating solutions for anisotropic α-effect (see their Table 3)[7].

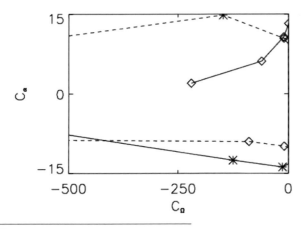

Figure 6.15: The torus dynamo embedded in vacuum for negative shear and both signs of the α-effect. Solutions with quadrupolar (dipolar) symmetry are given as solid (dashed) lines. Oscillating solutions are marked with $*$. The dominance of quadrupoles for positive α-effect changes to a dominance of dipoles for negative α-effect (Deinzer 1993).

---

[7] the oscillations of the magnetic field in the models of Brandenburg, Moss & Tuominen (1992, see their Table 5) and Ferrière & Schmitt (2000, see their Table 1) probably belong to the class of oscillating solutions of spherical $\alpha\Omega$-dynamos with too small pitch angles

A clear formulation of the kinematic dynamo problem has been presented by Deinzer (1993) and Deinzer, Grosser & Schmitt (1993) with their torus dynamo motivated by the geometrical torus-like structure of M 31 (Fig. 6.6). Outside the torus is vacuum, the boundary conditions are satisfied in full generality for axisymmetric solutions (Grosser 1988). The dynamo discussed by Deinzer, Grosser & Schmitt is of the $\alpha\Omega$-type with isotropic α-effect. The rotation law is linear. Deinzer (1993) gives the results for an $\alpha^2\Omega$-dynamo that can be interpreted with $C_\Omega < 0$ and both signs of C_α (Fig. 6.15). The main result is that the α^2-dynamo with the lowest C_α has a dipolar geometry and does not oscillate – in contrast to the disk solution. For moderate shear there is a stationary quadrupole for positive C_α and a stationary dipole for negative C_α. For very high C_Ω, i.e. in the true $\alpha\Omega$-regime, the torus dynamo always yields quadrupolar solutions, which are stationary for positive α-effect and oscillatory for negative α-effect.

6.5   Linear 3D Models

Consider now the complete dynamo equation (4.5) in cylindrical geometry and for axisymmetric disks. The α-tensor is basically antisymmetric with respect to the galactic midplane. This allows us to treat the $\boldsymbol{B}$-modes of Sm- and Am-type in the form

$$\bar{\boldsymbol{B}} = \Re\left\{\boldsymbol{B}_m(s, z, t)\, \mathrm{e}^{\mathrm{i}m\phi}\right\}, \qquad m = 0, 1, 2, \ldots, \tag{6.13}$$

where $\boldsymbol{B}_m$ are the complex axisymmetric field modes. The time evolution of the components of $\boldsymbol{B}_m$ in cylindrical coordinates is determined by

$$\frac{\partial B_s}{\partial t} = \frac{\mathrm{i}m}{s}\left\{-\Omega s B_s + \alpha_{zz}B_z - \eta_{\mathrm{T}}\left(\frac{1}{s}\frac{\partial}{\partial s}(sB_\phi) - \frac{\mathrm{i}m}{s}B_s\right)\right\} -$$
$$-\frac{\partial}{\partial z}\left\{\alpha_{\phi\phi}B_\phi - \eta_{\mathrm{T}}\left(\frac{\partial B_s}{\partial z} - \frac{\partial B_z}{\partial s}\right)\right\},$$

$$\frac{\partial B_\phi}{\partial t} = \frac{\partial}{\partial z}\left\{\Omega s B_z + \alpha_{ss}B_s - \eta_{\mathrm{T}}\left(\frac{\mathrm{i}m}{s}B_z - \frac{\partial B_\phi}{\partial z}\right)\right\} -$$
$$-\frac{\partial}{\partial s}\left\{-\Omega s B_s + \alpha_{zz}B_z - \eta_{\mathrm{T}}\left(\frac{1}{s}\frac{\partial}{\partial s}(sB_\phi) - \frac{\mathrm{i}m}{s}B_s\right)\right\} \tag{6.14}$$

and

$$\frac{\partial B_z}{\partial t} = \frac{1}{s}\frac{\partial}{\partial s}\left\{s\left(\alpha_{\phi\phi}B_\phi - \eta_{\mathrm{T}}\left(\frac{\partial B_s}{\partial z} - \frac{\partial B_z}{\partial s}\right)\right)\right\} -$$
$$-\frac{\mathrm{i}m}{s}\left\{s\Omega B_z + \alpha_{ss}B_s - \eta_{\mathrm{T}}\left(\frac{\mathrm{i}m}{s}B_z - \frac{\partial B_\phi}{\partial z}\right)\right\}. \tag{6.15}$$

The equations have to be completed by the correct number of boundary conditions, for which a local formulation for a galaxy embedded in vacuum has not yet been presented. In the ellipsoid-model of Stix (1975), and for the torus geometry used by Deinzer, Grosser & Schmitt (1993), the nonlocal vacuum condition could be exactly formulated as a local condition, but only for $m = 0$. Stepinski & Levy (1988) and Elstner, Meinel & Rüdiger (1990) therefore introduced the simulation concept 'without sharp boundaries', where the disk is embedded in a huge galactic halo where approximate boundary conditions such as pseudovacuum or perfectly conducting are applied.

We approximate the spatial operators in the induction equation by discrete difference operators, and follow the temporal evolution by means of Euler's formula

$$B(t + \delta t) = B(t) + \delta t \cdot \text{CURL}((u \times B + \mathcal{E}). \tag{6.16}$$

CURL denotes the difference scheme for vectors defined on a grid in the cylindrical region $0 \le s \le R_{\text{max}}$ and $0 \le z \le H$. This simple explicit algorithm is numerically stable because of the dissipation term.

In order to satisfy the divergence relation $\nabla \cdot B = 0$ a difference operator CURL is constructed, yielding for any discrete vector field c the relation DIV CURL $c \equiv 0$, for some finite difference approximation DIV of the divergence operator. This is almost trivial in Cartesian coordinates, and is also possible for curvilinear coordinates (Evans & Hawley 1988).

Conditions for B_m ensuring regularity of the magnetic field and its derivatives on the axis $s = 0$ are also needed. These conditions can be expressed as boundary conditions for $s = 0$, but they depend on m for $s = 0$ in the following way: $B_s = B_\phi = \partial B_z / \partial s = 0$ for $m = 0$, $B_z = \partial B_\phi / \partial s = 0$ for $m = 1$, and $B_s = B_\phi = B_z = 0$ in all other cases.

The code by Elstner, Meinel & Rüdiger allows one to model a galaxy of arbitrary (axisymmetric) geometrical shape. Such a model is defined by profiles of the angular velocity $\Omega(s, z)$, the turbulent magnetic diffusivity $\eta_{\text{T}}(s, z)$ and the α-effect $\alpha_{ij}(s, z)$. All functions are assumed to be axisymmetric. Because of the spiral structure of galaxies the assumption of axisymmetry of Ω, η and α is certainly not satisfied very well. This assumption seems to be justified though for galaxies showing an irregular flocculent spiral structure (Fig. 6.1). Often the simplified galactic rotation law

$$\Omega = \left\{ \begin{array}{ll} \Omega_0 & \text{for} \quad s < s_0 \\ \Omega_0 s_0 / s & \text{for} \quad s > s_0 \end{array} \right\} \tag{6.17}$$

is adopted instead of the smoother Brandt profile (5.68). Inside the turnover radius s_0 we have rigid rotation, and outside s_0 the flat rotation curve is realized. We assume $\alpha_{\phi\phi} = \alpha_0 f(z)\Omega(s)$, with $f(z) = \pm 2^{-(z/H)^2}$ for $z \gtrless 0$ and with positive α_0. For the radial variation of α we only consider the radial profile (6.17). As the magnetic field turbulence *increases* in the halo, we assume $\eta_{\text{T}} \propto 2^{(z/3H)^2}$ and neglect any s-dependence. The half-thickness of this profile is assumed to be three times larger than H. For the equatorial magnetic diffusivity $3 \cdot 10^{26}$ cm^2/s is used. H has been assumed as 1 kpc.

Table 6.5: Critical C_α for a series of model calculations (Elstner, Meinel & Beck 1992). The minimum C_α^{crit} are given in bold. They always belong to the S0-solution.

$\tilde{s}_0$	$\lvert C_\Omega \rvert$	S0	A0	S1	A1	$\tilde{s}_0$	$\lvert C_\Omega \rvert$	S0	A0	S1	A1
2	200	**0.09**	3.3	2.5	2.2	5	10	**0.64**	3.8	3.2	3.7
5	50	**0.13**	3.9	2.7	3.7	5	1	**3.7**	3.7	3.8	3.7
2	10	**1.6**	4.3	4.3	4.3	5	5	**1.2**	3.7	3.3	3.7

For a given galaxy model the marginal values C_α^{crit} as well as the corresponding field structure is computed by Elstner, Meinel & Beck (1992). Table 6.5 shows their results. The prediction of kinematic dynamo theory is a magnetic field configuration of the symmetry type that has the smallest C_α^{crit}. The final state, however, can only be determined from nonlinear simulations, where the backreaction of the field on the motions is fully taken into account.

In this linear regime the results in Table 6.5 always lead to magnetic fields of S0-type. Donner & Brandenburg (1990) obtained similar results. In no case is a preference for a non-axisymmetric S1 or A1 field obtained. Without special extras the galactic disk dynamo can only maintain axisymmetric magnetic fields. If the disk is embedded in vacuum then the preferred mode is S0, but if the halo has the same conductivity as the disk then the dynamo region changes to a solar-like one, and an oscillating dipole appears (see also Brandenburg, Moss & Tuominen 1992, their Table 5).

6.6 The Nonlinear Galactic Dynamo with Uniform Density

Let us now suppose that turbulence in galaxies is driven by a field of random SN explosions under the influence of global rotation. The turbulence intensity then has its maximum in the equatorial midplane. There would also be a strong correlation with the star formation rate and its time dependence during the galactic evolution (see Ko & Parker 1989, Ko 1993, Beck et al. 1996). Beck & Golla (1988) reported the observational background for this concept: For several spiral galaxies the radio continuum emission is correlated with their far-infrared emission. Our code allows the unified derivation of the complete turbulent EMF on the basis of a given turbulence field. The expansion (6.3) becomes fully tractable, and the internal rotation law in galaxies is well-known of course.

6.6.1 The Model

The α-effect only occurs in a stratified medium under the combined influence of turbulence and global rotation. Both density and intensity stratification of the turbulence field form the known α-sources. Unlike stellar convection zones, galactic disks are self-gravitating layers, hence their density stratification is weak.

The numerical simulations of SN explosions show that the influence of density stratification is extremely small. Our turbulence model is based on the anelastic approximation and also on the τ-approximation. The resulting expressions for uniform density are given by Eqs. (6.7), multiplied by the function (4.44) for the magnetic quenching. The main shortcoming of this approach consists in the fact that the equipartition value of the halo fields is overestimated. Our halo fields will thus simulate (magnetic) disks that are too thick.

Dynamo models including the back-reaction of the magnetic field on the turbulence and/or the mean flow yield magnetic field strengths depending on the feedback mechanism. Instabilities due to effects such as the Parker mechanism are ignored here. It is the equipartition value $B_{\mathrm{eq}} = \sqrt{\mu_0 \rho}\, u_{\mathrm{midplane}}$ that appears in the quenching functions, in the form of $\beta = B/B_{\mathrm{eq}}$.

The rotation law can be approximated by Eq. (6.17). We assume $V = 100$ km/s as the linear velocity for almost all of our models. For the eddy diffusivity we work with the scalar expression (4.70), which should be sufficiently accurate for slow rotation. Both turbulent and microscopic contributions have to be included in the magnetic diffusivity, so that

$$\eta = \eta_{\mathrm{T}} + \eta_{\mathrm{micro}}, \tag{6.18}$$

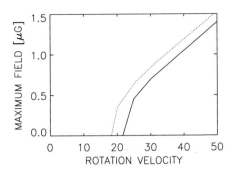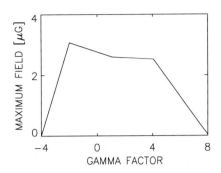

Figure 6.16: *Left*: The marginal state for slow rotation. Below $V \simeq 21.5$ km/s no dynamo is possible. The dotted curve represents a dynamo without advection terms. *Right*: Artificial amplification by the factor γ^* of the turbulent vertical advection quenches the dynamo action. The same effect appears for large 'inverse winds' with negative γ^*. $u_{\text{midplane}} = 15$ km/s, $u_{\text{halo}} = 8$ km/s. (Rüdiger, Elstner & Schultz 1993).

with $\eta_{\text{micro}} = 1/\mu_0\sigma$ and σ the microscopic conductivity. In the bulk of the galaxy, for $z < 0.5H$, the first term will always dominate. If, for greater vertical distances, $z > 0.5H$, the gas becomes neutral the second term dominates there. We shall consider two extreme cases, (i) $\eta_{\text{micro}} \to \infty$ (vacuum) and (ii) $\eta_{\text{micro}} = 0$ (hot plasma).

The dynamo only lives from the vertical stratification of the turbulence intensity. At the galactic midplane the turbulence velocity is u_{midplane}, and at the outer boundary, $z = 0.5H$, it is u_{halo}, with $u_{\text{halo}} < u_{\text{midplane}}$. Typical values for the observable midplane values are ≥ 10 km/s. In the model the upward decrease of the turbulence intensity enables the dynamo to operate. The critical eigenvalue is already reached for very small gradients ($u_{\text{midplane}} \simeq 5$ km/s).

Characteristic of our approach is the inclusion of the advection term $\boldsymbol{U}^{\text{dia}}$ simultaneously with the α-effect. It transports magnetic field to regions of lower turbulence intensity. If, in the galactic halo smaller velocities are present then the field is carried into the halo. If not, it is carried to the midplane (Brandenburg et al. 1993).

Only the global rotation produces an α-effect. The advection velocity $\boldsymbol{U}^{\text{dia}}$ does not vanish for vanishing rotation. The galactic dynamo thus possesses a minimum rotation rate for which dynamo generation is possible. In Fig. 6.16 (left) the dynamo regime is shown for the marginal case. The turbulence field with $u_{\text{midplane}} = 15$ km/s and $u_{\text{halo}} = 8$ km/s as well as the transition radius $s_0 = 2H$ are prescribed. The dynamo dies off for $V \leq 21.5$ km/s. Note here in particular that NGC 4449 with $V \simeq 30$ km/s possesses regular fields with ~ 14 μG amplitudes (Chyży et al. 2000).

According to Eq. (6.7) the diagonal and off-diagonal elements of the α-tensor are of the same order of magnitude. One may ask how long the dynamo tolerates an extremely high advection velocity. Therefore, we write

$$\alpha_{\phi s} = \gamma^* \, \alpha_{\phi\phi}/\Omega^*, \tag{6.19}$$

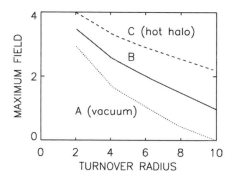

Figure 6.17: The maximum magnetic field in equipartition units vs. the geometry number $\tilde{s}_0$. The electrodynamical halo properties are varied (see text). $V = 100$ km/s, $u_{\mathrm{midplane}} = 15$ km/s, $u_{\mathrm{halo}} = 3$ km/s. Variation of the correlation time gives no significant effect. (Schultz, Elstner & Rüdiger 1994).

and vary the free parameter γ^*. For $\gamma^* > 1$ the vertical turbulent transport is amplified while for negative γ^* the field is advected to the galactic midplane. Figure 6.16 (right) shows the result. The maximum magnetic field is given for diverse amplification factors γ^*. Already for $\gamma^* > 8$ the dynamo breaks down. There is thus not too much freedom in the choice of the value of $\alpha_{\phi s}$. For strong equatorward turbulent winds the dynamo cannot operate either.

6.6.2   The Influences of Geometry and Turbulence Field

The dynamo regime may be studied in two ways. First the turbulent flow field is assumed to be given and the 'geometry' is varied. Only two parameters describe the geometry, i.e. $\tilde{s}_0 = s_0/H$ and η_{micro}. With the latter quantity we model the electrical properties of the halo. Vacuum is simulated if the value of η_{micro} exceeds the turbulent term (model A). In the case of turbulent 'warm' plasma (model B) the opposite is true. For a very hot plasma (C) the total halo dissipation is relatively small, simulating a nearly perfect conductor.

All these cases are considered in Fig. 6.17, which shows the maximum field strength created by the dynamo. The influence of the variations is small. Except for model C the midplane equipartition value B_{eq} is comparable with the induced magnetic fields. Dynamos with halos of very high conductivity (model C) create magnetic fields considerably stronger than the equipartition value in the disk.

Figure 6.18 exhibits the geometry of the dynamo-created magnetic fields. The greater the halo conductivity, the more distinct magnetic halo belts we obtain, whereas for vacuum-like boundary conditions (model A) they disappear completely. As no such belts are observed in real galaxies, the observations clearly favor this latter case. Highly intensive turbulent motions in the galactic halo lead to the same phenomenon. The characteristic properties of such models are

- the fields are confined to the galactic disk,
- the fields have a ring-like structure,
- the ring is attached at the turnover radius s_0,
- the pitch angles are rather small.

We now turn to the influence of special turbulence intensity profiles on the induced magnetism. The z-profile of the turbulent intensity, u_{T}, is characterized by two numbers: $u_{\mathrm{T}} = u_{\mathrm{midplane}}$ at the midplane and $u_{\mathrm{T}} = u_{\mathrm{halo}}$ in the halo. There is a smooth transition between the two

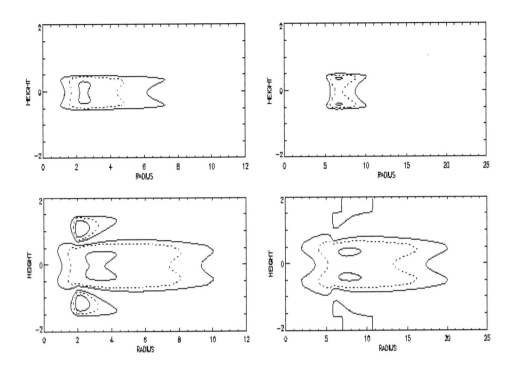

Figure 6.18: Characteristic toroidal field configurations for $\tilde{s}_0 = 2$ (*left*) and $\tilde{s}_0 = 6$ (*right*) for a vacuum halo (*top*, model A) and a hot halo (*bottom*, model C). Eddy lifetime 10 Myr, other parameters as in Fig. 6.17. Note that the maximal toroidal fields always occur close to $\tilde{s}_0$.

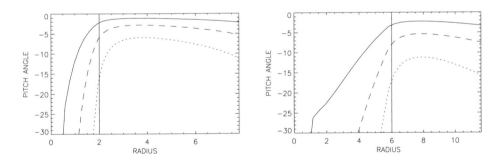

Figure 6.19: The pitch angles for diverse correlation times τ_{corr} (10 Myr (solid), 25 Myr (dashed) and 50 Myr (dotted) and two different turnover radii in kpc (marked by vertical lines). The velocities are $u_{\mathrm{midplane}} = 15$ km/s and $u_{\mathrm{halo}} = 3$ km/s. Note the absolute pitch angles outside the core increasing outward.

values. The α-effect exists only in this transition region. The α-effect also grows if, e.g., the midplane value increases, for a given halo velocity.

The pitch angle represents the ratio of radial to azimuthal component of the magnetic field. It vanishes if the field is purely azimuthal. Fields with globally large pitch angles form the observed patterns of the well-developed galaxies, so that we have to conclude that the α-effect is strong enough compared with the differential rotation to create considerable radial magnetic field components. The ratio of the two induction effects is given by Eq. (6.6)$_2$, i.e.

$$\frac{C_\alpha}{|C_\Omega|} \propto \Omega^{*2} \, \mathrm{Ma}_{\mathrm{T}}^2, \tag{6.20}$$

with the turbulent Mach number $\mathrm{Ma}_{\mathrm{T}} = u_{\mathrm{T}}/c_{\mathrm{ac}}$ and with $H\Omega \simeq c_{\mathrm{ac}}$, which is true not only for accretion disks but also for selfgravitating galaxies for $Q \lesssim 1$ and $H = c_{\mathrm{ac}}/\sqrt{G\rho}$. Obviously, for supersonic turbulence or longer correlation times the differential rotation no longer dominates. We thus expect larger pitch angles for longer correlation times. Figure 6.19 shows the results. Only for rather long correlation times do the pitch angles reach values that are compatible with observations. For large radii the pitch angles are independent of the chosen $\tilde{s}_0$. Generally, the pitch angles increase slightly going outward – contrary to observations, however. The observed radial decay of the pitch angles (Fig. 6.3) can only be explained with a radial decay of the turbulence.

6.7   Density Wave Theory and Swing Excitation

Most large disk galaxies are spiral systems that look very impressive if observed pole-on. They consist of bright oblate core domains with tangentially attached spiral arms (Fig. 6.4). The spirals are transient structures in a mixture of stars and molecular clouds. The density wave theory explains how the spirals are formed rotating around the galactic center and coexisting with the differential galactic rotation. It is the profile of the mass distribution that determines the resulting gravity-supported structures as rings or spirals.

6.7.1   Density Wave Theory

Studies of the general structure of self-gravitating cool disks are strongly stimulated by detections of exosolar planets. The idea to consider the solar system as the relic of a global instability is not new in the context of the density wave theory. Such a concept has its roots in the fascinating order of the Titius–Bode law, suggesting a global formation process rather than a local one. Modes with a wave number

$$k_{\mathrm{crit}} = \frac{1}{\pi} \frac{M_{\mathrm{c}}}{\Sigma_0 s^3} \tag{6.21}$$

become unstable in thin Keplerian disks if the Toomre parameter

$$Q = \frac{c_{\mathrm{ac}} \Omega}{\pi \, G \, \Sigma_0} \tag{6.22}$$

falls below unity (Toomre 1964, Safronov 1969, Goldreich & Ward 1973). M_{c} is the central mass and Σ_0 is the surface density of the disk. Numerical simulations with a large number of mass points (Cassen et al. 1981, Sellwood 1985, Mayer et al. 2003, Klessen 2003, Lufkin et al. 2004) generally confirm the stability criterion $Q > 1$.

Significant progress has also been made in explaining the spiral structure of galaxies with the density wave theory. The instability of global modes has been analyzed with considerable success (Meinel 1983, Binney & Tremaine 1987). Sellwood (1985) has shown how a global spiral instability occurred for galaxies with minimum $Q = 1$ but not for $Q = 1.5$.

It is an open question though whether axisymmetric or nonaxisymmetric modes are more unstable. Are rings or spirals excited once the cooling of the disk brings it to an unstable state? We also mention the suggestion by Morfill, Spruit & Levy (1993) that the existence of ring-like structures may help to overcome the main difficulty of the standard-accretion disk theory, i.e. the disagreement between the theoretically predicted and observed radial temperature profiles (Horne 1993, Beckwith 1994).

6.7.2   The Short-Wave Approximation

Small disturbances in an infinitesimally thin inviscid disk in nonuniform rotation are considered. We start with the linear relation between the pressure and density perturbations, $P' = c_{ac}^2 \rho'$, valid for isothermal as well as for isentropic disks. The linearized equations for the radial (u') and azimuthal (v') velocity disturbances, and also mass conservation, then become

$$\frac{\partial u'}{\partial t} + \Omega \frac{\partial u'}{\partial \phi} - 2\,\Omega\,v' + \frac{\partial \psi'}{\partial s} + \frac{\partial}{\partial s}\frac{c_{ac}^2 \Sigma'}{\Sigma_0} = 0,$$

$$\frac{\partial v'}{\partial t} + \Omega \frac{\partial v'}{\partial \phi} + \frac{\kappa^2}{2\Omega} u' + \frac{1}{s}\frac{\partial \psi'}{\partial \phi} + \frac{c_{ac}^2}{\Sigma_0 s}\frac{\partial \Sigma'}{\partial \phi} = 0,$$

$$\frac{\partial \Sigma'}{\partial t} + \Omega \frac{\partial \Sigma'}{\partial \phi} + \frac{1}{s}\frac{\partial (u' s \Sigma_0)}{\partial s} + \frac{\Sigma_0}{s}\frac{\partial v'}{\partial \phi} = 0 \qquad (6.23)$$

with κ being the epicyclic frequency $(5.24)_1$, which equals $\sqrt{2}\Omega$ for rotation laws with $s\Omega = $ const. The gravity potential ψ and the density fluctuations are related by the Poisson equation for an infinitely thin disk, $\Delta \psi' = 4\pi G \Sigma' \delta(z)$. The modes with different azimuthal wave numbers m can be considered independently. The Toomre (1964) solution

$$\psi' = -2\pi G \int_0^\infty S(k) J_m(ks) e^{im\phi - |kz|} dk, \quad \Sigma' = \int_0^\infty k S(k) J_m(ks) dk\, e^{im\phi} \qquad (6.24)$$

for the Poisson equation then applies, where $S(k)$ is the density spectrum in the Fourier–Bessel transform. The existence of the general solution of the Poisson equation demonstrates the convenience of the Fourier–Bessel transform. For a delta-like spectral function (only one mode in k) one can deduce from Eq. (6.24) that $\psi' = -2\pi G \Sigma'/k$, with the wave number k in the Fourier ansatz $\exp(i(ks - \omega t))$. The immediate consequence for ring structures is

$$\left(1 - \frac{kc_{ac}^2}{2\pi G \Sigma_0}\right)\frac{d^2 \psi'}{ds^2} + \cdots = -k\frac{\left(\kappa^2 - \omega^2\right)\psi'}{2\pi G \Sigma_0}, \qquad (6.25)$$

resulting in the dispersion relation $\omega^2 = \kappa^2 + c_{ac}^2 k^2 - 2\pi G \Sigma_0 k$ for the oscillation frequency ω. It is here only given in the short-wave approximation, i.e. $ks > 1$. Instability is indicated for

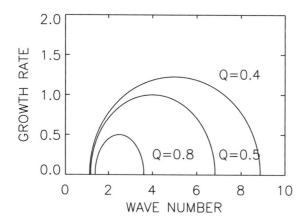

Figure 6.20: Dispersion relation for the Jeans instability in thin disks for various values of the Toomre parameter. For given Q all wave numbers between the intersections of a curve and the x-axis are unstable.

positive $\Im(\omega)$, which happens for $Q = c_{ac}\kappa/\pi G \Sigma_0 \leq 1$ starting at a wave number of $k_{crit} = \pi G \Sigma_0 / c_{ac}^2$. A reasonable value for Σ_0 for the galactic surface density is $\approx 75\ M_\odot/\text{pc}^2$ so that for $c_{ac} \simeq 10$ km/s the unstable wavelength is of order kpc. It is characteristic of the Jeans instability in thin disks that it only exists for wave numbers not too small and not too large (see Fig. 6.20).

From the above system of equations with $\omega_D = \omega - m\Omega$ we find the relations

$$u' = i\frac{\omega_D\left(\frac{d\psi'}{ds} + \frac{c_{ac}^2}{\Sigma_0}\frac{d\Sigma'}{ds}\right) - \frac{2m\Omega}{s}\left(\psi' + \frac{c_{ac}^2}{\Sigma_0}\Sigma'\right)}{\kappa^2 - \omega_D^2},$$

$$v' = \frac{\frac{\kappa^2}{2\Omega}\left(\frac{d\psi'}{ds} + \frac{c_{ac}^2}{\Sigma_0}\frac{d\Sigma'}{ds}\right) - \frac{m\omega_D}{s}\left(\psi' + \frac{c_{ac}^2}{\Sigma_0}\Sigma'\right)}{\kappa^2 - \omega_D^2}, \tag{6.26}$$

so that for $m < ks$

$$\frac{u'}{v'} \simeq \frac{2i\omega_D\Omega}{\kappa^2} \simeq \frac{i\omega_D}{\Omega}, \tag{6.27}$$

and a phase lag of $\pi/2$ exists between u' and v', e.g.

$$u' \propto -\frac{\omega_D}{\Omega}\cos(ks + m\phi - \omega t), \qquad\qquad v' \propto \sin(ks + m\phi - \omega t) \tag{6.28}$$

(Rohlfs 1977, Binney & Tremaine 1987). For trailing spirals the phase relation must ensure $ds/d\phi < 0$, so that $k > 0$ is required. The pattern velocity $\Omega_p = \omega/m$ must then be positive.

6.7.3 Swing Excitation in Magnetic Spirals

Density waves in galaxies form time-dependent mean motions that may also influence the dynamo mechanism. For a dynamo interacting with the galactic density waves Chiba & Tosa (1990) and Hanasz, Lesch & Krause (1991) suggested resonant solutions very close to the resonance phenomenon for the pendulum as described by Landau & Lifschitz (1969).

Dynamo Waves

We start with the dynamo equation applied to the simplest case of Parker's (1971) 1D dynamo in the Cartesian geometry with $\boldsymbol{B} = (B_x(z), B_y(z), 0)$ and $\boldsymbol{u} = (u_x(x), u_y(x), 0)$, where y denotes the azimuthal direction. All the eddy quantities are taken to be spatially uniform. The shear $\partial u_y/\partial x$ describes an extra differential rotation, whereas the shear $\partial u_x/\partial x$ represents a meridional flow in the radial direction. With uniform shears the magnetic field can be written as $\boldsymbol{B} = \hat{\boldsymbol{B}} e^{-ikz}$. With $\hat{B}_x = -\partial A/\partial z$ and $\hat{B}_y = B$ the dynamo equations become

$$\dot{A} + A = CB, \qquad\qquad \dot{B} + B = CA + iDA - EB, \qquad\qquad (6.29)$$

where

$$C = \frac{\alpha}{\eta k}, \qquad\qquad D = \frac{1}{\eta k^2}\frac{\partial u_y}{\partial x}, \qquad\qquad E = \frac{1}{\eta k^2}\frac{\partial u_x}{\partial x}. \qquad (6.30)$$

For a pure $\alpha\Omega$-dynamo and without meridional flow ($E = 0$) the resulting dispersion relation yields marginal modes with oscillation frequency $\omega = 1$ for $CD = 2$. The pure α^2-dynamo ($D = 0, E = 0$) starts at $C = 1$, but does not oscillate. As the shear D in galaxies is negative, also C in this model must be negative in order to ensure $CD = 2$.

We shall study the influence of oscillating shear flows on an oscillating $\alpha\Omega$-dynamo at the marginal state, i.e.,

$$CD = 2 + d\sin\gamma t, \qquad E = e\cos\gamma t, \qquad\qquad (6.31)$$

where d/C and e are the (positive) amplitudes of the shear flows oscillating with frequency γ. Note the phase lag between the oscillations of d and e. The dynamo equation then becomes

$$\ddot{A} + (2 + e\cos\gamma t)\dot{A} + (1 + e\cos\gamma t - i(2 + d\sin\gamma t))A = 0, \qquad (6.32)$$

which differs from the traditional Mathieu equation due to the complex coefficient of A. In contrast to the pendulum equation, the flow-free version of Eq. (6.32) has only one nondecaying solution, i.e. $A = e^{it}$ (Schmitt & Rüdiger 1992). The interaction of this mode with the sin- and cos-terms of the time-dependent shear in Eq. (6.31) leads to the excitation of the neighboring modes, $e^{i(1\pm\gamma)t}$, which themselves excite their outer neighbors, $e^{i(1\pm2\gamma)t}$, and feedback to their common inner neighbor, i.e. to e^{it}. Hence, we have to write

$$A = \sum_{n=0,\pm1,\dots} a_{2n+1}(t)e^{i(1+n\gamma)t}. \qquad\qquad (6.33)$$

The classical resonance frequency is $\gamma = 2$. Insertion of Eq. (6.33) into Eq. (6.32) together with the usual ansatz $a_{2n+1} \sim e^{\sigma t}$ provides an algebraic system of equations whose determinant must vanish to determine the eigenvalue σ. For $\Re(\sigma) > 0$ we have 'swing excitation'.

The solutions for various excitation frequencies γ are given in Fig. 6.21. The general result is that for a given amplitude d the absolute amplitude $|e|$ must exceed characteristic values $|e_{\mathrm{crit}}|$ in order to fulfill the constraints for swing excitation. i.e. $e > e_{\mathrm{crit},1}(d)$ or $e < e_{\mathrm{crit},2}(d)$. These critical values are defined by $\Re(\sigma) = 0$. Resonance is found primarily in the region $de > 0$. The axis $d = 0$ lies completely in the region of resonance. Except for large γ the axis $e = 0$ lies in the region of decay. Hence, the oscillation of the shear e produces resonance, while the oscillation of the shear d disturbs this effect. Whether excitation exists

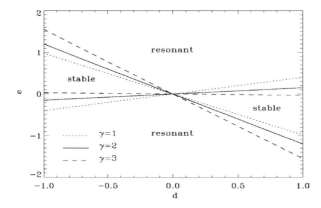

Figure 6.21:
Stability/resonance-diagram for varying shear amplitudes d and e, and three different excitation frequencies γ. (Schmitt & Rüdiger 1992).

depends on the relation between d and e. The classical resonance frequency $\gamma = 2$ does not play any particular role.

According to Eq. (6.31) the azimuthal stream and the radial stream are $90°$ out of phase. We thus find that the product de must have the same sign as C. With C negative (see above), galactic density waves are located in the second or fourth quadrant of Fig. 6.21, so that resonant behavior would require large amplitudes of the meridional shear. These findings are confirmed by Moss (1996) with a 2D code operating in the $s\phi$-space and modulating the spiral flow pattern as given below in Eq. (6.34).

3D Dynamo Model with Density Waves

After Rohde, Rüdiger & Elstner (1999) we now consider a spiral galaxy embedded in a plasma of given conductivity. Let the half-thickness of the galaxy be $H = 1.5$ kpc, and its radius 15 kpc. The differential rotation is described by a Brandt-type law with $q = 1$ and $s_0 = 3$ kpc. The generation of magnetic field in the model is described by the dynamo numbers with (say) $|C_\Omega| = 41.85$ and $C_\alpha = 4.65$. The value of C_α is high enough that nonaxisymmetric field modes are also excited. Their possible swing with the flow pattern drift can therefore be considered, although the axisymmetric field mode is the preferred one.

The galactic spiral arms are given by the expressions

$$u_s = \chi u_0 \cos\left(m_{\mathrm{p}}(\phi - \Omega_{\mathrm{p}}t) + K \frac{s}{R_{\max}} \right) f(s),$$

$$u_\phi = u_0 \sin\left(m_{\mathrm{p}}(\phi - \Omega_{\mathrm{p}}t) + K \frac{s}{R_{\max}} \right) f(s), \tag{6.34}$$

with $m_{\mathrm{p}} = 1, 2$ for a one-armed spiral and a two-armed spiral and χ negative. The frequency Ω_{p} gives the drift velocity of the spiral flow pattern, K its radial wave number[8]. Trailing spirals are obtained with a positive radial wave number K. In a trailing spiral the regions with inward-directed radial velocities are associated with enhanced density in the spiral arms. The velocity u_0 may be of order 10 km/s. The azimuthal flow of the outer (inner) edge of the spiral arm is positive (negative) in ϕ-direction. The computations are done with $\chi = -1$.

---

[8] $f(s)$ is introduced in order to suppress the flow pattern at maximal radius and at $s = 0$

m	C_α^{crit}	$\gamma\,[\mathrm{Gyr}^{-1}]$	$\omega\,[\mathrm{Gyr}^{-1}]$
0	0.2	3.58	–
1	2.5	2.14	13.6
2	3.6	1.24	13.0

Table 6.6: Critical dynamo numbers C_α^{crit}, growth rates γ of magnetic energy, and drift velocities ω_m of the magnetic field modes $m = 0, 1, 2$ in a model with axisymmetric flow pattern (Rohde, Rüdiger & Elstner 1999).

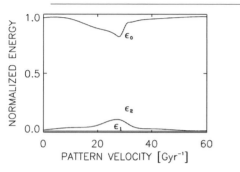
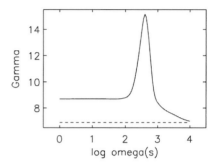

Figure 6.22: Two-armed spiral galaxy. *Left*: Normalized magnetic energies ϵ_0, ϵ_1 and ϵ_2 for magnetic field modes $m = 0, 1, 2$. The contribution of mode $m = 2$ is maximal for $\Omega_\mathrm{p} \approx 2\omega_m$. *Right*: Growth rate of the magnetic $m = 1$ mode for an $m = 2$ azimuthal modulation of the α-effect. The growth rate for axisymmetric α is given by the dashed line (Moss 1997).

In all the simulations a combination of field modes $m = 0$–5 is taken as the seed field. The ratio between the energy E_m of mode m and the total magnetic energy is

$$\epsilon_m = \frac{E_m}{E_{\mathrm{total}}} = \frac{E_m}{E_0 + E_1 + \ldots}. \tag{6.35}$$

The dynamo numbers used are supercritical for several magnetic field modes. The linear analysis gives the growth rates γ of the magnetic field energy. The magnetic field for modes with $m > 0$ exhibits a characteristic drift velocity ω_m in the azimuthal direction. For $m = 1$ and $m = 2$ the drift velocities are almost equal (see Table 6.6). Swing excitation can be simulated by tuning the pattern speed Ω_p of the large-scale velocity field (6.34).

A two-armed nonaxisymmetric velocity pattern with $K = 2\pi$ has been considered. The final solution is no longer a single mode. The solution can be described as a mixture of several modes, but in a two-armed spiral galaxy only the even modes contribute. For the magnetic field mode with $m = 2$ a (weak) resonance behavior is found: In the model with pattern speed $\Omega_\mathrm{p} = 26\,\mathrm{Gyr}^{-1}$ the energy of the magnetic $m = 2$ mode reaches its maximal value (Fig. 6.22, left).

It is quite another story to modulate the α-effect in the azimuthal direction with (say) $m = 2$, i.e.

$$\alpha \propto 1 + \hat{\alpha}\cos(2\phi - \omega_\mathrm{s}t - Ks),$$

which is a *leading* spiral (Moss 1997). Now a clear resonant behavior of the $m = 1$ mode is found (Fig. 6.22, right). There is a clear and broad resonant peak centered at $\omega_\mathrm{s} = 2\omega_0$, where ω_0 is the drift rate of the (here artificually excited) magnetic $m = 1$ mode[9].

---

[9] Bigazzi & Ruzmaikin (2004) even use a nonaxisymmetric α-effect in a solar dynamo to find 'preferred longitudes' for the magnetic fields

6.7.4 Nonlocal Density Wave Theory in Kepler Disks

The velocity components, u' and v' do *not* match transformations like Eq. (6.24). It is appropriate to describe the flow in terms of scalar potentials for the momentum disturbances, i.e.

$$\Sigma_0 u' = \frac{\partial \Phi}{\partial s} + \frac{1}{s}\frac{\partial V}{\partial \phi}, \qquad\qquad \Sigma_0 v' = \frac{1}{s}\frac{\partial \Phi}{\partial \phi} - \frac{\partial V}{\partial s} \tag{6.36}$$

(Rüdiger & Kitchatinov 2000b). Then the potentials Φ and V define the divergence and vorticity of the flow by

$$\nabla \cdot (\Sigma_0 \boldsymbol{u}') = \mathcal{D}_2 \Phi, \qquad\qquad (\nabla \times (\Sigma_0 \boldsymbol{u}'))_z = -\mathcal{D}_2 V, \tag{6.37}$$

where $\boldsymbol{u}'$ is the 2D velocity vector and $\mathcal{D}_2$ is the 2D Laplace operator

$$\mathcal{D}_2 = \frac{1}{s}\frac{\partial}{\partial s}s\frac{\partial}{\partial s} + \frac{1}{s^2}\frac{\partial^2}{\partial \phi^2}. \tag{6.38}$$

With these relations the system (6.23) can be reformulated in terms of the flow potentials. It then reads

$$\frac{\partial\,(\mathcal{D}_2\Phi)}{\partial t} + \Omega\frac{\partial\,(\mathcal{D}_2\Phi)}{\partial \phi} + \left(\frac{\kappa^2}{\Omega} - 2\Omega\right)\mathcal{D}_2 V - \left(\frac{\kappa^2}{\Omega} - 4\Omega\right)\frac{\partial}{\partial s}\left(\frac{\partial V}{\partial s} - \frac{1}{s}\frac{\partial \Phi}{\partial \phi}\right) +$$

$$+ \frac{\mathrm{d}\Sigma_0}{\mathrm{d}s}\frac{\partial \psi'}{\partial s} + \Sigma_0\,(\mathcal{D}_2\psi') + \mathcal{D}_2\left(c_{\mathrm{ac}}^2\Sigma'\right) + \frac{1}{s}\frac{\partial}{\partial s}\left(nc_{\mathrm{ac}}^2\Sigma'\right) = 0,$$

$$\frac{\partial\,(\mathcal{D}_2 V)}{\partial t} + \Omega\frac{\partial\,(\mathcal{D}_2 V)}{\partial \phi} - \frac{\kappa^2}{2\Omega}\mathcal{D}_2\Phi - \left(\frac{\partial \Phi}{\partial s} + \frac{1}{s}\frac{\partial V}{\partial \phi}\right)\frac{\mathrm{d}}{\mathrm{d}s}\frac{\kappa^2}{2\Omega} -$$

$$- \frac{1}{s}\frac{\mathrm{d}\Sigma_0}{\mathrm{d}s}\frac{\partial \psi'}{\partial \phi} + \frac{nc_{\mathrm{ac}}^2}{s^2}\frac{\partial \Sigma'}{\partial \phi} = 0,$$

$$\frac{\partial \Sigma'}{\partial t} + \Omega\frac{\partial \Sigma'}{\partial \phi} + \mathcal{D}_2\Phi = 0, \tag{6.39}$$

where $n(x) = -\mathrm{d}\log\Sigma_0/\mathrm{d}\log s$ is the radial slope of the surface density profile. The density profile used decreases as $1/s$ within the disk and falls more steeply beyond $s = R$ (see Boss 1998 for a detailed discussion). With the Fourier mode ansatz in the azimuthal coordinate ϕ all the perturbations become proportional to $\exp(\mathrm{i}(\omega t + m\phi))$. The basic dimensionless parameters of the normalized equation system are then the Mach number and fractional disk mass f,

$$\mathrm{Ma} = \frac{R\Omega_0}{c_{\mathrm{ac}}}, \qquad\qquad \mathrm{f} = \frac{\pi\Sigma_0 G}{R\Omega_0^2} \simeq \frac{\pi R^2 \Sigma_0}{M_{\mathrm{c}}} \tag{6.40}$$

with the central mass M_{c}. Ma and f can be combined into the global Toomre parameter (6.22), $Q = 1/(\mathrm{fMa})$.

   If the Fourier–Bessel expansions (6.24) are applied to the flow potentials Φ and V, a system of integral equations results that must be solved numerically. This was done for Kepler disks with a sharp edge at $s = R$, where the minimum of the local Toomre number also occurs. The resulting instability diagram for $m \le 2$ is given in Fig. 6.23. One can mimic

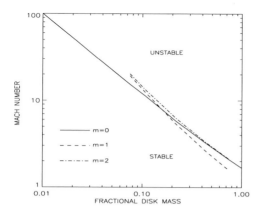

Figure 6.23: The marginal stability lines for rings ($m = 0$) and low-m spirals for various disk masses. The axisymmetry switches at f $\simeq 0.16$. For the massive disks on the right the nonaxisymmetric instabilities appear somewhat before the ring-like ones. (Rüdiger & Kitchatinov 2000b).

the cooling process by an upward movement in the plot. Note that the axisymmetric (ring-like) patterns only dominate up to disk masses of about 16%. For more massive disks the instability produces nonaxisymmetric spirals as displayed in Fig. 6.24. The stability diagram shows the one-armed spirals as the most easily excited nonaxisymmetric structures. This may be a special property of Keplerian disks. They possess a central mass concentration that acts as the second 'arm'.

The equations for global spirals possess a remarkable symmetry. If the functions Φ, V and Σ' constitute a solution of Eqs. (6.39) with eigenvalue $\omega = m\Omega_\mathrm{p} - \mathrm{i}\gamma$, then $-\Phi^*$, V^* and Σ'^* (where the asterisks denote the complex conjugate) is also a solution, with corresponding eigenvalue $\omega^* = m\Omega_\mathrm{p} + \mathrm{i}\gamma$. This means that any stable solution in the form of a trailing

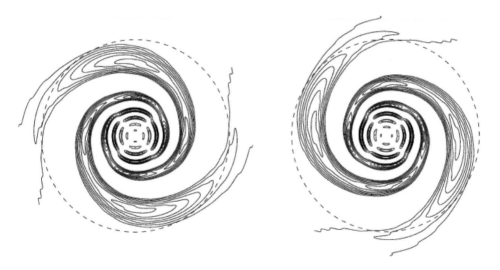

Figure 6.24: Decaying (*left*) and growing (*right*) two-arm spirals for f $= 0.3$. Only the positive density disturbances are shown. The dashed circle is the disk edge $s = R$. The background rotation is counter-clockwise. Note the high number of local density maxima, which also occur in the SPH code simulations given in Fig. 6.25.

spiral has a leading counterpart with exactly the same pattern speed Ω_p. This result is known as the antispiral theorem (Lynden-Bell & Ostriker 1967) for tightly-wound spiral waves. The equivalence of leading and trailing spirals does not hold, however, for unstable disturbances. If a leading spiral grows exponentially then a trailing one decays, and vice versa. The theorem does not show, however, whether the leading or the trailing patterns are growing. The numerical solutions always show excitation of trailing spirals similar to that shown in Fig. 6.24 (right), while the leading ones decay or have zero growth rates.

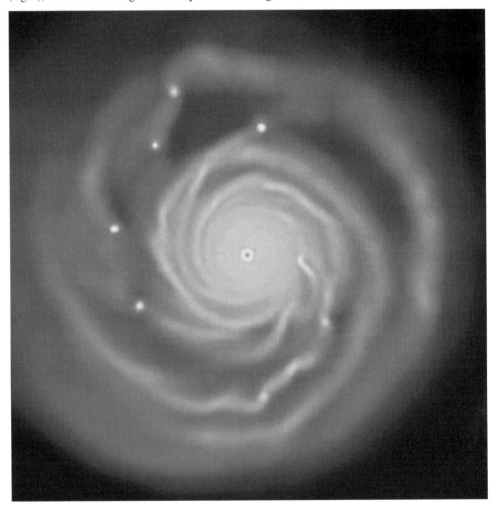

Figure 6.25: Planet formation in a disk of 20 AU, $T \simeq 50$ K, and a mass ratio f $\simeq$ 0.1 (Mayer et al. 2003). For the minimum solar nebula the mass ratio f is 0.01. Courtesy T. Quinn.

The next natural steps are the inclusion of the vertical stratification and the magnetic fields. Nakamura & Hanawa (1997) considered the gravitational instability in a very thin disk threaded by a vertical magnetic field that also in the disk has only a (nonhomogeneous) B_z-component. It is known that the Lorentz force

$$\boldsymbol{J} \times \boldsymbol{B} = -\frac{1}{2\mu_0}\nabla B^2 + \frac{1}{\mu_0}(\boldsymbol{B}\nabla)\boldsymbol{B} \tag{6.41}$$

for such a field is generally not zero, but it is a gradient since $(\boldsymbol{B}\nabla)\boldsymbol{B} = 0$ (Krause & Rüdiger 1975, Boss 2001). No MRI can thus develop, which is also true for the current-free fields that Nakamura & Hanawa postulate outside the disk. They consider the nonlinear evolution of a nonaxisymmetric finite isothermal perturbation during the protostellar collapse and find an instability (only) for $m = 2$ (their Fig. 2). Without rotation the instability produces bars during the protostellar collapse, while with rotation an elongated S-shape appears. The bars themselves are unstable and finally fragment (Burkert & Bodenheimer 1993).

6.8 Mean-Field Dynamos with Strong Halo Turbulence

Rohde & Elstner (1998) used models with strong halo ('high-z') turbulence to construct non-linear dynamos including the vertical magnetohydrostatic equilibrium of Eq. (6.10) in order to derive the turbulence parameters. All the coefficient functions in the turbulent EMF are consistently computed from the same turbulence model, which includes both density stratification and turbulence intensity stratification.

In the equation of the vertical equilibrium there is also the turbulent pressure, which itself is influenced by the induced magnetic fields. The back-reaction of the generated magnetic field on the turbulence pressure is involved in a general theory of the magnetic influence on the Reynolds stress (Moffatt 1966, Rüdiger 1974, Rädler 1974, Roberts & Soward 1975). For isotropic turbulence it can be summarized in the general formulation

$$Q_{ij} = \langle \boldsymbol{u}_0^2 \rangle \left(\psi(\beta)\delta_{ij} + \psi_1(\beta)\frac{\bar{B}_i \bar{B}_j}{\bar{B}^2} \right), \tag{6.42}$$

with

$$\psi = \frac{1}{8\beta^2}\left(\frac{\beta^2-1}{\beta^2+1} + \frac{\beta^2+1}{\beta}\tan^{-1}\beta \right), \quad \psi_1 = \frac{1}{8\beta^2}\left(\frac{\beta^2+3}{\beta^2+1} + \frac{\beta^2-3}{\beta}\tan^{-1}\beta \right).$$

For the vertical intensity one has $\langle u_z'^2 \rangle = \langle \boldsymbol{u}_0^2 \rangle(\psi + \psi_1 \bar{B}_z^2/\bar{B}^2)$. This effect must be included in the simulations when the vertical stratification alone may yield the turbulence-induced magnetic feedback. This is done in the nonlinear axisymmetric dynamo model of Rüdiger & Schultz (1997).

For strong fields the functions ψ and ψ_1 vary as $1/\beta$, so that for a vertically imposed magnetic field

$$\langle u_\parallel'^2 \rangle \simeq 2\langle u_\perp'^2 \rangle \simeq \frac{u_{\rm T}^2}{8\beta} \tag{6.43}$$

results. The turbulent velocity parallel to the magnetic field then exceeds the perpendicular component by a factor of 2. The magnetic quenching of all the turbulence intensities scales as $1/\beta$. In Fig. 10 of Ossendrijver, Stix & Brandenburg (2001) the quenching of the turbulence intensities for a vertical magnetic field with $\beta = 1$ is indeed of this order. With the original quenching expressions in Rüdiger (1974) one can show that in the high-conductivity limit no such catastrophic quenching exists for the turbulence intensity as it might exist for the α-effect and the magnetic eddy diffusivity (see Sect. 4.3.1).

6.8.1 Nonlinear 2D Dynamo Model with Magnetic Supported Vertical Stratification

If only the density stratification produces the α-tensor, it takes the form

$$\alpha_{ij} = -\alpha_{ij}^{(\rho)} \frac{\mathrm{d} \log \rho}{\mathrm{d}z} u_\mathrm{T}^2 \tau_\mathrm{corr}, \tag{6.44}$$

with

$$\alpha_{ij}^{(\rho)} = c_\alpha \begin{pmatrix} \frac{2}{5}\Omega^*\tilde{\psi} & -U^{\mathrm{buo}} & 0 \\ U^{\mathrm{buo}} & \frac{2}{5}\Omega^*\tilde{\psi} & 0 \\ 0 & 0 & -\frac{4}{5}\Omega^*\tilde{\psi}_z, \end{pmatrix}, \tag{6.45}$$

where $\tilde{\psi} = \left(\Psi + (15/8)\Psi_1 \bar{B}_z^2/\bar{B}^2\right)$ and $\tilde{\psi}_z = \left(\Psi_z - (15/16)\Psi_1 \bar{B}_z^2/\bar{B}^2\right)$. The α-quenching functions as well as the magnetic-advection velocity are taken from Eqs. (4.25) and (4.42), with $\Psi_z = -\Psi/2 + 3\Psi_2/4 + 3\Psi_3/4$.

The turnover radius s_0 is fixed to 2 kpc. The turbulence is assumed to be homogeneous in space, so $\langle u_0^2 \rangle = \mathrm{const.}$ and $\tau_\mathrm{corr} = \mathrm{const.}$, so only these two free parameters remain in the model. All other terms in the turbulent EMF are computed, the density profile included.

No Turbulence-Quenching

Dobler, Poezd & Shukurov (1996) formulated the question whether the 2D disk-dynamo model also saturates without any turbulence-quenching mechanism. The idea is to check whether the system is able to adjust itself without any magnetic influence upon the microscale. We find that the magnetic field flattens the vertical density profile so that the α-effect is reduced, and indeed a balance finally arises. The main question concerns the magnitude of the induced magnetic field. The equipartition value no longer occurs in the equations. The results of the calculations are given in Table 6.7. Only quadrupolar fields are excited. Stable solutions were not found for all parameters. For high turbulence intensities the magnetic fields are oscillatory.

Table 6.7: Characteristic magnetic fields for dynamo models with density profiles derived from the vertical stratification. Turbulence-quenching is completely ignored. $c_\alpha = 0.1$.

τ_corr [Myr]	$u_\mathrm{T} = 10$ km/s	$u_\mathrm{T} = 20$ km/s
10	27 μG	17 μG
20	unstable	15 μG

The fields are stronger (by a factor of 5) if α-quenching is absent. Hence, a system without microquenching appears to provide magnetic fields that are too strong. Nevertheless, by adjusting its structure the system is able to limit the magnetic field growth. The resulting density profiles, however, are extremely flat and far from realistic (Fig. 6.26). For strong fields it is thus not reasonable to ignore the magnetic influence upon the structure and evolution of the considered object.

Full Magnetic Feedback

The *complete* system is highly nonlinear. The magnetic influence in Eq. (6.10) is now two-fold: magnetic pressure and magnetic quenching of the turbulence pressure. The only fixed quantity remains the eddy diffusivity.

The results are presented in Table 6.8. All the various models produce nearly the same amplitude of the magnetic field of $\lesssim 10\ \mu G$. Figure 6.26 shows that the galactic disk becomes *flatter* as a result of the dynamo action. The models prove to be rather insensitive to variation of the free parameter c_α. Increasing this number from 0.1 to 1 increases the amplitude of the field only by a factor of 2 or 3. For $c_\alpha = 1$ the computed magnetic amplitudes agree with the observations (Table 6.8). The outer domain with $s > s_0$ is always in an almost stationary regime. In order to find the influence of the vertical stratification on the dynamo model it is necessary to compare the numbers of Table 6.8 with numbers for a model with pure α-quenching (same table, last two columns). They are quite characteristic of traditional dynamo models. The strength of the α-effect compared with the differential rotation grows both with the turbulence intensity and the correlation time. For correlation times of 20 Myr or more there are always oscillating parts in the magnetic field distributions. The equatorial parity remains quadrupolar.

Table 6.8: Magnetic fields in μG for dynamo models with extra magnetic influence on the turbulence pressure. In the last two columns values are given for a reference model with local α-quenching as the only nonlinearity. $c_\alpha = 1$, $\sim$ indicates oscillating modes.

τ_{corr} [Myr]	$u_T = 10$ km/s	$u_T = 20$ km/s	$u_T = 10$ km/s	$u_T = 20$ km/s
10	7.1	9.4	13	17
20	7.0($\sim$)	8.8($\sim$)	13	16($\sim$)

The results of the simulations are rather clear. The *maximal* amplitudes of the magnetic fields are, with $\leq 10\ \mu G$, close to the observations. The fields are stationary for short correlation times (≤ 10 Myr), and they have a time-dependent character for longer correlation times. It is also no problem to produce the observed magnetic amplitudes in timescales < 2 Gyr with only a few basic assumptions.

6.8.2 Nonlinear 3D Dynamo Models for Spiral Galaxies

Baryshnikova et al. (1987), Mestel & Subramanian (1991), Panesar & Nelson (1992), Moss (1996) and Hanasz & Lesch (1997) considered the question whether the dynamo theory for galaxies can also provide nonaxisymmetric magnetic modes (BSS). Baryshnikova et al. worked with an axisymmetric flow field, while Mestel & Subramanian took the α-effect, but not the eddy diffusivity, to depend on azimuth. In the first case the answer is clear: With an

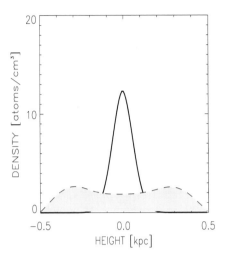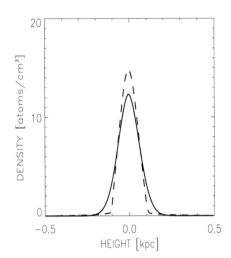

Figure 6.26: The vertical density stratification from Eq. (6.10) without (solid) and with (dashed) magnetic field (gray). Turbulence intensity 10 km/s, correlation time 10 Myr. *Left*: No magnetic influence on the turbulence. *Right*: Full magnetic feedback.

isotropic α-effect the axisymmetric magnetic field is always the mode with the lowest dynamo number. With anisotropic α-effect one can find S1 modes preferred, but only if the rigidly rotating inner core of the galaxy is unrealistically large, or the disk is very thin (Rüdiger, Elstner & Schultz 1993, their Fig. 3).

Studies of the consequences of azimuth-dependent turbulent EMF are much more challenging, as the various m-modes are then coupling. Two sorts of solutions exist, i.e. a series of the even m-modes and a series of the odd m-modes. In the model of Rohde & Elstner (1998) the radius of the galaxy $R_{\mathrm{max}} = 7.5$ kpc, the rigidly rotating core extends out to 2 kpc, and $V = 100$ km/s. In order to take into account the influence of the spiral arms in density and diffusivity the spiral profile in Eq. (6.2) is used for the density as well as for the turbulence intensity. Ω_{p} is the angular pattern speed of the $m = 2$ spiral, and is set to 13 Gyr^{-1} in all calculations. The pitch angle of the gaseous spirals is taken as $25°$.

With the given density stratification we take Eq. (6.10) to calculate the turbulence intensity. The midplane turbulent velocity is 10 km/s. The resulting vertical profile of the turbulent velocity is characterized by a saturation of the high-z velocities of about 40 km/s. The correlation time is now considered as a free parameter, which is shown to control the dynamo regime. A mixed seed field is used, having $P = 0.5$ (see Eq. (4.101)) and $\epsilon_0 = 0.5$ (see Eq. (6.35)).

The dynamo models with small correlation times ($\tau_{\mathrm{corr}} = 30$ Myr) lead to steady S0 solutions (Fig. 6.27). The growth time of the magnetic instability is about 1.5 to 2 Gyr. The magnetic pitch angle varies from $10°$ between the spiral arms to $30°$ within them. The field strength shows a strong concentration between the arms, in agreement with Hanasz & Lesch (1997), but in contrast to the results of Panesar & Nelson (1992), where the field lines concentrate within the spiral arms. In the latter model only the density is assumed to be nonax-

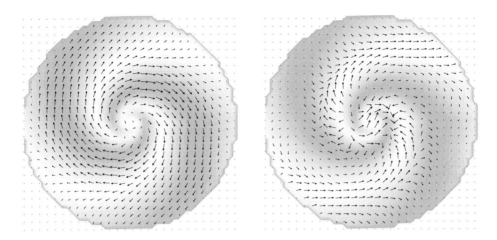

Figure 6.27: Magnetic field geometry for models with nonaxisymmetric α-effect. *Left*: $\tau_{\mathrm{corr}} = 30$ Myr (after 3.3 Gyr). *Right*: $\tau_{\mathrm{corr}} = 100$ Myr (after 2.0 Gyr). The optical spiral arms are shown in light gray.

isymmetric, which by the nonlinear quenching automatically leads to nonaxisymmetry of the α-effect. The diffusivity dissipates the field more intensively in the spiral arms. For larger correlation times (50 Myr, not shown) the solutions are much more complicated: The magnetic field in the inner part of the galaxy now has a nonaxisymmetric structure (the α-effect is anisotropic), drifting with a period of about 4 Gyr. In the outer parts a steady S0 field is excited. The magnetic pitch angle varies between $60°$ in the spirals and $20°$ between them.

For very large correlation time (100 Myr) the type of dynamo changes completely. This model leads to an S1 dynamo solution, where the magnetic field is clearly concentrated within the spiral arms (Fig. 6.27). The solution is oscillatory, and the magnetic pitch angle thus varies in time. The growth time in this model is extremely short. This very striking behavior is due to the strong α-effect, which works mostly in the spiral arms where the turbulent velocity is assumed to be large.

The dynamo model shows a great variety of solutions, even though it depends on only one free parameter. For short correlation times axisymmetric and steady solutions are found, of even parity ('quadrupoles') and concentrating the magnetic field between the spiral arms. For longer correlation times the field becomes nonaxisymmetric and oscillatory, still with even parity, but with the fields concentrated within the spiral arms. In both cases the pitch angles are so large that they could never be produced by a simple $\alpha\Omega$-dynamo.

6.9 New Simulations: Macroscale and Microscale

Dynamo theory is the search for self-excitation of magnetic fields for a given flow system, which often in astrophysics is a combination of large-scale and small-scale flow fields. For spiral galaxies the global flow system is controlled by the density wave theory. In order to understand the large-scale magnetic features of galactic fields one needs a detailed knowledge of the global flow system. The use of the analytical density wave theory that originally was developed for stellar disks only, has obvious strong defects. As it is a linear theory, the amplitude of the velocity field remains undetermined. On the other hand, all information about the vertical structures and vertical shears is lost; within the short-wave approximation the same holds for the azimuthal structures. A promising step toward better understanding is thus to use nonlinear, high-resolution numerical simulations. There are two possibilities. In the first the hydrodynamic flow $\bar{u}$ is derived with N-body simulations, and a code to solve the dynamo equation is then added. However, the resolution is not sufficient to obtain the small-scale motions, so that the turbulent EMF in the induction equation must still be parameterized in the traditional way, Eq. (4.7). The other possibility consists in the numerical simulation of the microscale, i.e. the theory of the interstellar matter under the influence of the basic rotation. Here we present the results regarding the α-effect in turbulence driven by SN explosions or networked explosions of many SN, also called a superbubble (SB, Fig. 6.11). The consequent question whether an ensemble of SN explosions is able to drive a dynamo is still open (see Korpi et al. 1999b).

6.9.1   Particle-Hydrodynamics for the Macroscale

Moss, Rautiainen & Salo (1999) and Elstner et al. (2000) simulated the large-scale flows of the interstellar gas with N-body codes. The galactic disk consists of collisionless stars and gaseous clouds moving in the gravitational potential of the stellar population. The whole system is embedded in an extra potential of the bulge and the dark matter halo (Otmianowska-Mazur et al. 1997). Collisions between clouds are highly inelastic and are accompanied by energy dissipation ('sticky particles'). The large-scale motions of the molecular gas are rotation, bar shocks and density-wave flows. The nonlinear feedback of the magnetic field upon the gas, however, is assumed to damp only the α-effect and the turbulent diffusion. The global galactic flow pattern itself remains unaffected by any magnetic feedback. The input parameters are those of Table 6.9. A total of 38,000 stellar particles and 19,000 gaseous clouds were used. The dynamo equation (4.5) is numerically solved with the flow field $\bar{u}$ from the particle code. From the α-tensor only the diagonal components are used. The distribution of the α-coefficient in z is approximated by a sine-profile. The α-components α_{ss} and $\alpha_{\phi\phi}$ are taken as 10 km/s and $\alpha_{zz} = -20$ km/s. The α-effect is magnetically quenched as in Eq. (4.44).

Table 6.9: Input parameters for the spiral galaxy model of Elstner et al. (2000).

	dark-halo	disk	gas	bulge
mass [$10^{10}\ M_\odot$]	9.6	6.0	0.8	6.0
scale length [kpc]	15.0	6.0	12.0	1.6

The magnetic diffusivity is assumed to be due to turbulent motions in the interstellar gas. If cloud collisions dominate, the turbulent diffusion is determined by the velocity dispersion u_T of the gas clouds, of order 10 km/s. With the standard cloud lifetime of 10 Myr this leads to a turbulent diffusion coefficient of about $5 \cdot 10^{26}$ cm^2/s. The enhancement of the small-scale turbulent motions in spiral arms is included by scaling the turbulent diffusivity with the density of the gas, which can be directly obtained from the simulation, and which leads to a

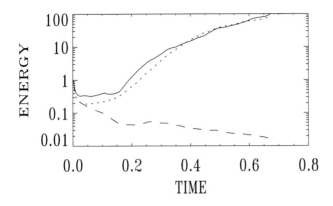

Figure 6.28: Time evolution (in Gyr) of the magnetic energy for the density-wave model, the reference model without spirals (dotted) and an artificial density-wave model without α-effect (dashed).

factor of 800 (!) for the peak-to-peak variation. In this way the model simulates the effects of density waves and the enhanced turbulent diffusion in spiral arms.

Figure 6.28 illustrates the time evolution of the total magnetic energy normalized to its initial value, compared with a reference model (without spirals) and a simulation without any α-effect. Without α-effect the magnetic energy decreases during the whole simulation time of several galactic rotations. The complete 3D large-scale flow system without α-effect does not provide magnetic field amplification! With α-effect, however, field amplification exists with or without density waves. The density waves thus have relatively little influence on the dynamo action. Note that the amplification of the magnetic field after several rotation periods (~ 1 Gyr) only amounts to a factor of 10.

In Fig. 6.29 (top) the magnetic field vectors and the gas density in the galactic plane after $t = 0.4$ Gyr are shown. The initial field was axisymmetric. The reference model without spirals represents the well-known S0 solution with a small pitch angle. If, on the other hand, the dissipation varies with the azimuth-dependent density, the magnetic field becomes enhanced *between* the optical arms. An increased turbulent diffusion in the star-forming spiral arms apparently advects the magnetic field into the interarm region. This is the main result of the calculations.

The magnetic field configuration one galactic rotation later is also shown (Fig. 6.29, bottom). With the density wave magnetic spiral arms again develop, but now within the gas arms. The magnetic arms become even longer than for $t = 0.4$ Gyr, extending to the outer disk where the gaseous arms are no longer visible. The high arm/interarm contrast of the field amplitude in the model nearly disappears. In all the models the magnetic arms drift into the interarm areas. The drift is slow though. There is not yet any indication of resonance between the magnetic drift and the density wave. Due to the radial outflow in the arms the pitch angle remains large, independent of the value of the eddy diffusivity (Otmianowska-Mazur et al. 2002).

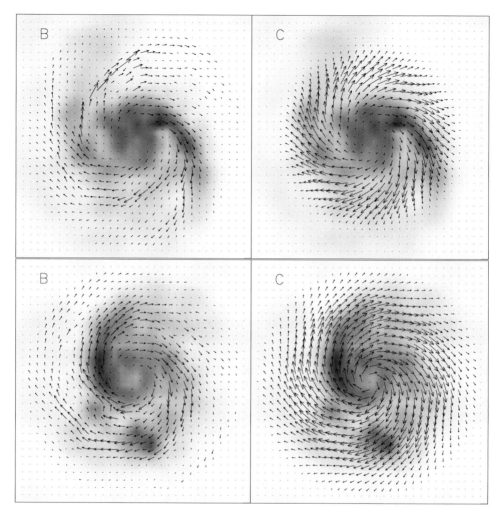

Figure 6.29: *Left*: Magnetic field vectors in the galactic plane overlaid upon the grayplot of the gas density after 0.4 Gyr (*top*) and after 0.65 Gyr (*bottom*). *Right*: The same for the reference model without spirals. From Elstner et al. (2000).

6.9.2   MHD for the Microscale

Ziegler (1996) numerically simulated SN explosions in the galactic disk (Fig. 6.30) and also calculated all components of the α-tensor for a random ensemble of explosions. Only isolated explosions under the influence of the galactic differential rotation have been simulated. From the time-dependent 3D MHD simulations the turbulent EMF is calculated and then averaged over an empirical SN distribution function to derive the α-tensor for a whole sample of explosions.

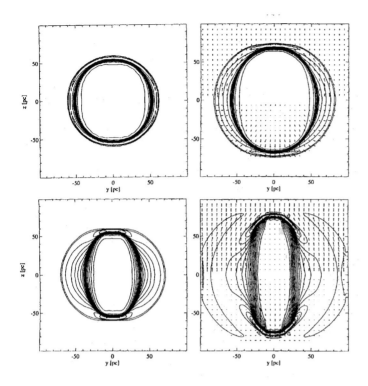

Figure 6.30: SN explosions under the influence of vertical magnetic fields at two different timesteps. *Top*: Weak magnetic field (2.5 μG). *Bottom*: Strong magnetic field (10 μG). From Ziegler (1996).

An empirical SN rate per unit volume, $\Phi(x)$, is introduced (see Kaisig, Rüdiger & Yorke 1993), implying that the evolution of every remnant is independent of the others. The spatial distribution of SN shows a strong dependence on z, but varies only slowly with radius. Φ is thus assumed to be a function of z alone. The total EMF is then

$$\mathcal{E}(z,t) = \int\limits_{0}^{\tau_{\mathrm{corr}}} \int\limits_{-z_0}^{z_0} \int\limits_{A'} \Phi(z'-z)\mathcal{E}^*(x',t')\,\mathrm{d}z'\,\mathrm{d}t'\,\mathrm{d}A', \tag{6.46}$$

where the third integral is to be taken over areas given by $z' = \mathrm{const}$. That is, $\mathcal{E}$ is simply the convolution of the EMF of one basic explosion and the distribution function of the SN. Since $\mathcal{E}^*$ is known from the simulations, the integrals can be evaluated. In all runs the interstellar medium is regarded as isothermal, with a uniform density of 10^{-24} g/cm^3 and a temperature of 5000 K, and the explosive energy release is 10^{51} erg. The uniform initial magnetic field is parallel to the coordinate axes. Figure 6.31 shows the results. The α-effect is weak and highly anisotropic. The results of only a few meters per second generally confirm the early analytical findings of Ferrière (1992a,b). The exact values of the amplitudes depend strongly on the scale height of the SN distribution. They vanish for very large scale heights. In the upper disk layer the $\alpha_{\phi\phi}$ is positive and the α_{zz} is negative. The off-diagonal component $\alpha_{s\phi}$ exceeds the

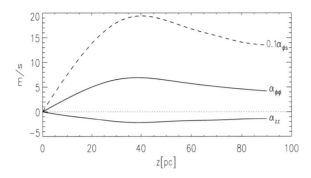

Figure 6.31: The α-effect on the upper disk plane for an ensemble of isolated SN explosions which are vertically distributed with a scale height of 90 pc: $\alpha_{\phi\phi}$ (5 m/s), α_{zz} (-2 m/s), the advection velocity goes upwards and reaches 200 m/s (Ziegler 1996).

diagonal elements by two orders of magnitude (Fig. 6.31). However, the pitch angle statistics requires a strong α-effect, and such a large escape velocity suppresses the dynamo action.

Ferrière (1996) suggested consideration of ensembles of 100–1000 SN explosions in clusters of OB stars[10]. The filling factor of the forcing is then reduced to values of (say) 0.2–0.4, but the correlation time is increased by a factor of 10 (see Eq. (6.8) for the consequences).

Korpi (1999) provides a detailed discussion of the filling factors of the warm (10^4 K) and hot (10^6 K) component of the ISM. However, the main consequence for the α-effect computation is the increase of the correlation time to 30 Myr. This important aspect is confirmed by the simulations of the interstellar turbulence with a local 3D MHD code including shear, SN heating and radiative cooling. The dynamics of the SN-regulated ISM is used to find the α-effect of the interstellar turbulence. The basic difference from former calculations is that the SN-created superstructures can interact in the model. A greatly enhanced α-effect of several km/s is the consequence (Fig. 6.32). The nonlocal character of the SN-driven turbulence seems indeed to resolve the contradiction between the strong α-effect in Sect. 6.3.2 and the weak α-effect that results from isolated SN-explosions.

Figure 6.30 shows two explosions under the influence of magnetic fields up to 10 μG. The shape of the explosions is strongly influenced by the magnetic field, and, as the field is frozen into the flow pattern, the components of the magnetic field are also strongly modified. The flows perpendicular to the magnetic field are damped.

MHD simulations of SN explosions also allow one to study the quenching of α. Ziegler (1996) derived the quenched coefficients $\alpha_{\phi\phi}$, α_{zz} and $\alpha_{\phi s}$ from the magnetically influenced flow pattern, and the magnetic fields of a single SN explosion and a vertically stratified ensemble of SN explosions. The stronger the field the smaller the α-effect. The $\alpha_{\phi\phi}$-component is much more strongly quenched in comparison to $\alpha_{\phi s}$. For strong fields $\alpha_{\phi\phi}$ scales as $\bar{B}^{-3}$, while the quenching of the advection velocity only scales as $\bar{B}^{-2}$. For both cases the best fit resulted for $B_{\mathrm{eq}} \simeq 4.5$ μG. While the dependence $\bar{B}^{-3}$ is known from quasilinear (SOCA) calculations (Moffatt 1972, Rüdiger 1974, Gilbert & Sulem 1990), the weak feedback of the magnetic field onto the advection velocity $\alpha_{\phi s}$ is a surprise. It leads to the puzzling situation that under the influence of a mean magnetic field of order of the equipartition value the diamagnetic advection term basically dominates the α-effect, so that the dynamo may easily

---

[10] Korpi et al. (1999) mainly consider $N = 50$

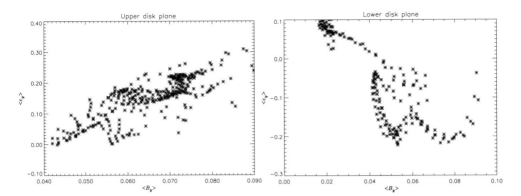

Figure 6.32: The $\alpha_{\phi\phi}$ for an ensemble of *coherent* SN explosions. *Left*: upper disk plane. *Right*: lower disk plane. The α-effect is the slope of the averaged lines. It is positive (negative) on the upper (lower) disk plane. Courtesy M. Korpi.

disappear. Perhaps here as well though the situation changes if more coherent structures are taken into account.

6.10 MRI in Galaxies

The question arises whether the MRI also exists in galaxies, with their differential rotation given by Eq. (5.68) and their aspect ratio $s_0/H \simeq 5$. See in Fig. 5.10 (left) the line with $q = 1$. The maximal amplitude of the vertical magnetic field is given on the right-hand branch in Fig. 5.10 (left), and according to Eq. (5.78) yields $B \simeq \sqrt{\mu_0 \rho}\, c_{\mathrm{ac}}$, with $c_{\mathrm{ac}} \simeq \Omega H$. With the characteristic values $\Omega = 10^{-15}$ s^{-1} and $\rho = 10^{-24}$ g/cm^3 one finds

$$B \lesssim 1\,\mu\mathrm{G}\,\frac{H}{100\ \mathrm{pc}}. \tag{6.47}$$

Sellwood & Balbus (1999) concluded from this that up to amplitudes of a few μG galaxies might be MRI-unstable. If so, the existence of interstellar turbulence can be explained without any stellar activity or SN explosions – in other words, stellar winds and SN explosions merely perturb the MRI. This idea agrees with the observations of the HI disk in NGC 1058, with its uniform distribution of interstellar turbulence without any star formation activity. In NGC 4414 (Fig. 6.1) one also finds ordered, large-scale magnetic fields without strong star formation activity.

A linear, global MRI model may clarify the situation. This is presented in Sect. 5.5.2 as the disk model of Kitchatinov & Mazur (1997) for $q = 1$, but now the magnetic Prandtl number must be much larger than unity due to the high values of the microscopic viscosity in galaxies ($\nu \sim 10^{18}$ cm^2/s, $\eta \sim 10^7$ cm^2/s, see Kulsrud & Anderson 1992). The neutral stability lines for Pm $\lesssim 10^4$ are given in Fig. 6.33 (left). As the real value of Pm for galaxies is much higher still, the results of the computations can only be used as the basis for a scaling procedure. From the computations Kitchatinov & Rüdiger (2004) find that

- the strong-field branch leads to $\Omega_0 H/V_{\mathrm{A}} \simeq 2.4$,

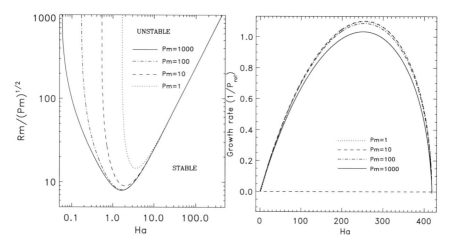

Figure 6.33: *Left*: Stability map for a galactic rotation law under the influence of a vertical magnetic field with the Hartman number Ha and for high magnetic Prandtl numbers. *Right*: The growth rates in units of the rotation time of the rigidly rotating core. (Kitchatinov & Rüdiger 2004).

- for large Pm the minima in Fig. 6.33 (left) converge to $\Omega_0 H^2 / \sqrt{\nu\eta} \simeq 8$ (galactic rotation is fast enough to fulfill this necessary condition),
- the minima for quadrupolar parity are smaller than for dipolar parity.

For the *minimum* magnetic field for which MRI is possible at all, Fig. 6.33 (left) leads to the condition

$$\mathrm{Ha}^* = \sqrt{\mathrm{Pm}}\,\mathrm{Ha} = 1.7 \tag{6.48}$$

for the Lundquist number $\mathrm{Ha}^* = V_A H/\eta$, also for large Pm. According to these results, and with the given microscopic viscosities, the minimum magnetic field for which the MRI starts is

$$B_{\min} \simeq 10^{-25}\mathrm{G}. \tag{6.49}$$

Sigl, Olinto & Jedamzik (1997) estimate the primordial magnetic field as up to 10^{-20}G on a (comoving) scale of 10 Mpc. Much higher magnetic amplitudes for cosmological fields are reported by Widrow (2002). Banerjee & Jedamzik (2003) present numerical models for the magnetic field evolution from the cosmic past to the present. They favor the surprisingly large cluster fields as of primordial origin. So far we do not finally know whether large-scale cosmological magnetic fields existed.

In Fig. 6.33 (right) the growth rates for $\mathrm{Rm} = 1000\sqrt{\mathrm{Pm}}$ are given in units of the rotation time. Between the left- and right-hand branches the growth time always, for all magnetic Prandtl numbers, equals the rotation time. For a rotation time of 70 Myr the amplification factor after 1 Gyr is thus 10^6. That is, in 1 Gyr the MRI can amplify a seed field of 10^{-12} G to 1 μG. Note though that if $\tau_{\mathrm{rot}} \simeq 90$ Myr then after (say) 500 Myr the amplification factor is only 100.

Nonlinear simulations of Dziourkevitch, Elstner & Rüdiger (2004) indeed indicate the occurrence of flow and field fluctuations due to MRI for runs with uniform density. The values

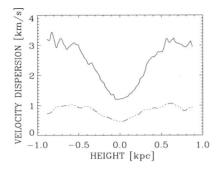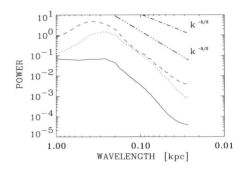

Figure 6.34: *Left*: The turbulent intensity u_{T} has a minimum at the disk midplane. The same holds for u_z (dashed). *Right*: Fourier spectrum of the magnetic field components (B_z solid, B_ϕ dashed, B_s dotted).

are given in Table 6.10. The initial field varies by a factor of 4. The typical wavelength λ of the disturbance and the resulting turbulence intensity u_{T} also grow with the magnetic field, but not by the same factor. It appears to be no problem to produce the observed values – also for the turbulence intensity of a few km/s. For too high initial vertical field (e.g. $B_z > 7\ \mu\mathrm{G}$) the disk proves to be stable.

$B_z\ [\mu\mathrm{G}]$	$\lambda\ [\mathrm{pc}]$	$u_{\mathrm{T}}\ [\mathrm{km/s}]$	$B_{\mathrm{T}}\ [\mu\mathrm{G}]$
0.11	130	2.3	1.0
0.44	400	17	4.3

Table 6.10: Two models with uniform density ($\rho = 5 \cdot 10^{-25}\ \mathrm{g/cm^3}$) and identical rotation laws.

The Fourier spectrum of the magnetic fluctuations is given in Fig. 6.34, which leads to the critical wavelengths in Table 6.10. The slope of the spectrum is much steeper than a Kolmogorov spectrum.

The model is then improved with the density stratification of Dickey & Lockman (1990) with a higher equatorial density of $10^{-24}\ \mathrm{g/cm^3}$. The initial vertical field for all models is $0.32\ \mu\mathrm{G}$. The magnetic energy is found to grow with the timescale τ. The results of two model runs are given in Table 6.11. The turbulence intensity u_{T} is of order 4 km/s, and the characteristic amplitude of the magnetic fluctuations is $1\ \mu\mathrm{G}$. The instability is very fast; its growth time (for magnetic energy) is only 20–40 Myr. The last two columns of Table 6.11 reveal the instability to be magnetically dominated. The magnetic energy exceeds the kinetic energy by a factor of 4, and the Maxwell stress exceeds the Reynolds stress by a factor of 50–100. The angular momentum transport is thus dominated by the magnetic-field fluctuations.

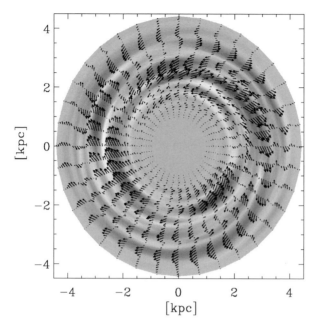

Figure 6.35: Density fluctuations (maximum dark, minimum white) and magnetic field vectors. Averaged over the entire disk the pitch angle is about $30°$. The turnover to differential rotation is at 2 kpc. Courtesy N. Dziourkevitch.

Table 6.11: The same as in Table 6.10, but for density-stratified models with different rotation laws. The last column gives the ratio of Maxwell to Reynolds stresses $B_0 = 0.32\ \mu$G, see Dziourkevitch, Elstner & Rüdiger (2004).

s_0 [kpc]	V [km/s]	u_T [km/s]	B_T [μG]	τ [Myr]	E_{mag}/E_{kin}	MS/RS
2.3	172	5.8	1.4	43	3.8	120
2.0	200	7.8	2.0	30	3.5	452

Figure 6.35 shows the density fluctuations and the magnetic field vectors. In the ring-like cells the Maxwell stress $B'_s \cdot B'_\phi$ is always negative, so that the angular momentum is transported outwards. The averaged pitch angle is about $-30°$.

7 Neutron Star Magnetism

7.1 Introduction

When a sufficiently massive star explodes in a SN at the end of its lifetime, the outermost layers are blown off into space (where the turbulence they produce in the interstellar medium may play a role in generating the galactic magnetic field, as discussed in Chapt. 6). In contrast, the core collapses to form a neutron star, in which some 1.3–1.4 solar masses are compacted into an object no more than 20 km across (Shapiro & Teukolsky 1983). Neutron stars are thus the densest objects known (excluding of course black holes, for which density is not properly defined at all).

Besides having the greatest densities, neutron stars also have the strongest magnetic fields, with field strengths of around 10^{8-9} G for millisecond pulsars, 10^{11-13} G for 'classical' radio and X-ray pulsars, and perhaps as much as 10^{14-15} G for the so-called magnetars. See for example Lyne (2000), Thompson (2000) or Reisenegger (2001) for recent reviews of some of the observational data.

Another interesting feature of these objects is the correlation between field strength and age, with magnetars and classical pulsars relatively young, 10^{3-7} yr old, but millisecond pulsars much older, 10^{8-10} yr old. The most plausible explanation for this is that all neutron stars start out with relatively strong fields, which then gradually decay away. This in turn raises two questions, namely (i) what is the origin of these very strong initial fields, and (ii) what is the mechanism responsible for the subsequent decay.

Regarding the origin of these fields, Thompson & Duncan (1993) and Bonanno, Rezzolla & Urpin (2003) consider the possibility of dynamo action, either in the very early stages of the neutron star itself, or else in the late stages of the precursor. Another possibility is the so-called thermoelectric effect (Blandford, Applegate & Hernquist 1983, Wiebicke & Geppert 1996), whereby temperature gradients (which are enormous in the first few thousand years of the neutron star's life) act as a battery, generating substantial fields. We note though that even fields as large as 10^{15} G do not require any dynamo or battery effects, but may have been amplified by nothing more exotic than simple compression of the precursor's magnetic field. In particular, Reisenegger (2001) points out that the most strongly magnetized neutron stars, White dwarfs (WDs) and MS stars all have astonishingly similar magnetic fluxes $\Phi = \pi R^2 B$, despite the fact that R varies by five orders of magnitude, and B by ten.

We will therefore not consider the origin of these fields further, but turn instead to our second question, namely what causes them to decay with age. There are again a number of possibilities. One suggestion is that accretion of mass from a binary companion is somehow responsible. What motivates this suggestion is the additional observation that the field strength

The Magnetic Universe: Geophysical and Astrophysical Dynamo Theory.
Günther Rüdiger, Rainer Hollerbach
Copyright © 2004 Wiley-VCH Verlag GmbH & Co. KGaA
ISBN: 3-527-40409-0

is correlated not only with age, but also with whether or not the neutron star has a companion. In particular, most weakly magnetic neutron stars have companions, whereas very few strongly magnetic ones do (e.g. Bhattacharya 1995). Furthermore, there are indeed a variety of mechanisms whereby accretion could plausibly cause field decay; see for example Blondin & Freese (1986), Romani (1990), or Urpin & Geppert (1995).

There is (at least) one difficulty with this field decay via accretion hypothesis though, namely why do all millisecond pulsars end up with field strengths in the relatively narrow range 10^{8-9} G? That is, if this process is so efficient that it can reduce the field by some four to five orders of magnitude, why should it suddenly – and consistently – stop once the field reaches 10^{8-9} G? We are motivated therefore to look for alternative mechanisms, and ideally mechanisms that naturally switch off once the field strength drops below a certain level.

The most plausible alternative, first proposed by Jones (1988), is the Hall effect, in which the magnetic field influences itself through a quadratic nonlinearity. As a result, the timescale on which it acts is inversely proportional to the field strength; Jones suggests it is given by

$$\tau \sim \frac{10^7}{B_{12}} \text{ yr},\tag{7.1}$$

where B_{12} is the field strength in units of 10^{12} G. See also Goldreich & Reisenegger (1992), who obtain a similar estimate. With this timescale, the observed correlation between field strength and age follows quite naturally: 10^{15} G fields would decay on 10^4-yr timescales, which is sufficiently short that there should indeed be very few such magnetars, whereas 10^{12} G fields would decay on 10^7-yr timescales, long enough for many more pulsars to be found in this evolutionary phase. And finally, once the field strength drops below 10^9 G the timescale becomes 10^{10} yr, at which point Ohmic decay becomes the dominant effect. We see therefore that this mechanism of field decay via Hall drift fits the observations rather well, potentially explaining not only the relative scarcity of magnetars, but also why the fields of millisecond pulsars do not continue to decay at the very rapid rates found earlier.

As elegant as it is, there is also one slight difficulty with this theory, namely that the Hall effect conserves magnetic energy (as we will see below), and so by itself cannot cause any field decay at all. Goldreich & Reisenegger (1992) suggested instead that it would induce a cascade of magnetic energy to ever shorter lengthscales, at which point Ohmic decay would take over.

7.2 Equations

As derived by Goldreich & Reisenegger (see also Sect. 5.4), the equation governing the evolution of a magnetic field under the influence of Hall drift and Ohmic decay is Eq. (5.34) with $\beta = c/4\pi n_e e$. Dimensional analysis then yields

$$\frac{B_0}{\tau} \sim \frac{c}{4\pi n_e e} \frac{B_0^2}{L^2} \qquad \Longrightarrow \qquad \tau \sim \frac{4\pi n_e e L^2}{c B_0}\tag{7.2}$$

as the natural timescale τ associated with Hall drift, where B_0 is a typical field strength, and L a typical lengthscale. This estimate, Eq. (7.1), thus amounts to nothing more complicated than inserting particular numbers into this result obtained purely by dimensional analysis.

Note though that the number density n_e varies by several orders of magnitude over the depth of the crust. It is therefore misleading to talk about a single Hall timescale; there is rather a range of timescales, perhaps $10^{5-9}/B_{12}$ yr. In Sect. 7.4 we will also see some less obvious consequences of variations in n_e.

Nondimensionalizing the field by B_0, length by the star's radius R, and time by the particular Hall timescale $4\pi n_e(R)eR^2/cB_0$, the governing equation becomes

$$\frac{\partial \boldsymbol{B}}{\partial t} = -\nabla \times \left(f(x)(\nabla \times \boldsymbol{B}) \times \boldsymbol{B}\right) + \mathrm{R_B}^{-1}\Delta\boldsymbol{B}, \tag{7.3}$$

where

$$f(x) = \frac{n_e(R)}{n_e(x)} \tag{7.4}$$

encapsulates the density profile, and

$$\mathrm{R_B} = \frac{\sigma B_0}{n_e(R)ec} \tag{7.5}$$

is the Hall parameter. One useful physical interpretation to associate with $\mathrm{R_B}$ is the ratio of the Ohmic timescale $4\pi\sigma R^2/c^2$ to this Hall timescale τ. Based on the values given above, that the Ohmic timescale is 10^{10} yr, whereas the Hall timescale is $\sim 10^{7-8}$ yr for a young pulsar, we conclude therefore that the limit we are interested in is $\mathrm{R_B} \gg 1$ (which turns out to be numerically very difficult). Finally, we note that we have taken the conductivity σ to be constant in Eq. (7.3). In real neutron stars this too varies, by perhaps two or three orders of magnitude over the depth of the crust. Because the Ohmic term in general is smaller than the Hall term though, one might expect that neglecting variations in σ would have relatively little effect.

We note that this Hall equation (7.3) bears certain similarities to the vorticity equation of ordinary fluid dynamics,

$$\frac{\partial \boldsymbol{\Omega}}{\partial t} = \nabla \times \left(\boldsymbol{u} \times \boldsymbol{\Omega}\right) + \mathrm{Re}^{-1}\Delta\boldsymbol{\Omega}, \tag{7.6}$$

where $\boldsymbol{u}$ and $\boldsymbol{\Omega} = \nabla \times \boldsymbol{u}$ are the velocity and vorticity, respectively, and Re the Reynolds number. It is on the basis of these similarities that Goldreich & Reisenegger suggested that the Hall effect would initiate a cascade to ever shorter lengthscales, analogous to the Kolmogorov cascade in ordinary fluid turbulence. By applying arguments from turbulence theory, they went on to suggest that the power spectrum of Hall turbulence should fall off as k^{-2}, where k is the wave number, and that the dissipation scale should be reached when $k = O(\mathrm{R_B})$.

However, there is also one crucial difference between Eq. (7.3) and Eq. (7.6). In particular, in Eq. (7.6) the nonlinear term contains only first derivatives of $\boldsymbol{\Omega}$, whereas in Eq. (7.3) the nonlinear term contains second derivatives of $\boldsymbol{B}$. As a result, in Eq. (7.6) one can always be certain that if one just goes to sufficiently short lengthscales, the diffusive term will eventually dominate the advective term, regardless of how large Re is. In contrast, in Eq. (7.3) one can go to arbitrarily short lengthscales, and still not be certain that the diffusive term will dominate the Hall term, because they both scale quadratically with the wave number. This means that the whole notion of a definite dissipation scale is much less clear in Eq. (7.3) than in Eq. (7.6). And indeed, we will find that, in general, there does not seem to be a definite dissipation scale associated with Eq. (7.3).

Ultimately we would like to consider fully 3D solutions of Eq. (7.3). In ordinary fluid turbulence it is known, for example, that two-dimensional results are very different from three-dimensional ones. If this analogy between Eqs. (7.3) and (7.6) has any basis at all, one might therefore expect the same to be true here. Unfortunately, Eq. (7.3) is such a difficult equation to solve – precisely because of this feature that the nonlinear term has just as many derivatives as the linear term – that even 2D solutions are quite challenging. We will therefore consider only axisymmetric solutions here.

For such axisymmetric fields, we can decompose $\boldsymbol{B}$ into toroidal and poloidal components

$$\boldsymbol{B} = \boldsymbol{B}_t + \boldsymbol{B}_p = B\hat{\boldsymbol{e}}_\phi + \nabla \times (A\hat{\boldsymbol{e}}_\phi), \tag{7.7}$$

yielding

$$\frac{\partial A}{\partial t} = -\hat{\boldsymbol{e}}_\phi \cdot f\left[(\nabla \times \boldsymbol{B}_t) \times \boldsymbol{B}_p\right] + \mathrm{R_B}^{-1}\Delta' A, \tag{7.8}$$

and

$$\frac{\partial B}{\partial t} = -\hat{\boldsymbol{e}}_\phi \cdot \nabla \times f\left[(\nabla \times \boldsymbol{B}_p) \times \boldsymbol{B}_p + (\nabla \times \boldsymbol{B}_t) \times \boldsymbol{B}_t\right] + \mathrm{R_B}^{-1}\Delta' B, \tag{7.9}$$

where $\Delta' = \Delta - 1/(r\sin\theta)^2$. There are then certain consequences one can deduce just from the general structure of these equations. First, if one starts out with a purely toroidal field, it will remain so, whereas if one starts out with a purely poloidal field, it will immediately induce a toroidal component as well, and once both components are present each will act back on the other. One might suppose therefore that considering purely toroidal solutions would be a sensible further simplification. However, doing so loses an essential part of the physics: while the term

$$-\hat{\boldsymbol{e}}_\phi \cdot \nabla \times f\left[(\nabla \times \boldsymbol{B}_t) \times \boldsymbol{B}_t\right] \tag{7.10}$$

may look like it also contains two derivatives of B, if one works through the algebra one finds that all the second derivative terms exactly cancel. This property that the Hall term has just as many derivatives as the Ohmic term is therefore associated entirely with the poloidal field.

It is of interest also to consider some of the symmetries associated with Eqs. (7.8) and (7.9). For example, the original equation (7.3) is clearly not invariant under $\boldsymbol{B} \to -\boldsymbol{B}$. If one considers Eqs. (7.8) and (7.9) though, one finds easily enough that it is only the sign of B that matters; since A appears as purely odd powers in Eq. (7.8) and as purely even powers in Eq. (7.9), its sign does not matter. Another symmetry worth mentioning is the equatorial symmetry, particularly as this is somewhat different from that usually encountered in stellar dynamos. One finds that solutions exist having A either symmetric or antisymmetric, but with B antisymmetric in both cases. In contrast, in stellar dynamo models the parity of B would also change, always being the opposite of A's (as in Chapt. 4). We can see easily enough though that this cannot be the case here, by noting this same property of Eqs. (7.8) and (7.9) already used above, that A enters as odd powers in Eq. (7.8), and as even powers in Eq. (7.9). Therefore, if either pure parity is allowed for A at all, the opposite one must also be allowed, but with B having the same parity in both cases.

Turning next to the boundary conditions associated with Eqs. (7.8) and (7.9), at $x = 1$ we have simply

$$\left(\frac{d}{dx} + \frac{l+1}{x}\right) A = 0, \qquad B = 0 \qquad \text{at } x = 1, \tag{7.11}$$

matching the poloidal field to an external potential field, and the toroidal field to zero. Note also how the condition on A involves the spherical harmonic degree l; since the numerical solution (Hollerbach 2000) involves spherical harmonics anyway, this causes no problems.

The boundary conditions at x_{in}, the inner edge of the crust, are not quite so straightforward, and depend on what assumptions we make about the core, about which little is known for certain. See for example Shapiro & Teukolsky (1983) for a discussion of various conjectures regarding the internal structure of neutron stars. However, one common assumption (e.g. Bhattacharya & Datta 1996) is that it is superconducting, in which case the magnetic field will be expelled from it. The boundary conditions are then that the normal component of the magnetic field and the tangential components of the associated electric field must vanish, which yield

$$A = 0, \qquad \frac{f}{x \sin \theta} \frac{\partial}{\partial \theta}\left(B \sin \theta\right) B + \mathrm{R_B}^{-1} \frac{1}{x} \frac{\partial}{\partial x}\left(Bx\right) = 0 \qquad \text{at } x = x_{\text{in}}, \quad (7.12)$$

respectively. At this point we realize also that stratified and unstratified calculations will require different inner boundary conditions on B. In particular, if we include density stratification $f(x_{\text{in}})$ is so small as to be essentially zero (indeed, the particular functional form we will implement in Sect. 7.4 has it identically zero), yielding the much simpler condition $\partial(xB)/\partial x = 0$. In contrast, without stratification $f(x_{\text{in}}) = 1$, yielding a very complicated, nonlinear boundary condition. We simplify this by noting that in the relevant $\mathrm{R_B} \gg 1$ limit the second term ought to be negligible (assuming $\partial B/\partial x$ does not increase with $\mathrm{R_B}$, that is, assuming that no boundary layers develop), in which case we are left with just $B = 0$.

Finally, we noted above that the Hall effect conserves magnetic energy. To obtain this result, simply take the dot product of Eq. (7.3) with $\boldsymbol{B}$, and apply various vector identities, yielding ultimately

$$\frac{\partial}{\partial t} \frac{\boldsymbol{B}^2}{2} = -\nabla \cdot \left[f(\boldsymbol{J} \times \boldsymbol{B}) \times \boldsymbol{B} + \mathrm{R_B}^{-1}(\boldsymbol{J} \times \boldsymbol{B})\right] - \mathrm{R_B}^{-1} \boldsymbol{J}^2, \qquad (7.13)$$

where here the dimensionless current $\boldsymbol{J} = \nabla \times \boldsymbol{B}$. When integrated over the crust therefore, the Hall term will contribute only surface integrals at x_{in} and 1, whereas the diffusive term will contribute both surface integrals and a negative-definite volume integral. The surface integrals at x_{in} turn out to vanish, for both the stratified and unstratified boundary conditions (and even though the unstratified $B = 0$ boundary condition is only an approximation). The surface integrals at $x = 1$ do not vanish, but instead turn out to be precisely what one needs to take into account the energy contained in the external field. The final result is that

$$\frac{\partial}{\partial t} \frac{1}{2} \int \boldsymbol{B}^2 \, \mathrm{d}V = -\mathrm{R_B}^{-1} \int \boldsymbol{J}^2 \, \mathrm{d}V, \qquad (7.14)$$

where the integral on the left extends over $x > x_{\text{in}}$, and the one on the right over $x_{\text{in}} \leq x \leq 1$. This is thus the desired energy balance, stating that the total magnetic energy decreases as a result of Ohmic decay only, with Hall drift rearranging the field, and hence also the energy, but neither creating nor destroying it. And once again, the hypothesis put forward by Goldreich & Reisenegger (1992) is that the Hall effect will induce sufficiently fine structures in the field that it decays on the $O(1)$ Hall timescale, rather than the $O(\mathrm{R_B})$ Ohmic timescale.

7.3 Without Stratification

In this section we briefly summarize the results obtained by Hollerbach & Rüdiger (2002) in the absence of stratification. In the next section we will then see some of the interesting and unexpected effects that stratification can have. The procedure we adopt to test this Goldreich-Reisenegger hypothesis is to start with some large-scale field and study the nature of the subsequent decay. In particular, consider the lowest order free-decay modes, $l = 1$ for A and $l = 2$ for B (to take advantage of this equatorial symmetry noted above). Our initial conditions will then consist of fields of the form $\boldsymbol{B}_{\mathrm{p}} + a\boldsymbol{B}_{\mathrm{t}}$, where the poloidal mode is normalized such that $B_r(1, 0) = 1$, and the toroidal such that $B_{\mathrm{max}} = 1$. See also Figs. 7.3 and 7.4 below.

We start with the simplest possible initial condition, namely $a = 0$, so just the poloidal mode by itself. Figure 7.1 shows how the first three harmonics b_1, b_3 and b_5 of the external field then evolve in time, where these b_l are defined by

$$B_r(1, \theta, t) = \sum_l b_l(t)\, P_l(\cos\theta). \tag{7.15}$$

That is, b_1 is nothing more than the coefficient of our $l = 1$ initial condition, etc. And indeed, we note how b_1 starts out at 1, and then slowly decays. It does not decay monotonically, but never deviates very much from the $\exp(-49\mathrm{R_B}^{-1}\,t)$ rate that Ohmic decay alone would have yielded. For these runs at least, the Hall effect has not significantly enhanced the decay rate (see Shalybkov & Urpin 1997).

Figure 7.2 shows what happens when we include some toroidal field in our initial condition, so $a \neq 0$ (and we remember that the sign of the toroidal field matters, so we have to consider positive/negative a separately). There are two reasons for including a toroidal component in the initial condition. First, all three of the above-mentioned mechanisms for generating the initial condition in the first place (a dynamo, the thermoelectric effect, simple compression of a pre-existing field) are likely to yield a considerable toroidal component, almost certainly at least as strong as the poloidal part. Secondly, if we return to Eq. (7.8), we note that without a toroidal field the poloidal field would not be affected at all. A stronger toroidal field might therefore be expected to have a greater effect. Indeed, seen in this light it is hardly surprising that Hall drift had relatively little effect in Fig. 7.1: the maximum induced toroidal field turns out to be no more than 0.25, so the poloidal field cannot be affected very much.

We see in Fig. 7.2 that increasing $|a|$ does indeed lead to increasingly large amplitudes for b_3. Otherwise we obtain much the same evolution, in particular the same rapid oscillation on the Hall timescale, and still the same slow decay on the Ohmic timescale. Another interesting point to note is that taking a positive or negative has relatively little effect after all; it merely determines where in the oscillation cycle the system starts off. Figures 7.3 and 7.4 show the full structure of the field through one of these oscillations.

Based on these results, one might expect that the solutions would exist for arbitrarily large a (and $\mathrm{R_B}$), and that eventually b_3 would even exceed b_1 for part of the cycle. Perhaps such solutions do exist, but we were unable to obtain them numerically. Figure 7.5 shows why; as a is increased the spectra get flatter and flatter, until eventually the numerical resolution fails. Increasing $\mathrm{R_B}$ instead of a has a similar effect. The other very interesting point to note about these spectra is that there is indeed no evidence of a dissipative cutoff. We conclude

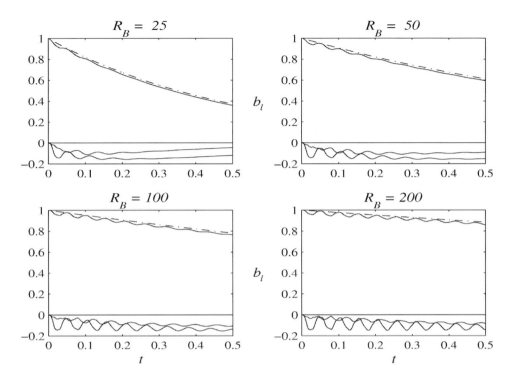

Figure 7.1: The harmonics b_1, b_3 and b_5 as functions of time, for the four indicated values of R_B. The solid line starting at 1 is b_1, with the dashed line being the $\exp(-49 R_B^{-1} t)$ decay rate of Ohmic decay alone. The next largest solid line is b_3, and the smallest b_5.

this section then by noting that the Hall effect does seem to generate something much like a cascade to short lengthscales, but that – in the absence of stratification at least – this does not appear to be sufficient to enhance the decay rate significantly beyond what Ohmic decay acting alone would have yielded.

7.4   With Stratification

From the way that this (inverse) density profile $f(x)$ enters into Eq. (7.3), it seems clear that including stratification will affect the Hall term in some way. The first to consider this effect were Vainshtein, Chitre & Olinto (2000), who suggested that instead of these oscillations found in the unstratified case, there would be a persistent motion of the field toward either the equator or the pole (depending on the sign of B), and that it would then decay much faster than if Ohmic decay alone were acting.

In order to understand this argument – but also some of its limitations – it is helpful to begin by recalling the more familiar induction equation

$$\frac{\partial \boldsymbol{B}}{\partial t} = \nabla \times (\boldsymbol{u} \times \boldsymbol{B}) + \mathrm{Rm}^{-1} \Delta \boldsymbol{B}. \tag{7.16}$$

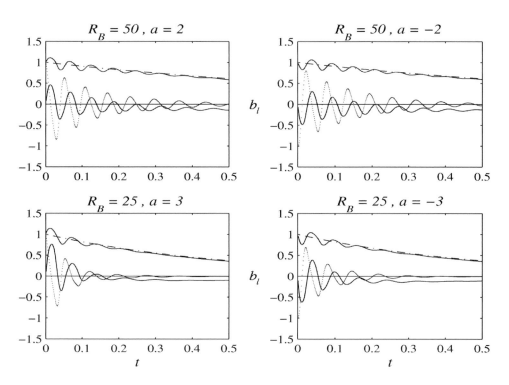

Figure 7.2: As in Fig. 7.1, the solid line starting at 1 is b_1 as a function of time. The solid line starting at 0 is b_3. The dotted line starting at ± 1 is the toroidal field at $x = 0.875$, $\theta = \pi/4$, divided by $|a|$. We see therefore that increasing a simply scales up the toroidal field, but otherwise has virtually no effect on it, only on b_3.

Comparing this with Eq. (7.3), we note that we can interpret Eq. (7.3) as if a flow $\boldsymbol{u} = -f(x)\nabla \times \boldsymbol{B}$ were advecting the field. A toroidal field B then corresponds to a meridional circulation having streamfunction $\Psi = f(x)Bx\sin\theta$. Figure 7.6 shows B and Ψ first without and then with stratification. Without stratification B and Ψ are identical except for this geometrical factor $x\sin\theta$. This 'flow' therefore has almost no effect, since it is mostly just advecting B around its own isolines. In contrast, with stratification B and Ψ are very different, and we see that depending on the sign of B (and hence Ψ) the flow will indeed be directed toward either the equator or the pole in the bulk of the interior.

The difficulty with this argument is that it only applies to toroidal fields – and indeed Vainshtein et al. did not allow for poloidal fields, which would have made their equations analytically intractable. If we do include poloidal fields though, the 'flow' we obtain corresponds to a differential rotation, which already makes the interpretation much more difficult, particularly with regard to the role that f might play. We therefore want to consider to what extent this Vainshtein et al. mechanism continues to operate when both toroidal and poloidal fields are present. See also Hollerbach & Rüdiger (2004), where these results are presented in more detail.

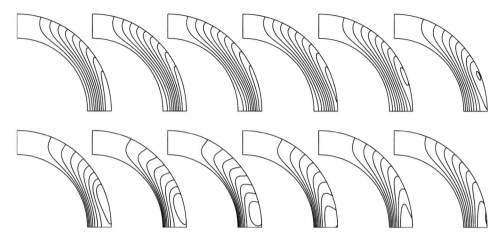

Figure 7.3: The structure of the poloidal field for $a = 3$, $R_B = 25$, and $t = 0.005, 0.01, \ldots, 0.06$. The solution at $t = 0.005$ also illustrates the free-decay-mode initial condition, as the field has not yet changed very much from $t = 0$.

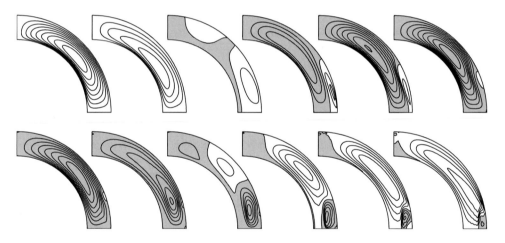

Figure 7.4: The same as in Fig. 7.3 but for the toroidal field.

Even though we are ultimately interested in mixed fields, we begin by considering purely toroidal fields, to verify that we can reproduce the basic mechanism. Figure 7.7 shows the results, which we see are indeed in perfect agreement with the Vainshtein et al. prediction; if B is negative/positive in the northern/southern hemispheres, the evolution is toward the equator, whereas if the signs are reversed it is toward the poles. In both cases very fine structures, scaling as $O(R_B^{-1})$, are then formed, in which the field reconnects sufficiently rapidly that the energy decays on the $O(1)$ Hall timescale rather than the $O(R_B)$ Ohmic timescale.

Figure 7.8 shows what happens when we start off with a purely poloidal initial condition. As noted above, this will immediately induce a toroidal field as well, which turns out to be

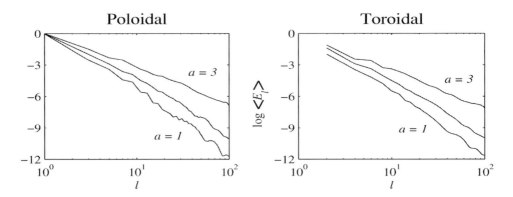

Figure 7.5: Power spectra of the solutions for $R_B = 25$ and $a = 1$, 2 and 3, and averaged over $t = 0$ to 0.1 (long enough to average over roughly two of these cycles, but short enough to avoid the subsequent decay).

negative/positive in the northern/southern hemispheres. According to Fig. 7.7, the evolution should therefore be toward the equator. We see that this is indeed the case, but unlike in Fig. 7.7, this time it never reaches the equator; once the poloidal field has been 'compressed' by a certain amount, it resists further advection of either field component. The toroidal field still sets up a current sheet (although not quite as fine as before), but because it is away from the equator it does not bring together fields of opposite sign, and hence does not lead to rapid dissipation.

That is, for a purely toroidal initial condition we obtain exactly the rapid dissipation predicted by Vainshtein et al., but for a purely poloidal initial condition we do not. To assess which is more relevant to real neutron stars, we again consider initial conditions containing both toroidal and poloidal fields. Figure 7.9 shows the results for the initial condition $-B_t + 0.3B_p$. We see that initially the poloidal field again manages to prevent the

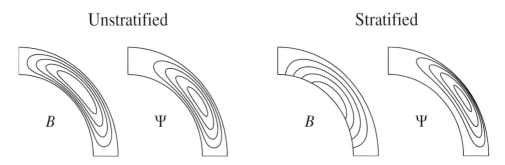

Figure 7.6: *Left*: B and $\Psi = Bx\sin\theta$ without stratification. *Right*: B and $\Psi = f(x)Bx\sin\theta$ with stratification, where $f(x) = ((x - x_{in})/(1 - x_{in}))^2$. Note also the different inner boundary conditions with and without stratification.

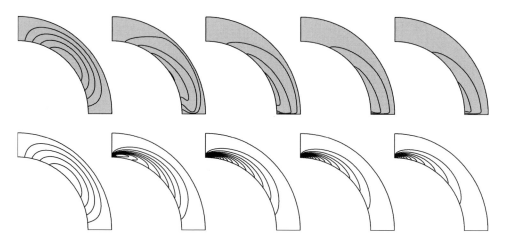

Figure 7.7: The upper row shows how negative B advects itself toward the equator, the lower row how positive B advects itself toward the pole. As in all these results, the solutions are antisymmetric about the equator. $R_B = 200$, and from left to right $t = 0$ (that is, the initial condition itself), 0.25, 0.5, 0.75 and 1.

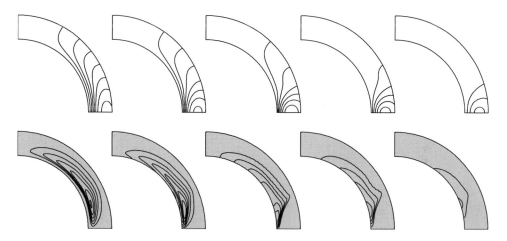

Figure 7.8: The upper row shows the poloidal field, the lower the toroidal. $R_B = 200$, and from left to right $t = 0.05, 0.1, 0.2, 0.4$ and 0.8.

toroidal field from reaching the equator. Eventually, however, the toroidal field overwhelms the poloidal – we note, for example, how the poloidal field has been almost completely annihilated by $t = 1$ – and thereafter does manage to establish a current sheet on the equator. Not surprisingly, the subsequent evolution is then much like the purely toroidal case, with the energy decaying on the Hall timescale. For values larger than 0.3 though, the toroidal field never manages to overwhelm the poloidal, and the evolution is as in Fig. 7.8 rather than Fig. 7.7 (upper row).

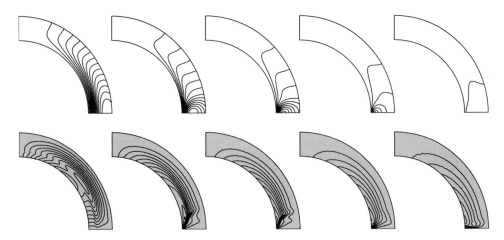

Figure 7.9: The upper row shows the poloidal field, the lower the toroidal. $R_B = 200$, and from left to right $t = 0.05, 0.25, 0.5, 0.75$ and 1.

Next, Fig. 7.10 shows the results for the initial condition $B_t + 0.05B_p$. Because the poloidal field is so weak, the negative toroidal field induced by B_p is overwhelmed by the positive contribution from B_t itself. We would expect the evolution to be toward the pole therefore, as in Fig. 7.7 (lower row). And indeed, that is exactly what happens, except that it never reaches the pole. As the poloidal field gets compressed into the polar regions, its effect becomes enhanced, to the point where it is able to induce a small region of negative B, which then blocks any further migration toward the pole. And as before, the details of these fine structures are sufficiently altered that they are no longer as effective in dissipating the field, with the result that the decay is again on the Ohmic rather than the Hall timescale. We see therefore that even a very weak poloidal field can already be sufficient to disrupt this Vainshtein et al. fast dissipation mechanism.

Finally, as interesting as this contrast between fast versus slow dissipation is, that between Figs. 7.9 and 7.10 is perhaps even more remarkable. In particular, this most basic feature, that the sign of the toroidal field in the two hemispheres dictates whether the migration (of both field components) is toward the poles or the equator, could be quite important, for the following reason: The strength of a pulsar's signal is determined at least in part by how narrowly focused its beam is, which in turn is determined by the structure of the magnetic field. So if the field is concentrated at the poles, rather than dispersed around the equator, the signal is also likely to be stronger. How visible a pulsar is may therefore be affected by something as seemingly unrelated as the orientation of its internal toroidal field.

7.5   Magnetic-Dominated Heat Transport

In the preceding sections we considered the evolution of the magnetic field alone, without regard to temperature. The two are in fact coupled though, through a variety of mechanisms (and even after the temperature gradients are too small for the thermoelectric effect to be sig-

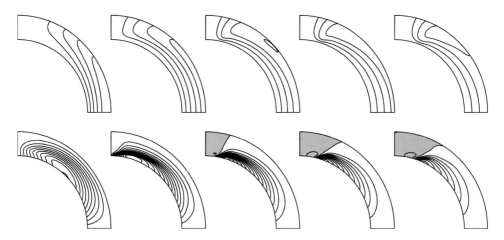

Figure 7.10: The upper row shows the poloidal field, the lower the toroidal. $R_B = 200$, and from left to right $t = 0.05, 0.25, 0.5, 0.75$ and 1. The figure differs from Fig. 7.9 by the initial condition.

nificant). For example, the field will obviously affect T via Ohmic heating, particularly in these current sheets that we saw above. Similarly, T will affect B through the temperature-dependence of the conductivity σ. (This is in fact one of the mechanisms whereby accretion could also cause field decay: the infalling matter heats up the surface, which reduces σ, causing the field to decay faster.)

In addition to these general theoretical considerations, there is also observational evidence suggesting a link between B and T. X-ray emissions (0.1–10 keV) indicate an inhomogeneous temperature distribution on the surfaces of some neutron stars (Becker & Pavlov 2001). The hot spots are located at the polar caps, and are quite concentrated ($\lesssim 1$ km), strongly suggesting a link with the field (Geppert et al. 2003). The particular mechanism proposed is the influence of the field on the heat flux (Schaaf 1990, Potekhin & Yakovlev 2001, Heyl & Hernquist 2001). In particular, both heat and magnetic flux are transported mainly by the electrons, so sufficiently strong magnetic fields will organize the heat flux preferentially along the field lines. If the field has a generally dipolar structure, the poles will then indeed be more strongly coupled to the hot interior than the equatorial regions.

To test whether this mechanism is sufficient, we write $F_i = -\rho C_p \chi_{ij} T_{,j}$, with

$$\chi_{ij} = \chi_1 \delta_{ij} + \chi_2 B_i B_j, \tag{7.17}$$

(compare with Eq. (3.79)). Obviously, χ_1 represents the heat flux perpendicular to the field, which we agreed should be quenched for sufficiently strong fields, so we take $\chi_1 = \chi_0/(1 + a^2 B^2)$. If we similarly take $\chi_2 = a^2 \chi_0/(1 + a^2 B^2)$, then the heat flux along the field remains finite even for $B \to \infty$. From Eq. (5.54) we have $aB = \omega_B \tau$, given in Fig. 7.11 (left) as 1–1000 for 10^{12} G at different stages in the neutron star's life.

For some given spatial structure of the field (see Fig. 7.11 as an example) and an appropriate boundary condition for the temperature ($F_r = \sigma T^4$) the heat equation can then be solved numerically. Figure 7.11 (right) gives the results for a core temperature 10^6 K. We find that the poles are indeed always warmer than the equator. Within the crust the temperature difference between pole and equator exceeds 100%, but at the surface this value is reduced to 26%. This particular value obviously only holds for this particular field structure, but it does already illustrate that this magnetically channeled heat flux, Eq. (7.17), can indeed have a significant effect. It would be of considerable interest then to repeat this calculation for more complicated field structures, such as some of those computed in the previous sections. And, of course, ultimately one might want to include some of the other mechanisms mentioned above, and consider the self-consistently coupled evolution of B and T together.

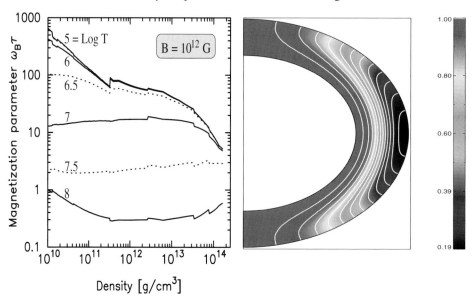

Figure 7.11: *Left*: The magnetization parameter $a_e = \omega_B \tau$ for various temperatures as a function of the density. *Right*: Field lines and temperature distribution in the crust for a magnetic field of $3 \cdot 10^{12}$ G. (Geppert, Küker & Page 2004).

7.6   White Dwarfs

Five per cent of the total white dwarf population possess magnetic fields from 10^5 to 10^9 G. Remarkably, their mean mass of 0.9 $M_\odot$ exceeds the mass of the nonmagnetic WDs. No other correlation with stellar parameters is known, not even with the rotation period (which is only known for the magnetic WDs). The shortest periods are 10 min, but much larger values are also observed. See the reviews by Wickramasinghe & Ferrario (2000) or Schmidt (2001).

Under flux conservation the contraction of a star of 1 $R_\odot$ to the WD size of only 10^4 km provides an amplification factor for the magnetic field of 10^4, so that 1 kG is amplified to 10^7 G. It is thus possible that the magnetic WDs are remnants of the magnetic Ap stars (Angel, Borra & Landstreet 1981, also Moss 2003b).

Indeed, Wickramasinghe (2001) argues that the magnetic field geometry of magnetic WDs is as complicated as it is for Ap stars. Muslimov, van Horn & Wood (1995) discuss the magnetic field decay of WDs under application of realistic conductivity values. Typically the Ohmic dissipation reduces the magnetic amplitudes of a poloidal dipolar field after 1 Gyr by a factor of (say) 2, while the higher modes decay faster ($M_{\text{WD}} = 1 \, M_\odot$, their Figs. 1 and 2). The decay of the toroidal field is very similar.

However, for strong fields, of order 10^9 G, the magnetic evolution is also influenced by the Hall effect. A relation

$$R_B \simeq (0.05 - 0.7) \frac{B}{10^9 \, \text{G}} \tag{7.18}$$

is suggested. Hence, the consequences of the Hall effect are only important for magnetic fields of order $B > 10^9$ G. Muslimov, van Horn & Wood (1995) only consider three magnetic modes, which from the experience of Shalybkov & Urpin (1997) is almost certainly too low. The reader is referred to the corresponding results of Sect. 7.3. The 'helicoidal' oscillations that are characteristic for the magnetic decay under the influence of the Hall effect cannot appear if the number of neighboring modes is too small.

8 The Magnetic Taylor–Couette Flow

8.1 History

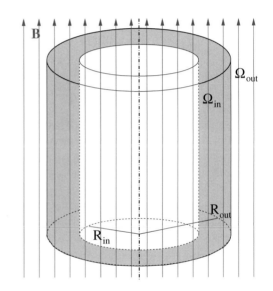

Figure 8.1: *Left*: Taylor (1923) vortices at the onset of the hydrodynamic instability. Note that the aspect ratio of the vortices is close to unity (from Koschmieder 1993). *Right*: Cylinder geometry of the magnetic Taylor–Couette flow.

The Taylor–Couette flow between concentric rotating cylinders (Fig. 8.1) is a classical problem of hydrodynamic stability (Couette 1890, Taylor 1923). For viscous flows in the absence of any transverse pressure gradient the most general form of the rotation law in the container is

$$\Omega(s) = a + \frac{b}{s^2}, \tag{8.1}$$

where a and b are two constants related to the angular velocities $\Omega_{\rm in}$ and $\Omega_{\rm out}$ with which the inner and the outer cylinders rotate. With $R_{\rm in}$ and $R_{\rm out}$ ($R_{\rm out} > R_{\rm in}$) being the radii of the two cylinders, one finds the coefficients a and b

$$a = \frac{\hat{\mu} - \hat{\eta}^2}{1 - \hat{\eta}^2}\,\Omega_{\rm in}, \qquad\qquad b = \frac{1 - \hat{\mu}}{1 - \hat{\eta}^2}\,\Omega_{\rm in} R_{\rm in}^2, \tag{8.2}$$

The Magnetic Universe: Geophysical and Astrophysical Dynamo Theory.
Günther Rüdiger, Rainer Hollerbach
Copyright © 2004 Wiley-VCH Verlag GmbH & Co. KGaA
ISBN: 3-527-40409-0

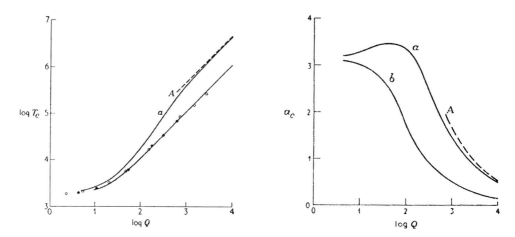

Figure 8.2: Chandrasekhar's (1961) results for small gap and small Pm. *Left*: The influence of the magnetic field (here Q is the Hartmann number) on the critical Taylor number. The magnetic field suppresses the instability. *Right*: The same for the wave number. The magnetic field expands the flow pattern in the vertical direction. Boundary conditions: (a) conducting walls, (b) insulating walls.

where $\hat{\mu} = \Omega_{\rm out}/\Omega_{\rm in}$ and $\hat{\eta} = R_{\rm in}/R_{\rm out}$. According to the Rayleigh criterion (5.9) the ideal flow is stable whenever the specific angular momentum increases outwards,

$$\hat{\mu} > \hat{\eta}^2. \tag{8.3}$$

Viscosity, however, has a stabilizing effect, so that a flow with $\hat{\mu} < \hat{\eta}^2$ becomes unstable only if the Reynolds number of the inner rotation exceeds some critical value, or in other words, if the inner cylinder rotates sufficiently rapidly.

If the fluid is electrically conducting and an axial magnetic field is applied then the critical value grows with growing amplitude of the magnetic field. Figure 8.2 shows the results of Chandrasekhar (1961) together with the experimental results of Donnelly & Ozima (1960), working with mercury. The Reynolds numbers of these first MHD Taylor–Couette experiments are also given in Fig. 8.16. These basic data were obtained for very narrow gaps and very small magnetic Prandtl numbers. Theory and observation are in nearly perfect agreement, but there is no indication of any magnetic-induced instability such as MRI. Ji, Goodman & Kageyama (2001) supposed that this absence of MRI is due to the use of the small magnetic Prandtl number limit. The magnetic Prandtl number is really very small under laboratory conditions (Table 8.1). For future dynamo experiments based on MHD Taylor–Couette flows the role of the material parameters must be known in detail. With the small-gap approximation but with Pm as a free parameter Kurzweg (1963) found that for *weak* magnetic fields and sufficiently large magnetic Prandtl number the critical Taylor number becomes smaller than in the hydrodynamic case (Fig. 8.3). If the field is not too strong it can play a destabilizing role and can lead to MRI via the Lorentz force. For the ideal magnetic Taylor–Couette flow this was first discovered by Velikhov (1959). In the MHD regime the Rayleigh criterion for stability, Eq. (8.3), changes to

$$\hat{\mu} > 1, \tag{8.4}$$

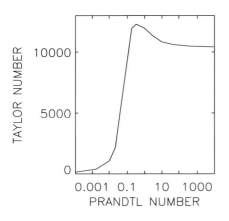 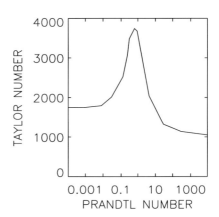

Figure 8.3: The variation of the critical Taylor number as a function of the magnetic Prandtl number for strong (*left*) and weak (*right*) axial magnetic field in the small-gap approximation. For weak magnetic fields and large magnetic Prandtl number the excitation is subcritical. For $B = 0$ is $\text{Ta}_{\text{crit}} = 1750$. After Kurzweg (1963).

i.e. only flows with superrotation are stable in the MHD regime (Velikhov, his Fig. 1). He found a growth rate along the Rayleigh line (i.e. $a = 0$) of $2\Omega_{\text{in}}\hat{\eta}$ and a critical ('dangerous') wave number of $k \leq 2\Omega_{\text{in}}\hat{\eta}/V_{\text{A}}$, with V_{A} the Alfvén velocity of the given axial field. He also suggested the stabilizing effect of an applied toroidal field decreasing outwards[1]. These results were derived via a dispersion relation of second order for ω^2 with the Fourier frequency ω that is only negative (indicating instability) if its absolute term is negative. This is only true if V_{A} is smaller than the shear $-s^2 d\Omega/ds$ multiplied with some positive factor, i.e. it is a weak-field instability. Chandrasekhar (1960) confirmed these results.

The hydrodynamic Taylor–Couette flow is stable if its angular momentum increases with radius, but the hydromagnetic Taylor–Couette flow is only stable if the angular velocity itself increases with radius. This remains true also for nonideal fluids. The MRI decreases the critical Reynolds number for weak magnetic field strengths for hydrodynamically unstable flow (Fig. 8.4) and it destabilizes the hydrodynamically stable flow for $\hat{\eta}^2 < \hat{\mu} < 1$. The MRI depends only on the amplitude of the external magnetic field and not on its direction. It exists in hydrodynamically unstable situations ($\hat{\mu} < \hat{\eta}^2$) only if Pm is not very small, as first shown by Kurzweg (1963, Fig. 8.3).

As we shall demonstrate here, the critical Reynolds numbers vary as 1/Pm for hydrodynamically stable flows ($\hat{\eta}^2 < \hat{\mu} < 1$), so that it is the magnetic Reynolds number Rm that controls the instability (see Fig. 8.7). According to the numbers given in Table 8.1 the magnetic diffusivity is at least 1000 cm^2/s (similar to the solar plasma, see Fig. 3.32); it is therefore not easy to reach magnetic Reynolds numbers of the required $O(10)$. This is the main reason why the MRI has never been observed experimentally in the laboratory.

---

[1] azimuthal fields were later considered in detail by Knobloch (1992) and Pringle (1996)

If the Hall effect is included in the induction equation then the new situation appears that the Taylor–Couette flow becomes unstable for *any ratio* of the angular velocities of the inner and the outer cylinder. This new instability, however, does not exist for both signs of the axial magnetic field B_0. For positive shear $d\Omega/ds$ the Hall instability exists for negative Hartmann number and for negative shear it exists for positive Hartmann number. For negative shear, of course, the Hall instability combines with the MRI.

Table 8.1: Parameters of the fluids suitable for MHD experiments taken from Chandrasekhar (1961) and Noguchi et al. (2002).

	ρ [g/cm^3]	ν [cm^2/s]	η [cm^2/s]	Pm
Mercury	5.4	$1.1 \cdot 10^{-3}$	7600	$1.4 \cdot 10^{-7}$
Gallium	6.0	$3.2 \cdot 10^{-3}$	2060	$1.5 \cdot 10^{-6}$
Sodium	0.92	$7.1 \cdot 10^{-3}$	810	$0.88 \cdot 10^{-5}$

8.2   The Equations

A viscous electrically conducting incompressible fluid between two rotating infinite cylinders in the presence of a uniform axial magnetic field B_0 admits the basic solution $B_z = B_0$, $u_\phi = as + b/s$, and all other vector components vanishing. We are interested in the stability of this solution. By developing the disturbances u' and B' into normal modes, the solutions of the linearized MHD equations are considered in the form

$$u' = u(s)\mathrm{e}^{\mathrm{i}(\omega t + kz + m\phi)}, \qquad\qquad B' = B(s)\mathrm{e}^{\mathrm{i}(\omega t + kz + m\phi)}. \qquad (8.5)$$

The equations have been derived by Chandrasekhar (1961) and Roberts (1964). Here a different Ohm's law is used and also different normalizations.

The general form of the induction equation with Hall effect is Eq. (5.34). The electric field for which the induction equation results is

$$E = \frac{J}{\sigma} - u \times B + \beta(\nabla \times B) \times B. \qquad (8.6)$$

The Navier-Stokes equation is used in its standard form, Eq. (4.14). $H = \sqrt{R_{\mathrm{in}}(R_{\mathrm{out}} - R_{\mathrm{in}})}$ is used as the unit of length, η/H as the unit of the perturbed velocity, ν/H^2 as the unit of frequencies, B_0 as the unit of the magnetic field fluctuations, H^{-1} as the unit of the wave number and Ω_{in} as the unit of Ω. The dimensionless numbers of the problem are the magnetic Prandtl number, the Hartmann number and the Reynolds number, i.e.

$$\mathrm{Pm} = \frac{\nu}{\eta}, \qquad\qquad \mathrm{Ha} = \frac{B_0 H}{\sqrt{\mu_0 \rho \nu \eta}}, \qquad\qquad \mathrm{Re} = \frac{\Omega_{\mathrm{in}} H^2}{\nu}. \qquad (8.7)$$

We only consider marginal stability, i.e. $\Im(\omega) = 0$. Using the same symbols for normalized quantities as before, the equations can be written as a system of 10 equations of first order, given by Goodman & Ji (2002) and Rüdiger & Shalybkov (2004b). Here only the relations representing the induction equation including the Hall effect are given, i.e.

$$\frac{\mathrm{d}B_s}{\mathrm{d}s} = -\frac{B_s}{s} - \mathrm{i}\frac{m}{s}B_\phi - \mathrm{i}kB_z,$$

$$\frac{\mathrm{d}B_z}{\mathrm{d}s} = \mathrm{i}\left(\frac{m^2}{ks^2} + k\right)B_s - \frac{\mathrm{Pm}}{k}(\omega + m\,\mathrm{Re}\,\Omega)B_s + u_s - \frac{m}{ks}X -$$

$$-\mathrm{i}R_B\frac{m}{s}B_z + \mathrm{i}R_B\,k\,B_\phi \tag{8.8}$$

and

$$\frac{\mathrm{d}X}{\mathrm{d}s} = \left(\frac{m^2}{s^2} + k^2\right)B_\phi + \mathrm{i}\mathrm{Pm}(\omega + m\,\mathrm{Re}\,\Omega)B_\phi - 2\mathrm{i}\frac{m}{s^2}B_s - \mathrm{i}ku_\phi +$$

$$+2\mathrm{Pm}\cdot\mathrm{Re}\frac{b}{s^2}B_s + R_B\frac{m^2}{s^2}B_s - R_B^2\frac{km}{s}B_z + R_B^2k^2B_\phi +$$

$$+\mathrm{i}R_B(\omega + m\mathrm{Re}\,\Omega)\,\mathrm{Pm}\,B_s - \mathrm{i}R_B\,ku_s + \mathrm{i}R_B\,\frac{m}{s}X, \tag{8.9}$$

with $X = \mathrm{d}B_\phi/\mathrm{d}s + B_\phi/s$ and the Hall parameter (7.5), i.e.

$$R_B = \frac{\beta B_0}{\eta}. \tag{8.10}$$

For electrons we have $R_B = a_e = \omega_B\tau$ after Eq. (5.54). The influence of the Hall effect is indicated by the R_B-terms. The ratio $\beta_0 = R_B/\mathrm{Ha}^*$ with the Lundquist number after Eq. (6.48) does not depend on the magnetic field; it is

$$\beta_0 = \frac{\sqrt{\mu_0\rho}}{H}\beta \tag{8.11}$$

a normalized Hall effect parameter. An appropriate set of 10 boundary conditions is needed to solve this system of equations. It is easy to see that the Hall effect leaves the boundary conditions unchanged, i.e. the no-slip conditions for the velocity $u_s = u_\phi = u_z = 0$, and for the conducting walls $\mathrm{d}B_\phi/\mathrm{d}s + B_\phi/s = B_s = 0$. The boundary conditions are valid at both R_{in} and R_{out}[2]. For insulating walls the magnetic boundary conditions are different at R_{in} and R_{out}, i.e. for $s = R_{\mathrm{in}}$

$$B_s + \mathrm{i}\frac{B_z}{I_m(ks)}\left(\frac{m}{ks}I_m(ks) + I_{m+1}(ks)\right) = 0, \tag{8.12}$$

and for $s = R_{\mathrm{out}}$

$$B_s + \mathrm{i}\frac{B_z}{K_m(ks)}\left(\frac{m}{ks}K_m(ks) - K_{m+1}(ks)\right) = 0, \tag{8.13}$$

where I_m and K_m are the modified Bessel functions. The condition for the toroidal field is one and the same at both locations, namely $k_sB_\phi = mB_z$.

---

[2] note that in this case $B_\phi \propto 1/s$ forms a free solution of the system

The homogeneous set of linear equations with the boundary conditions determines the eigenvalue problem for given Pm. The real part, $\Re(\omega)$, of ω describes a drift of the pattern along the azimuth. $\Re(\omega)$ is the second quantity that is fixed by the complex eigenequation. For a fixed Hartmann number, a fixed Prandtl number and a given vertical wave number we find the eigenvalue Re of the equation system. They are always minimal for a certain wave number that by itself defines the marginally unstable mode. The corresponding value is the desired Reynolds number which does not belong to a *prescribed* wave number.

8.3   Results without Hall Effect

In this section $R_B = 0$ is used in Eqs. (8.8) and (8.9).

8.3.1   Subcritical Excitation for Large Pm

Figure 8.4 shows the stability lines for axisymmetric modes for containers with both conducting and insulating walls with resting outer cylinder, for fluids of various magnetic Prandtl number. Only the vicinity of the classical hydrodynamic solution with $\mathrm{Re} = 68$ is shown. There is a strong difference of the bifurcation lines for $\mathrm{Pm} \gtrsim 1$ and $\mathrm{Pm} < 1$. For fluids with low electrical conductivity the magnetic field only suppresses the instability, so that all the critical Reynolds numbers exceed the value 68, and higher the stronger the magnetic field is.

For sufficiently small magnetic Prandtl number the stability lines hardly differ, which is the situation previously considered by Chandrasekhar (1961) without regard for the MRI. The opposite is true for $\mathrm{Pm} \gtrsim 1$. In Fig. 8.4 for materials with high electrical conductivity the resulting critical Reynolds numbers are smaller than $\mathrm{Re} = 68$. The magnetic field with small Hartmann numbers enhance the instability rather than suppress it. This effect becomes more effective for increasing Pm, but it vanishes for stronger magnetic fields. Obviously, the MRI only exists for weak magnetic fields and high enough electrical conductivity and/or microscopic viscosity (when the fields can be considered as frozen in and/or enough viscosity prevents the action of the Taylor–Proudman theorem).

In order to find a minimum due to the MRI the magnetic Prandtl number must exceed some critical value, $\mathrm{Pm_{min}}$, for hydrodynamically unstable flow ($\hat{\mu} < \hat{\eta}^2$). The critical magnetic Prandtl numbers lie in the narrow interval 0.25–1.75 for all $\hat{\mu}$ and $\hat{\eta}$. Thus, if the electrical conductivity is as small as it is for liquid metals then the MRI cannot be observed by corresponding experiments with hydrodynamically unstable flows (Rüdiger & Shalybkov 2002).

8.3.2   The Rayleigh Line ($a = 0$) and Beyond

There is a universal scaling with Pm for the special case for $\hat{\mu} = \hat{\eta}^2$, i.e. with $a = 0$ in the basic flow profile of Eq. (8.1). Then the terms with $\partial(s^2 \Omega)/\partial s$ vanish in the equations and for $m = \omega = 0$ one finds that the quantities u_s, u_z, B_s and B_z scale as $\mathrm{Pm}^{-1/2}$, while u_ϕ, B_ϕ, k and Ha scale as Pm^0. The Reynolds number for the axisymmetric modes then scales as

$$\mathrm{Re} \propto \mathrm{Pm}^{-1/2}. \tag{8.14}$$

This scaling does not depend on the boundary conditions, as these also comply with these relations for $m = 0$. The result (8.14) has also been found numerically for vacuum boundary

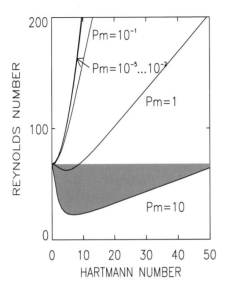 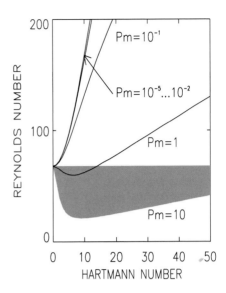

Figure 8.4: Marginal stability lines for axisymmetric modes with resting outer cylinder of conducting material (*left*) and vacuum (*right*). The shaded areas denote subcritical excitations of unstable axisymmetric modes by the external magnetic field. It only appears for $\mathrm{Pm} \gtrsim 1$ (Rüdiger, Schultz & Shalybkov 2003).

conditions by Willis & Barenghi (2002). However, for $a > 0$ Rüdiger & Shalybkov (2002) found the much steeper scaling $\mathrm{Re} \propto \mathrm{Pm}^{-1}$ (see Fig. 8.7), resulting in the surprisingly simple relation

$$\mathrm{Rm} = \frac{\Omega_{\mathrm{in}} R_{\mathrm{in}} (R_{\mathrm{out}} - R_{\mathrm{in}})}{\eta} \propto \mathrm{const.} \tag{8.15}$$

for the magnetic Reynolds number Rm, and $\mathrm{Ha} \propto \mathrm{Pm}^{-1/2}$, resulting in

$$\mathrm{Ha}^* = \frac{B_0 \sqrt{R_{\mathrm{in}} (R_{\mathrm{out}} - R_{\mathrm{in}})}}{\sqrt{\mu_0 \rho \eta}} \propto \mathrm{const.} \tag{8.16}$$

for the Lundquist number Ha^*. For small magnetic Prandtl numbers the exact value of the microscopic viscosity is not relevant for the excitation of the instability. In consequence, however, the corresponding Reynolds numbers for the MRI seem to differ by 2 orders of magnitude, i.e. 10^4 and 10^6. Experiments with $\hat{\mu} = \hat{\eta}^2$ thus seem to look much more promising than experiments with $\hat{\mu} > \hat{\eta}^2$.

However, this possibility cannot be utilized. The critical Reynolds number for $\hat{\mu} = \hat{\eta}^2$ and $\mathrm{Pm} = 1$ as a function of $\hat{\eta}$ has the overall minimum 54.4 for $\hat{\eta} = 0.27$, so that according to Eq. (8.14) one expects the value $1.7 \cdot 10^4$ for the Reynolds number for $\mathrm{Pm} = 10^{-5}$. Figure 8.5 shows the behavior of this result in the vicinity of $\hat{\mu} = \hat{\eta}^2$. There is a vertical jump from 10^4 to 10^6 in an extremely small interval of the abscissa. This sharp transition does not exist for

Pm $= 1$; it only exists for very small values of Pm. For this case the coexistence of both hydrodynamic and hydromagnetic instability is also presented in Fig. 8.5.

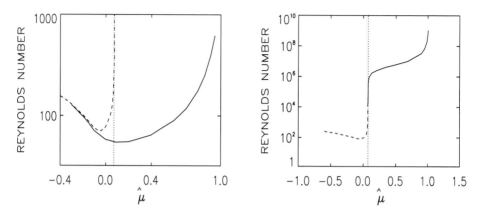

Figure 8.5: Critical Reynolds numbers for the Taylor–Couette flow versus $\hat{\mu}$ for $\hat{\eta} = 0.27$ and Pm $= 1$ (*left*) and Pm $= 10^{-5}$ (*right*). From all $\hat{\eta}$, $\hat{\eta} = 0.27$ yields the lowest minimum. The curve for the hydrodynamic instability (Ha $= 0$) is dashed and the hydromagnetic curve (Ha > 0) is solid. The dotted lines denote the location of $a = 0$.

The jump profile for Pm $= 10^{-5}$ in Fig. 8.5 (right) makes it clear that experiments with $\hat{\mu} = \hat{\eta}^2$ are not possible. Even the smallest deviation from the condition $\hat{\mu} = \hat{\eta}^2$ drastically changes the excitation condition. For $\hat{\mu}$ smaller than $\hat{\eta}^2$ (negative deviations) the hydrodynamic instability sets in and for $\hat{\mu}$ (slightly) exceeding $\hat{\eta}^2$ (positive deviations) the Reynolds number suddenly jumps by two orders of magnitudes.

Another situation occurs if the outer cylinder rotates so fast that the rotation law no longer fulfills the Rayleigh criterion and a solution for Ha $= 0$ cannot exist. The nonmagnetic eigenvalue along the vertical axis moves to infinity but a minimum remains. Figure 8.6 presents the results for Pm $= 1$ and Pm $= 10^{-5}$. There are always minima of Re for certain Hartmann numbers (also Goodman & Ji 2002). The minima and the critical Hartmann numbers increase for decreasing magnetic Prandtl numbers. For $\hat{\eta} = 0.5$ and $\hat{\mu} = 0.33$ the critical Reynolds numbers together with the critical Hartmann numbers are plotted in Fig. 8.7. Table 8.2 gives the exact coordinates of the absolute minima for experiments with rotating outer cylinder, for Pm $= 10^{-5}$. They are characterized by magnetic Reynolds numbers of order 10, very similar to the values of the existing dynamo experiments (Stieglitz & Müller 2001, Gailitis et al. 2001).

For liquid sodium in a container with insulating walls we find the critical numbers

$$f_{\text{in}} = 64 \frac{\text{Re}/10^6}{(R_{\text{out}}/10\text{cm})^2} \text{ Hz}, \qquad B = \frac{2.2 \text{ Ha}}{R_{\text{out}}/10\text{cm}} \text{ G}, \qquad (8.17)$$

with f_{in} as the real frequency of the inner cylinder and B as the necessary external field. A container with an outer radius of 22 cm (and an inner radius of 11 cm) filled with liquid

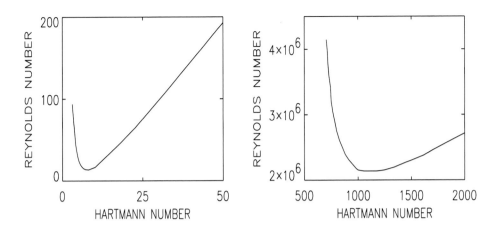

Figure 8.6: Marginal stability lines for axisymmetric modes in containers with rotating outer cylinder of conducting material for $\text{Pm} = 1$ (*left*) and $\text{Pm} = 10^{-5}$ (*right*). $\hat{\eta} = 0.5, \hat{\mu} = 0.33$.

Table 8.2: Coordinates of the absolute minima in Fig. 8.6 for rotating outer cylinder with $\hat{\mu} = 0.33$, $\hat{\eta} = 0.5$ and $\text{Pm} = 10^{-5}$.

	conducting walls	insulating walls
Reynolds number	$2.13 \cdot 10^6$	$1.42 \cdot 10^6$
mag. Reynolds number	21	14
Hartmann number	1100	1400
Lundquist number	3.47	4.42

sodium, therefore, requires a *rotation of about 19 Hz* in order to find the MRI. Following Eq. (8.17) and the values of Table 8.2 the required magnetic field is about 1400 G.

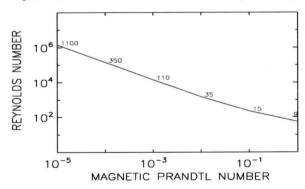

Figure 8.7: The critical Reynolds numbers vs. magnetic Prandtl numbers marked with those Hartmann numbers where the Reynolds number is minimum. $\hat{\eta} = 0.5$, $\hat{\mu} = 0.33$. From Rüdiger & Shalybkov (2002).

The results for containers with *conducting* walls are also given in Table 8.2. Note that the minimal Reynolds numbers are higher than for insulating cylinder walls. The influence of the boundary conditions is not small.

8.3.3   Excitation of Nonaxisymmetric or Oscillatory Modes

After the Cowling theorem only nonaxisymmetric magnetic fields can be maintained by a dynamo process. With respect to future dynamo experiments it is thus important to know the excitation conditions of nonaxisymmetric modes. We start with the results for containers with insulating walls and outer cylinders at rest and for $Pm = 10^{-5}$ (Fig. 8.8, left). There are linear instabilities even without magnetic fields. For $Ha = 0$ solutions for $m = 0$ ($Re = 68$), $m = 1$ ($Re = 75$) and $m = 2$ ($Re = 127$) are known. The axisymmetric mode possesses the lowest Reynolds number. This is also true in the MHD regime: we do not find any crossover of the instability lines for axisymmetric and nonaxisymmetric modes. The same is true for containers with rotating outer cylinder (Fig. 8.8, right). For growing $\hat{\mu}$ the Reynolds number for the hydrodynamic solution moves upwards, vanishing to infinity for $\hat{\mu} \geq \hat{\eta}^2 = 0.25$ (in this case). The MRI, however, is hardly influenced. It is represented by characteristic minima, in our case for $\hat{\mu} = 0.33$ at Hartmann numbers of order 10^3 and Reynolds numbers of order 10^6 (see Table 8.2).

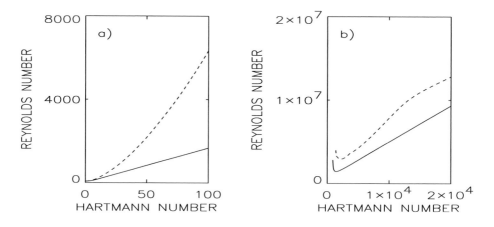

Figure 8.8: Insulating walls (vacuum): Stability lines for axisymmetric ($m = 0$, solid lines) and nonaxisymmetric instability modes with $m = 1$ (dashed). *Left*: Resting outer cylinder. *Right*: Rotating outer cylinder ($\hat{\mu} = 0.33$). $Pm = 10^{-5}$, $\hat{\eta} = 0.5$.

The main difference between the two sorts of boundary conditions is the existence of crossovers of the instability lines for $m = 0$ and $m = 1$ in the case of conducting walls (Fig. 8.9). For both resting and rotating outer cylinders Hartmann numbers exist above which the nonaxisymmetric mode possesses a lower Reynolds number than the axisymmetric mode. The occurrence of nonaxisymmetric solutions as the preferred modes is a rather general phe-

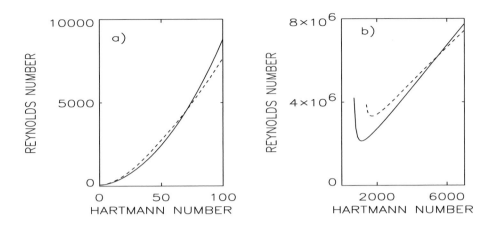

Figure 8.9: The same as in Fig. 8.8 but for conducting walls. For stronger magnetic fields the $m = 1$ mode is preferred.

nomenon for containers with conducting walls, which can become important for the design of future dynamo experiments.

Not only are stationary solutions possible; the instability can also occur in the form of oscillatory solutions ('overstability'). In the case of rotating convection between two layers heated from below the onset of instability in the form of oscillating solutions even possesses the lowest eigenvalues for certain Prandtl numbers (see Sect. 2.3.4). We find a very similar behavior for the MHD Taylor–Couette flow between conducting cylinders for resting outer cylinder (Fig. 8.10). It is a pair of waves traveling in positive and negative z-direction. Note that the cylinder considered here has no bound in the vertical direction. If the cylinder is finite, however, the possibility exists that traveling waves might be combined into standing waves.

8.3.4 Wave Number and Drift Frequencies

The unstable magnetic Taylor–Couette flow forms Taylor vortices. With our normalizations the vertical extent δz of a Taylor vortex is given by the expression

$$\frac{\delta z}{R_{\mathrm{out}} - R_{\mathrm{in}}} = \frac{\pi}{k} \sqrt{\frac{\hat{\eta}}{1 - \hat{\eta}}}. \tag{8.18}$$

The dimensionless vertical wave numbers k associated with the critical Reynolds numbers are given in Fig. 8.11. In the case of hydrodynamically unstable flows we have $\delta z \simeq R_{\mathrm{out}} - R_{\mathrm{in}}$ for small magnetic fields (Ha $\simeq 0$), independently of gap size and boundary conditions. The cell therefore has the same vertical extent as it has in radius (see Koschmieder 1993).

As all our results demonstrate, the influence of strong magnetic fields on turbulence consists of suppression and deformation. The deformation consists in a prolongation of the cell structure in the vertical direction, so that δz is expected to become larger and larger (the wave number becomes smaller and smaller) for increasing magnetic field. This is true for Pm $\simeq 1$,

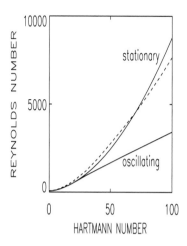

Figure 8.10: The same as in Fig. 8.9 (left) but with the inclusion of oscillating axisymmetric modes (overstability) appearing here for lower Reynolds numbers than the stationary modes do.

but for smaller Pm the vertical cell size has a minimum for an intermediate value of the magnetic field.

The cell size is minimum for the critical Reynolds number for all calculated examples of hydrodynamically stable flows with a conducting boundary. This is not true, however, for containers with insulating walls, for which the cell size grows with increasing magnetic field. For experiments with the critical Reynolds numbers the vertical cell size is generally 2–3 times larger than the radial one. The smaller the magnetic Prandtl number the longer are the cells in the vertical direction.

The influence of boundary conditions on the cell size disappears, of course, for sufficiently wide gaps. For the small and medium gaps, however, one finds the cells vertically more elongated for containers with insulating walls (Fig. 8.11).

For nonaxisymmetric modes the drift velocity $\Re(\omega)$ is always positive, i.e. the pattern drifts in the direction of the rotation (eastward). It is given by $\dot\phi = (\Re(\omega)\Omega_{\mathrm{in}})/(m\mathrm{Re})$ so that for $m = 1$ the drift period in units of the rotation period changes as $\mathrm{Re}/\Re(\omega)$. A typical value for this ratio is 2.

8.4   Results with Hall Effect

Equations (8.8) and (8.9) are now considered with $\beta_0 = 1$. The linearized induction equation (5.37) with Hall effect is invariant against the transformation $B_0 \to -B_0$, $u_0 \to -u_0$, so that the simultaneous change of the signs of $\mathrm{d}\Omega/\mathrm{d}s$ and B_0 leads to the same instability. After the splitting of the induction equation into poloidal and toroidal components one finds the scheme

$$B_{\mathrm{tor}} \xrightarrow[\mathrm{Hall}]{} B'_{\mathrm{pol}} \xrightarrow[\mathrm{Hall}]{\mathrm{shear}} B'_{\mathrm{tor}}, \tag{8.19}$$

hence the shear must also be changed if the Hall effect is changed. If this is true then it is impossible to find a magnetohydrodynamic instability with Hall effect but without shear.

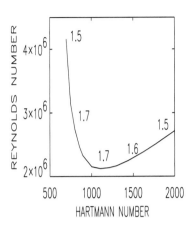
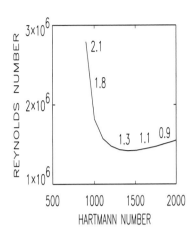

Figure 8.11: The stability line for $\hat{\eta} = 0.5$, $\hat{\mu} = 0.33$ and $\mathrm{Pm} = 10^{-5}$. There is no hydrodynamic instability. The line is marked with those wave numbers for which the Reynolds number is minimum. *Left*: Conducting walls. *Right*: Insulating walls. In both cases the variation of the wave numbers with the magnetic field differs.

In our scenario the shear is necessary for the existence of an instability, which can never be provided by the Hall effect alone as the shear is the energy source of the instability. The computations confirm these expectations.

The magnetic field needed for the Hall effect to become important is rather high. For both positive and negative shear the smallest value of Re occurs for $R_B \sim 1$. The corresponding value of the magnetic field is $B_0 \simeq \eta/\beta$. With the Hall coefficient ($\mu_0\beta$ in our notation) of 10^{-10} m^3/C for liquid metals and the magnetic diffusivity for sodium (see Table 8.1) one obtains 10^7 G as the critical magnetic fields for Taylor–Couette experiments (Rüdiger & Shalybkov 2004b).

8.4.1 Hall Effect with Positive Shear

The following instability only exists with Hall effect. It destabilizes flows with $\hat{\mu} > 1$ (i.e. with positive shear $d\Omega/ds$), for which no other instability is known. Figure 8.12 illustrates the instability for both conducting and nonconducting boundary conditions for a container with a medium gap. The flow is unstable, but only for negative Hartmann number, i.e. if the angular velocity and magnetic field have opposite directions[3]. The fact that the Hall effect destabilizes flows with angular velocity increasing outwards was already noted by Wardle (1999) and Balbus & Terquem (2001) (see Sect. 5.4).

For $\hat{\mu} \gg 1$ the angular velocity profile hardly depends on the inner angular velocity, Ω_{in}. In this case, the Reynolds number of the outer rotation, $\mathrm{Re}_{\mathrm{out}}$, is the real parameter of the

---

[3] for negative Hall resistivity the orientation is opposite

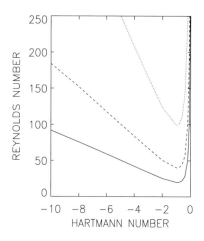 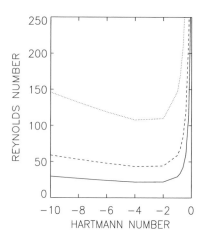

Figure 8.12: The line of marginal stability for magnetic Taylor–Couette flow with Hall effect ($\beta_0 = 1$) for $\hat{\eta} = 0.5$, Pm $= 1$, and positive $d\Omega/ds$: $\hat{\mu} = 2$ (solid line), 1.5 (dashed line) and 1.2 (dotted line). Boundary conditions for conducting cylinder walls (*left*) and insulating (*right*).

problem, with $\mathrm{Re_{out}} = \hat{\mu}\mathrm{Re} = \Omega_{out}H^2/\nu$. We have indeed numerically confirmed a Re $\propto 1/\hat{\mu}$ behavior for large $\hat{\mu}$. The value of $\mathrm{Re_{out}}$ corresponding to minimal Re (which is for Hartmann number of order unity) is about 20. For vacuum boundary conditions and not too small magnetic fields there is only a rather weak dependence of the critical Reynolds number on the Hartmann number (see Fig. 8.12).

In Fig. 8.13 the lines for both axisymmetric and nonaxisymmetric modes are given for both sorts of boundary conditions. A crossover of the lines (again) only exists for conducting cylinder walls. At the minimum the mode $m = 0$ dominates, but for stronger magnetic fields the mode with $m = 1$ dominates.

8.4.2   Hall Effect with Negative Shear

The Hall effect also modifies the critical Reynolds numbers for magnetohydrodynamically unstable flows ($\hat{\mu} < 1$), resulting in a rather complex situation illustrated with Fig. 8.14. The dashed line is the MRI without Hall effect. The combination of MRI and Hall instability is given as the solid line. A deep minimum of the Reynolds number is produced for weak magnetic fields – much deeper than the MRI minimum resulting without Hall effect. On the other hand, for increasing Hartmann numbers the solid line has a very weak slope, so that the magnetic field dependence of the combined instability is rather weak, as already shown by Wardle (1999, see his Fig. 1c). And once again, the Hall effect is only important for one orientation of the magnetic field.

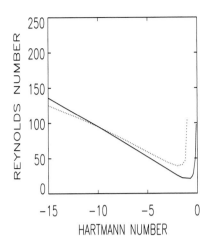
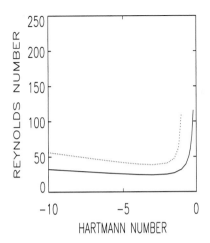

Figure 8.13: The same as in Fig. 8.12 but only for $\hat{\mu} = 2$ and for $m = 0$ (solid) and $m = 1$ (dashed). Note the crossover of the lines again for conducting boundary conditions (*left*). *Right:* No crossover exists for insulating boundary conditions.

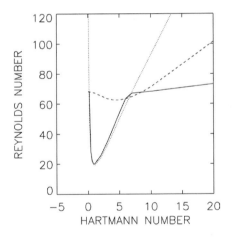

Figure 8.14: The same as in Fig. 8.12 but for resting outer cylinder. $\beta_0 = 0$ (dashed line) and $\beta_0 = 1$ (solid line). The dotted line is for $\beta_0 = 1$, but for $u = 0$ (velocity fluctuations neglected, i.e. kinematic case). The minimum of the dashed line indicates the Lorentz-force-induced MRI.

8.4.3   A Hall-Driven Disk-Dynamo?

The MRI is now widely accepted as the reason for turbulence in hot accretion disks. However, there may be difficulties in weakly ionized protoplanetary disks but the Hall effect has been considered as the source of instability in such disks (Wardle 1999, Balbus & Terquem 2001, Sano & Stone 2002a,b). In protoplanetary disks the critical magnetic field amplitude at $R = 1$ AU is about 0.1 G (Sano & Stone 2002a). Such high values can hardly be imagined as due to a magnetized central object. Polar field strengths of order $\sim 10^5$ G at the surface of a (windless) protosun are needed in order to produce 0.1 G at a distance of 1 AU.

Magnetic fields with amplitudes of (say) 1 G at 1 AU should be generated by the action of a (turbulent) dynamo. In this case, however, the microscopic conductivities cannot be used to estimate the values of the plasma parameters. The turbulent magnetic diffusivity may increase, for example, by several orders of magnitude *and* the turbulent magnetic Prandtl number approaches unity (see Yousef, Brandenburg & Rüdiger 2003, for the case of forced turbulence). Consideration of the effect of turbulence on the Hall diffusivity (Helmis 1968, Mininni, Gómez & Mahajan 2002) shall become important in the future.

Let us demonstrate the influence of the Hall effect on the disk stability with a simple 1D kinematic model. The half-thickness of the disk is H in the z-direction and a vertical magnetic field B_z is given and will be unchanged by the special geometry. With time in units of diffusion time and length in units of H the induction equation simply reads

$$\frac{\partial B_x}{\partial t} = \frac{\partial^2 B_x}{\partial z^2} + \mathrm{R_B}\frac{\partial^2 B_y}{\partial z^2}, \qquad \frac{\partial B_y}{\partial t} = \frac{\partial^2 B_y}{\partial z^2} + \mathrm{Rm}B_x - \mathrm{R_B}\frac{\partial^2 B_x}{\partial z^2}, \qquad (8.20)$$

with

$$\mathrm{Rm} = \partial u_y/\partial x \cdot H^2/\eta. \qquad (8.21)$$

The radial and azimuthal magnetic fields are confined to the disk by the boundary conditions $B_x(\pm 1) = B_y(\pm 1) = 0$. The solution fulfilling the boundary conditions is then

$$b_{x,y} = \sum_{n=1,3,5,\ldots} a_{x,y}^n \cos\left(\frac{\pi n}{2}z\right) e^{\gamma_n t} + \sum_{n=2,4,6,\ldots} b_{x,y}^n \sin\left(\frac{\pi n}{2}z\right) e^{\gamma_n t}, \qquad (8.22)$$

with

$$\gamma_n = -\frac{n^2\pi^2}{4} \pm \sqrt{-\frac{n^2\pi^2}{4}\mathrm{R_B}\left(\frac{n^2\pi^2}{4}\mathrm{R_B} + \mathrm{Rm}\right)}, \qquad (8.23)$$

where the coefficients a^n and b^n are defined by the initial conditions. One finds that an instability exists if

$$|\mathrm{Rm}| \geq \frac{\pi^2}{4}\frac{1 + \mathrm{R_B^2}}{|\mathrm{R_B}|}, \qquad (8.24)$$

with the minimum $|\mathrm{Rm}| = \pi^2/2$ at $|\mathrm{R_B}| = 1$ (Rüdiger & Shalybkov 2004a).

The critical shear $|\mathrm{Rm}|$ is the same for very small as for very large $\mathrm{R_B}$. The growth rates, however, differ significantly for the two branches of the stability map (Fig. 8.15, left). One may estimate the growth rate γ as $\gamma \sim |\mathrm{Rm}|/(2\tau_{\mathrm{diff}})$ for $|\mathrm{R_B}| \gg 1$ and $\gamma \sim |\mathrm{RmR_B}|/(2\tau_{\mathrm{diff}})$ for $|\mathrm{R_B}| \ll 1$, where $|\mathrm{R_B}| = \omega_B\tau$. An unstable solution, therefore, develops rapidly with the rotation scale only for a strongly magnetized plasma ($\omega_B\tau \gg 1$). Again the sign of the magnetic field is important for the shear-Hall instability. This means that the direction of the vertical magnetic field B_z should be opposite to the angular velocity Ω for positive β.

As no findings about the nonaxisymmetric modes are possible with the presented model we cannot ask whether it works as a dynamo. In order to find their role in the interplay of shear, dissipation and Hall effect the model of Kitchatinov & Mazur (1997) has been used with a Keplerian rotation law. The aspect ratio is $\tilde{s}_0 = 5$. The results *without* Lorentz force (i.e. only the induction equation is considered no MRI exists) are given in Fig. 8.15 (right). For

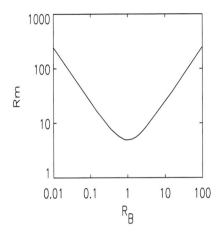
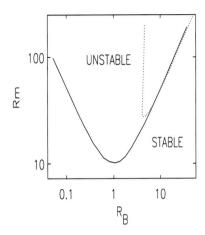

Figure 8.15: *Left*: Above the line of marginal stability after Eq. (8.24) the shear flow becomes unstable influenced by the Hall effect. *Right*: Induction equation (5.34) with Hall effect for the disk model of Kitchatinov & Mazur (1997) for $m = 0$ (solid) and $m = 1$ (dashed).

the axisymmetric mode the characteristic minimum at $|R_B| = 1$ again occurs for $Rm \simeq 10$. A minimum also exists for $m = 1$ but it lies higher. No difference exists for the axisymmetric and nonaxisymmetric solution at the strong-field branch of the instability map, but for the weak-field branch this is not true. There is obviously a threshold value of the Hall number for the excitation of modes with $m = 1$.

8.5   Endplate effects

We note that all of the previous discussion assumes a cylinder of infinite extent in z. Any actual laboratory apparatus would necessarily involve a finite cylinder though, so before concluding that an MRI experiment could be built, we must consider the effects of the endplates, which for solid plates turn out to be substantial. The difficulty is caused by the fact that both the inner and outer cylinders must be rotated, and extremely rapidly, as discussed above. As a result, the Taylor–Proudman theorem applies, stating that the flow will be almost independent of z. That implies though that if the endplates are rigidly rotating (with either $\Omega_{\rm in}$ or $\Omega_{\rm out}$), then the angular velocity throughout the interior will be the same. That is, instead of the desired profile $a + b/s^2$, one will have essentially solid-body rotation almost everywhere, with the entire difference between $\Omega_{\rm in}$ and $\Omega_{\rm out}$ accommodated in a so-called Stewartson layer, situated on whichever cylinder is rotating differently from the endplates.

More quantitatively, the relevant parameter is the Ekman number $E = Re^{-1}$ (so based on the above Reynolds numbers we would need to achieve $E \leq O(10^{-6})$ to have any chance of observing the MRI. There are then two aspects where E enters. First, the Taylor–Proudman theorem makes itself felt an $O(E^{-1/2})$ distance in z. That is, if one wanted to make a cylinder so long that the fluid in the middle no longer feels the effects of the end-plates, at $E = O(10^{-6})$ it would have to be some 1000 times as long as it is wide.

Since that is clearly not feasible, we conclude that the profile will not be the desired $a + b/s^2$, but instead this essentially solid-body interior plus a Stewartson layer at one or the other end. The second point where E enters is therefore in the thickness of this layer; Stewartson (1957) showed that the angular velocity adjusts over an $O(E^{1/4})$ distance. Instead of the desired smooth profile, essentially all of the adjustment would therefore occur over a very small fraction, on the order of 1/30, of the gap width.

The narrowness of this layer indicates also how difficult it would be to overcome this problem by splitting the endplates into several rings, and rotating each ring at some rate intermediate between Ω_{in} and Ω_{out}. Unless one had at least 10 rings or so, one would still not end up with a smooth profile, but rather with a step profile, with individual Stewartson layers separating the different rings. We conclude therefore that there are still considerable technical difficulties to be overcome before the MRI can be obtained in a real cylinder in the lab.

8.6   Water Experiments

It has been shown that for hydrodynamically stable Taylor–Couette flows the MRI should be visible for liquid sodium for Reynolds numbers of order 10^6. For an experimental realization of this effect it is important to know up to what Reynolds number hydrodynamically linearly stable Taylor–Couette flows are also nonlinearly stable. According to Eq. (8.3) the flow should be maximally stable for resting inner cylinder, i.e. $\hat{\mu} \to \infty$. Richard & Zahn (1999) focused attention on the experimental results of Wendt (1933), who found *nonlinear* instability for this case for Reynolds numbers of order 10^5. However, later experiments demonstrated that the results of Wendt were due to imperfections in the container, and the flow remained laminar for the same order of the Reynolds number (Schultz-Grunow 1959). In Fig. 8.16 the known hydrodynamical ('water') experiments are reported with resting inner cylinder. The numerical details are given in Table 8.3. Nevertheless, the possibility of finite-amplitude instabilities of hydrodynamically linearly stable rotation laws should remain in the astrophysical discussion (Richard & Zahn 1999, Richard 2003, Hersant, Dubrulle & Huré 2004).

Table 8.3: Details of the experiments reported in Fig. 8.16.

	R_{out} [cm]	ΔR [cm]	ν [10^{-2} cm^2/s]	$\hat{\eta}$	Re
Wendt	14.7	4.7	H$_2$O	0.68	42 100
Schultz-Grunow	2.5	0.2	0.65	0.92	40 100
Taylor	4.05	0.85		0.79	34 300
Richard et al.	5.0	1.5	H$_2$O	0.7	35 000

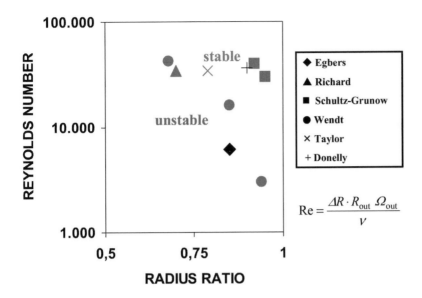

Figure 8.16: The existing water experiments with resting inner cylinder do not exceed Reynolds numbers of order 10^5. Only the rotation law of Schultz-Grunow (1959) proved to be stable. The experiments of Taylor (1923), Donnelly & Ozima (1960, MHD), Richard et al. (2001) and Egbers & Pfister (2000) are given for comparison.

8.7   Taylor–Couette Flow as Kinematic Dynamo

Taylor vortices and the rotation law (8.1) form a flow pattern that can be probed for its ability to work as a kinematic dynamo. Similar patterns were tested by Dudley & James (1989), with the result that for certain combinations of circulation and shear a kinematic dynamo can indeed operate. Laure, Chossat & Daviaud (2000) and Willis & Barenghi (2002) have considered this problem, which might be relevant for technical experiments. Here we only refer to the case of a vertically unbounded container considered by Willis & Barenghi. Only for infinite containers can the vacuum boundary conditions be exactly formulated. The dynamo-excited fields are small-scaled in the sense that their size is determined by the cell structure of the flow (two flow-cells exactly form a single field-cell).

   The standard case with $\hat{\mu} = 0$ and $\hat{\eta} = 0.5$ is not encouraging. Growing magnetic modes with $m = 1$ are observed only for $Pm \geq 1.5$, and only for a minimum Reynolds number of the inner rotation of 109, which leads to a (rather high) magnetic Reynolds number of about 163. No solution exists for smaller Pm. A slightly different situation holds for $\hat{\mu} \simeq 0.19$. Here Willis & Barenghi find that Rm does not vary for a wide range of Pm. For their minimum Pm of 0.06 a kinematic dynamo works but only for $Rm \geq 230$. If, on the other hand, experiments with $Re \simeq 10^6$ for liquid sodium are possible then the resulting magnetic Reynolds number is of order 10, so that the kinematic dynamo is still far away. Also for randomly forced turbulence small-scale dynamos seem to exist only for not too small magnetic Prandtl numbers (Schekochihin et al. 2004).

9 Bibliography

Abramenko VI, Wang T, Yurchishin VB, 1996, *Solar Physics* 168, 75

Abramowicz MA, Livio M, Piran T, Wiita PJ, 1984, *The Astrophysical Journal* 279, 367

Abramowicz MA, Livio M, Soker N, Szuszkiewicz E, 1990, *Astronomy & Astrophysics* 239, 399

Abramowicz MA, Brandenburg A, Lasota JP, 1996, *Month. Not. Roy. Astr. Soc.* 281, L21

Abt HA, Willmarth DW, 2000, in *Stellar Astrophysics*, L. S. Cheng et al. (Eds.), Kluwer, p. 175

Acheson DJ, 1972, *J. Fluid Mechanics* 52, 529

Acuña MH, Connerney JEP, Wasilewski P, et al., 2001, *J. Geophysical Research* 106, 23403

Ahrens B, Stix M, Thorn M, 1992, *Astronomy & Astrophysics* 264, 673

Ak T, Ozkan MT, Mattei JA, 2001, *Astronomy & Astrophysics* 369, 882

Alfè D, Gillan MJ, Vocadlo L, et al., 2002, *Phil. Trans. Royal Society London A* 360, 1227

Amado PJ, Cutispoto G, Lanza AF, Rodonò M, 2001, in *Cool Stars, Stellar Systems and the Sun*, R.J. García López et al. (Eds.), ASP, p. 895

Anantharamaiah KR, Radhakrishnan V, Shaver PA, 1984, *Astronomy & Astrophysics* 138, 131

Angel JRP, Borra EF, Landstreet JD, 1981, *Astrophys. J. Suppl.* 45, 457

Antia HM, Basu S, Chitre SM, 1998, *Month. Not. Roy. Astr. Soc.* 298, 543

Antia HM, 2002, in *SOLMAG 2002: Magnetic Coupling of the Solar Atmosphere*, H. Sawaya-Lacoste (Ed.), ESA Publications Division, p. 71

Anzer U, Börner G, Monaghan JJ, 1987, *Astronomy & Astrophysics* 176, 235

Anzer U, Börner G, 1992, *Comments on Astrophysics* 16, 31

Applegate JH, 1992, *The Astrophysical Journal* 385, 621

Arlt R, Rüdiger G, 2001, *Astronomy & Astrophysics* 374, 1035

Arlt R, Hollerbach R, Rüdiger G, 2003, *Astronomy & Astrophysics* 401, 1087

Arlt R, Urpin VA, 2004, *Astronomy & Astrophysics*, subm.

Armitage PJ, 1998, *The Astrophysical Journal* 501, 189

Athreya RM, Kapahi VK, McCarthy PJ, van Breugel W, 1998, *Astronomy & Astrophysics* 329, 809

Aurnou J, Olson P, 2000, *Physics Earth Planetary Interiors* 117, 111

Aurnou J, Olson P, 2001, *J. Fluid Mechanics* 44, 441

Backus G, 1958, *Annals of Physics* 4, 372

Baglin A, Morel PJ, Schatzman E, 1985, *Astronomy & Astrophysics* 149, 309

Balbus SA, 1991, *The Astrophysical Journal* 372, 25

Balbus SA, Hawley JF, 1991, *The Astrophysical Journal* 376, 214

Balbus SA, Hawley JF, 1994, *Month. Not. Roy. Astr. Soc.* 266, 769

Balbus SA, 1995, *The Astrophysical Journal* 453, 380

Balbus SA, Hawley JF, Stone JM, 1996, *The Astrophysical Journal* 467, 76

Balbus SA, Terquem C, 2001, *The Astrophysical Journal* 552, 235

Baliunas SL, Vaughan AH, 1985, *Ann. Rev. Astronomy Astrophysics* 23, 379

Baliunas SL, Donahue RA, Soon WH, et al., 1995, *The Astrophysical Journal* 438, 269

Baliunas SL, Nesme-Ribes E, Sokoloff DD, Soon WH, 1996, *The Astrophysical Journal* 460, 848

Baliunas SL, Frick P, Sokoloff DD, Soon WH, 1997, *Geophysical Research Letters* 24, 1351

van Ballegooijen AA, 1982, *Astronomy & Astrophysics* 113, 99

van Ballegooijen AA, 1998, in *Synoptic Solar Physics*, K. S. Balasubramaniam et al. (Eds.), ASP, p. 17

Balthasar H, Vázquez M, Wöhl H, 1986, *Astronomy & Astrophysics* 155, 87

Banerjee R, Jedamzik K, 2003, *Phys. Rev. Lett.* 91, 251301

Bao S, Zhang H, 1998, *The Astrophysical Journal* 496, L43

Bardou A, v. Rekowski B, Dobler W, et al., 2001, *Astronomy & Astrophysics* 370, 635

Barenghi CF, Jones CA, 1991, *Geophysical Astrophysical Fluid Dyn.* 60, 211

Barker T, 1993, in *Planetary Nebulae*, R. Weinberger et al. (Eds.), Kluwer, p. 181

Barnes JR, CollierCameron A, James DJ, et al., 2000, in *Stellar Clusters and Associations: Convection, Rotation, and Dynamos*, R. Pallavicini et al. (Eds.), ASP, p. 345

Baryshnikova I, Shukurov A, Ruzmaikin A, Sokoloff DD, 1987, *Astronomy & Astrophysics* 177, 27

Basu S, Antia HM, 1994, *Month. Not. Roy. Astr. Soc.* 269, 1137

Basu S, Mouschovias TC, 1994, *The Astrophysical Journal* 432, 720

Basu S, 1997, *Month. Not. Roy. Astr. Soc.* 288, 572

Basu S, Antia HM, 2001, *Month. Not. Roy. Astr. Soc.* 324, 498

Battaner E, Florido E, 1995, *Month. Not. Roy. Astr. Soc.* 277, 1129

Beck R, 1982, *Astronomy & Astrophysics* 106, 121

Beck R, Golla G, 1988, *Astronomy & Astrophysics* 191, L9

Beck R, 1993, in *The Cosmic Dynamo*, F. Krause et al. (Eds.), Kluwer, p. 283

Beck R, Poezd AD, Shukurov A, Sokoloff DD, 1994, *Astronomy & Astrophysics* 289, 94

Beck R, 1996, in *High-Sensitivity Radio Astronomy*, D. Jackson (Ed.), University of Manchester, p. 117

Beck R, Hoernes P, 1996, *Nature* 379, 47

Beck R, Brandenburg A, Moss D, et al., 1996, *Ann. Rev. Astronomy Astrophysics* 34, 155

Beck R, 2000, *Phil. Trans. Royal Society London A* 358, 777

Beck R, 2002, in *Highlights of Astronomy*, H. Rickman (Ed.), ASP, p. 712

Becker W, Pavlov G, 2001, in *The Century of Space Science*, J. Bleeker et al. (Eds.), Kluwer, Chapt. 8

Beckwith SVW, 1994, in *Theory of Accretion Disks – 2*, W. J. Duschl et al. (Eds.), Kluwer, p. 1

Beer J, Tobias SM, Weiss NO, 1998, *Solar Physics* 181, 237

Belvedere G, Paternò L, Stix M, 1980, *Geophysical Astrophysical Fluid Dyn.* 14, 209

Belvedere G, Lanzafame G, Proctor MRE, 1991, *Nature* 350, 481

Benevolenskaya EE, 1995, *Solar Physics* 161, 1

Berdyugina SV, Korhonen H, Tuominen I, 2001, in *Magnetic Fields Across the Hertzsprung-Russell Diagram*, G. Mathys et al. (Eds.), ASP, p. 243

Berdyugina SV, Usoskin IG, 2003, *Astronomy & Astrophysics* 405, 1121

Bergman MI, 1997, *Nature* 389, 60

Bhattacharya D, 1995, *J. Astrophysics & Astronomy* 16, 217

Bhattacharya D, Datta B, 1996, *Month. Not. Roy. Astr. Soc.* 282, 1059

Biermann L, 1951, *Z. Astrophysik* 28, 304

Bigazzi A, Ruzmaikin A, 2004, *The Astrophysical Journal* 604, 944

Binney J, Tremaine S, 1987, *Galactic Dynamics*, Princeton University Press

Birk GT, Wiechen H, Lesch H, 2002, *Astronomy & Astrophysics* 393, 685

Blackman EG, Field GB, 2000, *The Astrophysical Journal* 534, 984

Blackman EG, Brandenburg A, 2002, *The Astrophysical Journal* 579, 359

Blandford RD, Payne DG, 1982, *Month. Not. Roy. Astr. Soc.* 199, 883

Blandford RD, Applegate JH, Hernquist L, 1983, *Month. Not. Roy. Astr. Soc.* 204, 1025

Blondin JM, Freese K, 1986, *Nature* 323, 786

Bloxham J, Gubbins D, Jackson A, 1989, *Phil. Trans. Royal Society London A* 329, 417

Bloxham J, Jackson A, 1991, *Rev. Geophysics* 29, 97

Bloxham J, 2000a, *Nature* 405, 63

Bloxham J, 2000b, *Phil. Trans. Royal Society London A* 358, 1171

Bloxham J, 2002, *Geophysical Research Letters* 29, art. no. 1854

Bloxham J, Zatman S, Dumberry M, 2002, *Nature* 420, 65

Bodenheimer P, 1995, *Ann. Rev. Astronomy Astrophysics* 33, 199

Böhmer S, Rüdiger G, 1998, *Astronomy & Astrophysics* 338, 295

Bogdan TJ, 2000, *Solar Physics* 192, 373

Bogdan TJ, Hansteen M, Carlsson V, et al., 2003, *The Astrophysical Journal* 599, 626

Bonanno A, Elstner D, Rüdiger G, Belvedere G, 2002, *Astronomy & Astrophysics* 390, 673

Bonanno A, Rezzolla L, Urpin V, 2003, *Astronomy & Astrophysics* 410, L33

Boss AP, 1998, *The Astrophysical Journal* 503, 923

Boss AP, 2001, in *The Formation of Binary Stars*, H. Zinnecker et al. (Eds.), ASP, p. 371

Bouvier J, Cabrit S, Fernandez M, et al., 1993, *Astronomy & Astrophysics* 272, 176

Braginsky SI, 1963, *Soviet Physics Doklady* 149, 8

Braginsky SI, 1964, *Soviet Physics JETP* 20, 726

Braginsky SI, Roberts PH, 1987, *Geophysical Astrophysical Fluid Dyn.* 38, 327

Braginsky SI, Meytlis VP, 1990, *Geophysical Astrophysical Fluid Dyn.* 55, 71

Braginsky SI, Roberts PH, 1995, *Geophysical Astrophysical Fluid Dyn.* 79, 1

Brandenburg A, Krause F, Meinel R, et al., 1989, *Astronomy & Astrophysics* 213, 411

Brandenburg A, Tuominen I, Moss D, 1989, *Geophysical Astrophysical Fluid Dyn.* 49, 129

Brandenburg A, Tuominen I, Moss D, Rüdiger G, 1990, *Solar Physics* 128, 243

Brandenburg A, Tuominen I, 1991, in *The Sun and Cool Stars. Activity, Magnetism, Dynamos*, I. Tuominen et al. (Eds.), Springer, p. 223

Brandenburg A, Moss D, Rüdiger G, Tuominen I, 1991, *Geophysical Astrophysical Fluid Dyn.* 61, 179

Brandenburg A, Moss D, Tuominen I, 1992, *Astronomy & Astrophysics* 265, 328

Brandenburg A, Donner KJ, Moss D, et al., 1993, *Astronomy & Astrophysics* 271, 36

Brandenburg A, 1994a, in *Lectures on Solar and Planetary Dynamos*, M. R. E. Proctor et al. (Eds.), Cambridge University Press, p. 117

Brandenburg A, 1994b, in *Advances in Solar Physics*, G. Belvedere et al. (Eds.), Springer, p. 73

Brandenburg A, Nordlund Å, Stein RF, Torkelsson U, 1995, *The Astrophysical Journal* 446, 741

Brandenburg A, Jennings RL, Nordlund Å, et al., 1996, *J. Fluid Mechanics* 306, 325

Brandenburg A, Donner KJ, 1997, *Month. Not. Roy. Astr. Soc.* 288, L29

Brandenburg A, 1998, in *Theory of Black Hole Accretion Disks*, M. A. Abramowicz et al. (Eds.), Cambridge University Press, p. 61

Brandenburg A, Schmitt D, 1998, *Astronomy & Astrophysics* 338, L55

Brandenburg A, 2000, *Phil. Trans. Royal Society London A* 358, 759

Brandenburg A, 2001, *The Astrophysical Journal* 550, 824

Brandenburg A, Subramanian K, 2004, *Astrophysical Magnetic Fields and Nonlinear Dynamo Theory, Phys. Reports*, subm.

Brandt PN, Scharmer GB, Ferguson SH, et al., 1988, *Nature* 335, 238

Braun DC, Fan Y, 1998, *The Astrophysical Journal* 508, L105

Bray RJ, Loughhead RE, 1964, *Sunspots*, Chapman & Hall

Brown TM, 1984, *Science* 226, 687

Brummell NH, Hurlburt NE, Toomre J, 1996, *The Astrophysical Journal* 473, 494

Brun AS, Turck-Chièze S, Zahn JP, 1999, *The Astrophysical Journal* 525, 1032

Brun AS, Toomre J, 2002, *The Astrophysical Journal* 570, 865

Brun AS, 2003, in *Magnetism and Activity of the Sun and Stars*, J. Arnaud et al. (Eds.), EAS, EDP Sciences, p. 179

Buffett BA, Huppert HE, Lister JR, Woods AW, 1992, *Nature* 356, 329

Buffett BA, 1996, *Geophysical Research Letters* 23, 3803

Buffett BA, Huppert HE, Lister JR, Woods AW, 1996, *J. Geophysical Research* 101, 7989

Buffett BA, 1997, *Nature* 388, 571

Buffett BA, Glatzmaier GA, 2000, *Geophysical Research Letters* 27, 3125

Buffett BA, Wenk HR, 2001, *Nature* 413, 60

Buffett BA, 2003, *Geophysical J. International* 153, 753

Bumba V, 1963, *Bulletin Astronomical Institute Czechoslovakia* 14, 91

Burkert A, Bodenheimer P, 1993, *Month. Not. Roy. Astr. Soc.* 264, 798

Burrows CJ, Stapelfeldt KR, Watson AM, et al., 1996, *The Astrophysical Journal* 473, 437

Busse FH, 1970, *The Astrophysical Journal* 159, 629

Busse FH, Miin SW, 1979, *Geophysical Astrophysical Fluid Dyn.* 14, 167

Busse FH, 2000, *Ann. Rev. Fluid Mechanics* 32, 383

Busse FH, 2002a, *Physics of Fluids* 14, 1301

Busse FH, 2002b, *Geophysical Research Letters* 29, art no 1105

Cabot W, Pollack JB, 1992, *Geophysical Astrophysical Fluid Dyn.* 64, 97

Caligari P, Moreno-Insertis F, Schüssler M, 1995, *The Astrophysical Journal* 441, 886

Caligari P, Schüssler M, Moreno-Insertis F, 1998, *The Astrophysical Journal* 502, 481

Camenzind M, Lesch H, 1994, *Astronomy & Astrophysics* 284, 411

Campbell CG, 1997, *Magnetohydrodynamics in Binary Stars*, Kluwer

Campbell CG, Papaloizou JCB, Agapitou V, 1998, *Month. Not. Roy. Astr. Soc.* 300, 315

Campbell CG, 2002, *Month. Not. Roy. Astr. Soc.* 336, 999

Canuto VM, Minotti FO, Schilling O, 1994, *The Astrophysical Journal* 425, 303

Canuto VM, 1998, *The Astrophysical Journal* 497, L51

Carilli CL, Taylor GB, 2002, *Ann. Rev. Astronomy Astrophysics* 40, 319

Cassen PM, Smith BF, Miller RH, Reynolds RT, 1981, *Icarus* 48, 377

Cattaneo F, Hurlburt NE, Toomre J, 1990, *The Astrophysical Journal* 349, L63

Cattaneo F, Hughes DW, Weiss NO, 1991, *Month. Not. Roy. Astr. Soc.* 253, 479

Cattaneo F, Vainshtein SI, 1991, *The Astrophysical Journal* 376, L21

Chan KL, 2001, *The Astrophysical Journal* 548, 1102

Chandrasekhar S, Fermi E, 1953, *The Astrophysical Journal* 118, 113

Chandrasekhar S, 1960, *Proc. Nat. Acad. Sci.* 46, 53

Chandrasekhar S, 1961, *Hydrodynamic and Hydromagnetic Stability*, Oxford University Press

Charbonneau P, MacGregor KB, 1993, *The Astrophysical Journal* 417, 762

Charbonneau P, MacGregor KB, 1997, *The Astrophysical Journal* 486, 502

Charbonneau P, Dikpati M, Gilman PA, 1999, *The Astrophysical Journal* 526, 523

Charbonneau P, 2001, *Solar Physics* 199, 385

Charbonneau P, MacGregor KB, 2001, *The Astrophysical Journal* 559, 1094

Chiba M, Tosa M, 1990, *Month. Not. Roy. Astr. Soc.* 244, 714

Childress S, Soward AM, 1972, *Physical Review Letters* 29, 837

Choudhuri AR, 1989, *Solar Physics* 123, 217

Choudhuri AR, 1990, *The Astrophysical Journal* 355, 733

Choudhuri AR, 1992, *Astronomy & Astrophysics* 253, 277

Choudhuri AR, Schüssler M, Dikpati M, 1995, *Astronomy & Astrophysics* 303, L29

Choudhuri AR, 1998, *The Physics of Fluids and Plasmas. An Introduction for Astrophysicists*, Cambridge University Press

Christensen UR, Olson P, Glatzmaier GA, 1999, *Geophysical J. International* 138, 393

Christensen UR, Aubert J, Cardin P, et al., 2001, *Physics Earth Planetary Interiors* 128, 25

Christensen UR, Olson P, 2003, *Physics Earth Planetary Interiors* 138, 39

Christensen-Dalsgaard J, Monteiro MJPFG, Thompson MJ, 1995, *Month. Not. Roy. Astr. Soc.* 276, 283

Christensen-Dalsgaard J, Thompson MJ, 1997, *Month. Not. Roy. Astr. Soc.* 284, 527

Chyży KT, Beck R, Kohle S, et al., 2000, *Astronomy & Astrophysics* 355, 128

Clarke TE, Norman ML, Fiedler RA, 1994, *ZEUS-3D User Manual*, University of Illinois

CollierCameron A, 2002, *Astronomische Nachrichten* 323, 336

Connerney JEP, 1993, *J. Geophysical Research* 98, 18659

Couette M, 1890, *Ann. Chem. Phys.* 21, 433

Couvidat S, García RA, Turck-Chièze S, et al., 2003, *The Astrophysical Journal* 597, L77

Cowling TG, 1934, *Month. Not. Roy. Astr. Soc.* 94, 768

Creager KC, 1997, *Science* 278, 1284

Crutcher RM, 1999, *The Astrophysical Journal* 520, 706

Curry C, Pudritz RE, Sutherland PG, 1994, *The Astrophysical Journal* 434, 206

Curry C, Pudritz RE, 1995, *The Astrophysical Journal* 453, 697

Dame TM, Thaddeus P, 1994, *The Astrophysical Journal* 436, L173

Davidson PA, Siso-Nadal F, 2002, *Geophysical Astrophysical Fluid Dyn.* 96, 49

DeCampli WM, Baliunas SL, 1979, *The Astrophysical Journal* 230, 815

Deinzer W, 1993, in *The Cosmic Dynamo*, F. Krause et al. (Eds.), Kluwer, p. 185

Deinzer W, Grosser H, Schmitt D, 1993, *Astronomy & Astrophysics* 273, 405

Deiss BM, Reich W, Lesch H, Wielebinski R, 1997, *Astronomy & Astrophysics* 321, 55

DeLuca EE, Gilman PA, 1991, in *Solar Interior and Atmosphere*, A. N. Cox et al. (Eds.), Arizona University Press, p. 275

DeRosa ML, Toomre J, 2001, in *Helio- and Asteroseismology at the Dawn of the Millennium*, A. Wilson (Ed.), ESA Publications Division, p. 595

De Wijs GA, Kresse G, Vocadlo L, et al., 1998, *Nature* 392, 805

Dickey JM, Lockman FJ, 1990, *Ann. Rev. Astronomy Astrophysics* 28, 215

Dikpati M, Charbonneau P, 1999, *The Astrophysical Journal* 518, 508

Dikpati M, Gilman PA, 2000, *The Astrophysical Journal* 528, 552

Dikpati M, Gilman PA, 2001, *The Astrophysical Journal* 559, 428

Dobler W, Poezd AD, Shukurov A, 1996, *Astronomy & Astrophysics* 312, 663

Donahue RA, Baliunas SL, 1992, *The Astrophysical Journal* 393, L63

Donahue RA, 1996, in *Stellar Surface Structure*, K. G. Strassmeier et al. (Eds.), Kluwer, p. 261

Donahue RA, Saar SH, Baliunas SL, 1996, *The Astrophysical Journal* 466, 384

Donati JF, CollierCameron A, 1997, *Month. Not. Roy. Astr. Soc.* 291, 1

Donati JF, Mengel M, Carter BD, et al., 2000, *Month. Not. Roy. Astr. Soc.* 316, 699

Donati JF, CollierCameron A, Petit P, 2003, *Month. Not. Roy. Astr. Soc.* 345, 1187

Donati JF, CollierCameron A, Semel M, et al., 2003, *Month. Not. Roy. Astr. Soc.* 345, 1145

Donnelly RJ, Ozima M, 1960, *Physical Review Letters* 4, 497

Donner KJ, Brandenburg A, 1990, *Astronomy & Astrophysics* 240, 289

Dorch SBF, Nordlund Å, 2001, *Astronomy & Astrophysics* 365, 562

Dorfi E, 1982, *Astronomy & Astrophysics* 114, 151

Dormy E, Cardin P, Jault D, 1998, *Earth Planetary Science Letters* 160, 15

Dormy E, Valet JP, Courtillot V, 2000, *Geochem. Geophys. Geosys.* 1, 62

Draine BT, Roberge WG, Dalgarno A, 1983, *The Astrophysical Journal* 264, 485

Drecker A, Hollerbach R, Rüdiger G, 1998, *Month. Not. Roy. Astr. Soc.* 298, 1030

Drecker A, Rüdiger G, Hollerbach R, 2000, *Month. Not. Roy. Astr. Soc.* 317, 45

Dubrulle B, 1993, *Icarus* 106, 59

Dubrulle B, Morfill G, Sterzik M, 1995, *Icarus* 114, 237

Dudley ML, James RW, 1989, *Proc. Royal Society London A* 425, 407

Durisen RH, 2001, in *The Formation of Binary Stars*, H. Zinnecker et al. (Eds.), ASP, p. 381

Durney BR, Latour J, 1978, *Geophysical Astrophysical Fluid Dyn.* 9, 241

Durney BR, Spruit HC, 1979, *The Astrophysical Journal* 234, 1067

Durney BR, 1987, in *The Internal Solar Angular Velocity*, B. R. Durney et al. (Eds.), Reidel, p. 235

Durney BR, 1989, *The Astrophysical Journal* 338, 509

Durney BR, 1995, *Solar Physics* 160, 213

Duschl WJ, Strittmatter PA, Biermann PL, 2000, *Astronomy & Astrophysics* 357, 1123

Duvall TL, D'Silva S, Jefferies SM, et al., 1996, *Nature* 379, 235

Duvall TL, Gizon L, 2000, *Solar Physics* 192, 177

Dziembowski WA, Goode PR, 1991, *The Astrophysical Journal* 376, 782

Dziembowski WA, Kosovichev A, 1987, *Acta Astronomica* 37, 341

Dziourkevitch N, Elstner D, Rüdiger G, 2004, *Astronomy & Astrophysics*, subm.

Eddington AS, 1929, *Month. Not. Roy. Astr. Soc.* 90, 54

Eddy JA, 1976, *Science* 192, 1189

Egbers C, Pfister G, 2000, *Physics of Rotating Fluids*, Springer

Egorov P, Rüdiger G, Ziegler U, 2004, *Astronomy & Astrophysics*, subm.

Elstner D, Rüdiger G, Tschäpe R, 1989, *Geophysical Astrophysical Fluid Dyn.* 48, 235

Elstner D, Meinel R, Rüdiger G, 1990, *Geophysical Astrophysical Fluid Dyn.* 50, 85

Elstner D, Meinel R, Beck R, 1992, *Astronomy & Astrophysics Suppl.* 94, 587

Elstner D, Golla G, Rüdiger G, Wielebinski R, 1995, *Astronomy & Astrophysics* 297, 77

Elstner D, Rüdiger G, Schultz M, 1996, *Astronomy & Astrophysics* 306, 740

Elstner D, Otmianowska-Mazur K, von Linden S, Urbanik M, 2000, *Astronomy & Astrophysics* 357, 129

Eltayeb IA, Kumar S, 1977, *Proc. Royal Society London A* 353, 145

Evans CR, Hawley JF, 1988, *The Astrophysical Journal* 332, 659

Fan Y, Fisher GH, DeLuca EE, 1993, *The Astrophysical Journal* 405, 390

Fautrelle Y, Childress S, 1982, *Geophysical Astrophysical Fluid Dyn.* 22, 235

Fearn DR, 1979, *Geophysical Astrophysical Fluid Dyn.* 14, 103

Fearn DR, Loper DE, Roberts PH, 1981, *Nature* 292, 232

Fearn DR, Proctor MRE, 1983, *J. Fluid Mechanics* 128, 21

Fearn DR, 1984, *Geophysical Astrophysical Fluid Dyn.* 30, 227

Fearn DR, Weiglhofer WS, 1991, *Geophysical Astrophysical Fluid Dyn.* 60, 275

Fearn DR, Lamb CJ, McLean DR, Ogden RR, 1997, *Geophysical Astrophysical Fluid Dyn.* 86, 173

Fearn DR, 1998, *Reports on Progress in Physics* 61, 175

Feigelson ED, Gaffney JA, Garmire G, et al., 2003, *The Astrophysical Journal* 584, 911

Fendt C, Elstner D, 1999, *Astronomy & Astrophysics* 349, L61

Fendt C, Elstner D, 2000, *Astronomy & Astrophysics* 363, 208

Ferrière K, 1992a, *The Astrophysical Journal* 391, 188

Ferrière K, 1992b, *The Astrophysical Journal* 389, 286

Ferrière K, 1996, *Astronomy & Astrophysics* 310, 438

Ferrière K, 1998, *Astronomy & Astrophysics* 335, 488

Ferrière K, Schmitt D, 2000, *Astronomy & Astrophysics* 358, 125

Ferriz-Mas A, Schüssler M, 1993, *Geophysical Astrophysical Fluid Dyn.* 72, 209

Ferriz-Mas A, Schüssler M, 1995, *Geophysical Astrophysical Fluid Dyn.* 81, 233

Fisher GH, Longcope DW, Linton MG, et al., 1999, in *Stellar Dynamos: Nonlinearity and Chaotic Flows*, M. Núñez et al. (Eds.), ASP, p. 35

Fleming TP, Stone JM, Hawley JF, 2000, *The Astrophysical Journal* 530, 464

Frank-Kamenezki DA, 1967, *Vorlesungen über Plasmaphysik*, VEB Deutscher Verlag der Wissenschaften

Freytag B, Ludwig HG, Steffen M, 1996, *Astronomy & Astrophysics* 313, 497

Frick P, Galyagin D, Hoyt DV, et al., 1997, *Astronomy & Astrophysics* 328, 670

Frick P, Grossmann A, Tchamitchian P, 1998, *J. Math. Phys.* 39, 4091

Fricke K, 1969, *Astronomy & Astrophysics* 1, 388

Fricke K, Smith RC, 1971, *Astronomy & Astrophysics* 15, 329

Fröhlich HE, Schultz M, 1996, *Astronomy & Astrophysics* 311, 451

Fröhlich HE, Tschäpe R, Rüdiger G, Strassmeier KG, 2002, *Astronomy & Astrophysics* 391, 659

Gabriel M, Carlier F, 1997, *Astronomy & Astrophysics* 317, 580

Gailitis A, 1970, *Magnitnaya Gidrodinamika* 6, 19

Gailitis A, 1994, in *Energy Transfer in Magnetohydrodynamic Flows*, A. Alemany et al. (Eds.), Plenum Press, p. 177

Gailitis A, Rüdiger G, 1982, *Astrophysical Letters* 22, L89

Gailitis A, Lielausis O, Platacis E, et al., 2001, *Physical Review Letters* 86, 3024

Galli D, Shu FH, 1993, *The Astrophysical Journal* 417, 243

Gammie CF, 1996, *The Astrophysical Journal* 457, 355

Garaud P, 2001, *Month. Not. Roy. Astr. Soc.* 324, 68

Garaud P, 2002, *Month. Not. Roy. Astr. Soc.* 329, 1

Geppert U, Gil J, Rüdiger G, Zub M, 2003, in *Young Neutron Stars and Their Environment*, F. Camilo et al. (Eds.), IAU Symp. 218, p. 12

Geppert U, Küker M, Page D, 2004, *Astronomy & Astrophysics*, subm.

Gilbert AD, Sulem PL, 1990, *Geophysical Astrophysical Fluid Dyn.* 51, 243

Gilman PA, 1977, *Geophysical Astrophysical Fluid Dyn.* 8, 93

Gilman PA, Miller J, 1981, *Astrophys. J. Suppl.* 46, 211

Gilman PA, Howard RF, 1984, *Solar Physics* 93, 171

Gilman PA, 1986, in *Physics of the Sun*, P. A. Sturrock (Ed.), Reidel, p. 95

Gilman PA, Miller J, 1986, *Astrophys. J. Suppl.* 61, 585

Gilman PA, 1992, in *The Solar Cycle*, K. L. Harvey (Ed.), ASP, p. 241

Gilman PA, Fox PA, 1997, *The Astrophysical Journal* 484, 439

Gilman PA, Howe R, 2003, in *Local and Global Helioseismology: The Present and Future*, H. Saways-Lacosta (Ed.), ESA Publications Division, p. 283

Glatzmaier GA, 1984, *J. Computational Physics* 55, 461

Glatzmaier GA, 1985, *The Astrophysical Journal* 291, 300

Glatzmaier GA, Roberts PH, 1995, *Nature* 377, 203

Glatzmaier GA, Roberts PH, 1996, *Science* 274, 1887

Glatzmaier GA, Coe RS, Hongre L, Roberts PH, 1999, *Nature* 401, 885

Glatzmaier GA, 2002, *Ann. Rev. Earth Planetary Science* 30, 237

Goldman I, Wandel A, 1995, *The Astrophysical Journal* 443, 187

Goldreich P, Schubert G, 1967, *The Astrophysical Journal* 150, 571

Goldreich P, Ward WR, 1973, *The Astrophysical Journal* 183, 1051

Goldreich P, Reisenegger A, 1992, *The Astrophysical Journal* 395, 250

Golla G, Hummel E, 1993, in *The Cosmic Dynamo*, F. Krause et al. (Eds.), Kluwer, p. 299

Golla G, Hummel E, 1994, *Astronomy & Astrophysics* 284, 777

Golla G, Wielebinski R, 1994, *Astronomy & Astrophysics* 286, 733

Goode PR, Dziembowski WA, 1993, in *Gong 1992: Seismic Investigation of the Sun and Stars*, T. M. Brown (Ed.), ASP, p. 229

Goodman J, Ji H, 2002, *J. Fluid Mechanics* 462, 365

Goodson AP, Böhm KH, Wingley RM, 1999, *The Astrophysical Journal* 524, 142

Gough DO, 1976, *Lecture Notes in Physics* 71, 15

Gough DO, 1978, in *Solar Rotation*, G. Belvedere et al. (Eds.), University of Catania, p. 87

Gough DO, Toomre J, 1991, *Ann. Rev. Astronomy Astrophysics* 29, 627

Gough DO, Kosovichev AG, 1995, in *Helioseismology*, J. T. Hoeksema et al. (Eds.), ESA, p. 47

Gough DO, McIntyre ME, 1998, *Nature* 394, 755

Granzer T, 2002, *Astronomische Nachrichten* 323, 395

Greenspan HP, 1968, *The Theory of Rotating Fluids*, Cambridge University Press

Grosser H, 1988, *Astronomy & Astrophysics* 199, 235

Grossmann-Doerth U, Schüssler M, Steiner O, 1998, *Astronomy & Astrophysics* 337, 928

Grote E, Busse FH, Tilgner A, 1999, *Physical Review E* 60, 5025

Grote E, Busse FH, 2000, *Physical Review E* 62, 4457

Grote E, Busse FH, Tilgner A, 2000a, *Geophysical Research Letters* 27, 2001

Grote E, Busse FH, Tilgner A, 2000b, *Physics Earth Planetary Interiors* 117, 259

Gubbins D, 1981, *J. Geophysical Research* 86, 1695

Gubbins D, Zhang K, 1993, *Physics Earth Planetary Interiors* 75, 225

Gubbins D, 1994, *Rev. Geophysics* 32, 61

Gubbins D, 1997, *Astron. Geophysics* 38, 15

Gubbins D, 1999, *Geophysical J. International* 137, 1

Gubbins D., 2001, *Physics Earth Planetary Interiors* 128, 3

Gubbins D, Alfè D, Masters G, et al., 2003, *Geophysical J. International* 155, 609

Guenther EW, Emerson JP, 1996, *Astronomy & Astrophysics* 309, 777

Guenther EW, Lehmann H, Emerson JP, Staude J, 1999, *Astronomy & Astrophysics* 341, 768

Guinan E, Giménez A., 1992, in *The Realm of Interacting Binary Stars*, J. Sahade et al. (Eds.), Kluwer, p. 51

Gullbring E, Barwig H, Schmitt JHMM, 1997, *Astronomy & Astrophysics* 324, 155

Guzik JA, Swenson FJ, 1997, *The Astrophysical Journal* 491, 967

Haber DA, Hindman BW, Toomre J, et al., 2002, *The Astrophysical Journal* 570, 855

Hale GE, 1927, *Nature* 119, 708

Hall DS, 1976, in *Multiple Periodic Variable Stars*, W. S. Fitch (Ed.), Reidel, p. 287

Hall DS, 1989, *Space Science Review* 50, 219

Hall DS, 1990, in *Active Close Binaries*, C. Ibanoglu (Ed.), Kluwer, p. 95

Hall DS, 1991, in *The Sun and Cool Stars: Activity, Magnetism, Dynamos*, I. Tuominen et al. (Eds.), Springer, p. 353

Hanasz M, Lesch H, Krause M, 1991, *Astronomy & Astrophysics* 243, 381

Hanasz M, Lesch H, 1997, *Astronomy & Astrophysics* 321, 1007

Hathaway DH, Somerville RCJ, 1983, *J. Fluid Mechanics* 126, 75

Hathaway DH, Nandy D, Wilson RM, Reichmann EJ, 2003, *The Astrophysical Journal* 589, 665

Hatzes AP, 1995, *The Astrophysical Journal* 451, 784

Hawley JF, Balbus SA, 1991, *The Astrophysical Journal* 376, 223

Hawley JF, Gammie CF, Balbus SA, 1995, *The Astrophysical Journal* 440, 742

Hawley JF, 2000, *The Astrophysical Journal* 528, 462

Hayashi MR, Shibata K, Matsumoto R, 1996, *The Astrophysical Journal* 468, 37

Helmis G, 1968, *Monatsbericht der deutschen Akademie der Wissenschaften zu Berlin* 10, 280

Hempelmann A, Kurths J, 1990, *Astronomy & Astrophysics* 232, 356

Hempelmann A, Schmitt JHMM, Stępień K, 1996, *Astronomy & Astrophysics* 305, 284

Hempelmann A, Schmitt JHMM, Baliunas SL, Donahue RA, 2003, *Astronomy & Astrophysics* 406, L39

Henning T, Wolf S, Launhardt R, Waters R, 2001, *The Astrophysical Journal* 561, 871

Henry GW, Eaton JA, Hamer J, Hall DS, 1995, *Astrophys. J. Suppl.* 97, 513

Hersant F, Dubrulle B, Huré JM, 2004, *Astronomy & Astrophysics*, subm.

Herzenberg A, 1958, *Phil. Trans. Royal Society London A* 250, 543

Heyl JS, Hernquist L, 2001, *Month. Not. Roy. Astr. Soc.* 324, 292

Heyvaerts J, Norman C, 1989, *The Astrophysical Journal* 347, 1055

Hill GM, Bohlender DA, Landstreet JD, et al., 1998, *Month. Not. Roy. Astr. Soc.* 297, 236

Hodgson LS, Brandenburg A, 1998, *Astronomy & Astrophysics* 330, 1169

Hollerbach R, Ierley GR, 1991, *Geophysical Astrophysical Fluid Dyn.* 56, 133

Hollerbach R, Barenghi CF, Jones CA, 1992, *Geophysical Astrophysical Fluid Dyn.* 67, 3

Hollerbach R, Jones CA, 1993, *Nature* 365, 541

Hollerbach R, Proctor MRE, 1993, in *Solar and Planetary Dynamos*, M. R. E. Proctor et al. (Eds.), Cambridge University Press, p. 145

Hollerbach R, 1994a, *Proc. Royal Society London A* 444, 333

Hollerbach R, 1994b, *Physics of Fluids* 6, 2540

Hollerbach R, 1997, *Geophysical Astrophysical Fluid Dyn.* 84, 85

Hollerbach R, 2000, *Int. J. Numer. Meth. Fluids* 32, 773

Hollerbach R, Rüdiger G, 2002, *Month. Not. Roy. Astr. Soc.* 337, 216

Hollerbach R, 2003, in *Earth's Core: Dynamics, Structure, Rotation*, V. Dehant et al. (Eds.), AGU, p. 181

Hollerbach R, Rüdiger G, 2004, *Month. Not. Roy. Astr. Soc.*, 347, 1273

Holme R, Bloxham J, 1996, *J. Geophysical Research* 101, 2177

Hood LL, Jirikowic JL, 1990, in *Climate Impact of Solar Variability*, NASA, p. 98

Horne K, 1993, in *Accretion Disks in Compact Stellar Systems*, J. C. Wheeler (Ed.), World Scientific, p. 117

Howard RF, LaBonte BJ, 1980, *The Astrophysical Journal* 239, L33

Howard RF, LaBonte BJ, 1983, in *Solar and Stellar Magnetic Fields: Origins and Coronal Effects*, J. O. Stenflo (Ed.), Reidel, p. 101

Howe R, Hill F, Komm RW, et al., 2001, in *Recent Insights into the Physics of the Sun and Heliosphere: Highlights from SOHO and Other Space Missions*, P. Brekke et al. (Eds.), ASP, p. 40

Howe R, Komm RW, Hill F, 2002, *The Astrophysical Journal* 580, 1172

Hoyng P, 1988, *The Astrophysical Journal* 332, 857

Hoyng P, 1993, *Astronomy & Astrophysics* 272, 321

Hoyng P, Schmitt D, Teuben LJW, 1994, *Astronomy & Astrophysics* 289, 265

Hoyt DV, Schatten KH, Nesme-Ribes E, 1994, in *The Solar Engine and Its Influence on the Terrestrial Atmosphere and Climate*, E. Nesme-Ribes (Ed.), Springer, p. 57

Hubrig S, North P, Mathys G, 2000, *The Astrophysical Journal* 539, 352

Hubrig S, North P, Medici A, 2000, *Astronomy & Astrophysics* 359, 306

Hujeirat A, Camenzind M, Yorke HW, 2000, *Astronomy & Astrophysics* 354, 1041

Hujeirat A, Myers P, Camenzind M, Burkert A, 2000, *New Astronomy* 4, 601

Hummel E, Dettmar RJ, 1990, *Astronomy & Astrophysics* 236, 33

Hummel E, Beck R, Dahlem M, 1991, *Astronomy & Astrophysics* 248, 23

Huré JM, Richard D, Zahn JP, 2001, *Astronomy & Astrophysics* 367, 1087
Hurlburt NE, Toomre J, Massaguer JM, 1986, *The Astrophysical Journal* 311, 563
Hurlburt NE, Toomre J, Massaguer JM, Zahn JP, 1994, *The Astrophysical Journal* 421, 245
Hutcheson KA, Fearn DR, 1996, *Physics Earth Planetary Interiors* 97, 43
Igumenshchev IV, Abramowicz MA, Narayan R, 2000, *The Astrophysical Journal* 537, L271
Jackson A, Jonkers ART, Walker MR, 2000, *Phil. Trans. Royal Society London A* 358, 957
Jardine M, Wood K, CollierCameron A, et al., 2002, *Month. Not. Roy. Astr. Soc.* 336, 1364
Jault D, Gire C, Le Mouel JL, 1988, *Nature* 333, 353
Jault D, 1996, *CR Acad. Sci. II A* 323, 451
Jennings RL, Weiss NO, 1991, *Month. Not. Roy. Astr. Soc.* 252, 249
Jetsu L, Huovelin J, Tuominen I, Savanov I, 1991, *Astronomy & Astrophysics* 248, 574
Ji H, Goodman J, Kageyama A, 2001, *Month. Not. Roy. Astr. Soc.* 325, L1
Jiménez-Reyes SJ, Régulo C, Pallé PL, Roca Cortés T, 1998, *Astronomy & Astrophysics* 329, 1119
Jin L, 1996, *The Astrophysical Journal* 457, 798
Johns-Krull CM, 1996, *Astronomy & Astrophysics* 306, 803
Johns-Krull CM, Christopher M, Valenti JA, et al., 1999, *The Astrophysical Journal* 510, L41
Joncour I, Bertout C, Ménard F, 1994, *Astronomy & Astrophysics* 285, L25
Jones CA, 2000, *Phil. Trans. Royal Society London A* 358, 873
Jones CA, Roberts PH, 2000, *J. Fluid Mechanics* 404, 311
Jones CA, Soward AM, Mussa AI, 2000, *J. Fluid Mechanics* 405, 157
Jones CA, Mussa AI, Worland SJ, 2003, *Proc. Royal Society London A* 459, 773
Jones PB, 1988, *Month. Not. Roy. Astr. Soc.* 233, 875
Kaisig M, Rüdiger G, Yorke HW, 1993, *Astronomy & Astrophysics* 274, 757
Karato S, 1993, *Science* 262, 1708
Karato S, 1999, *Nature* 402, 871
Keinigs RK, 1983, *Physics of Fluids* 76, 2558
Kempf S, Pfalzner S, Henning T, 1999, *Icarus* 141, 388
Keppens R, Casse F, Goedbloed JP, 2002, *The Astrophysical Journal* 569, L121
Kim EJ, 1997, *The Astrophysical Journal* 477, 183
Kim EJ, Diamond PH, 2002, *The Astrophysical Journal* 578, L113
Kim KT, Kronberg PP, Dewdney PE, Landecker TL, 1990, *The Astrophysical Journal* 355, 29
Kippenhahn R, 1954, *Z. Astrophysik* 35, 165
Kippenhahn R, 1958, *Z. Astrophysik* 46, 26
Kippenhahn R, 1963, *The Astrophysical Journal* 137, 664
Kitchatinov LL, 1986, *Geophysical Astrophysical Fluid Dyn.* 35, 93
Kitchatinov LL, 1987, *Geophysical Astrophysical Fluid Dyn.* 38, 273
Kitchatinov LL, 1991, *Astronomy & Astrophysics* 243, 483
Kitchatinov LL, Rüdiger G, 1992, *Astronomy & Astrophysics* 260, 494
Kitchatinov LL, Rüdiger G, 1993, *Astronomy & Astrophysics* 276, 96
Kitchatinov LL, Pipin VV, Rüdiger G, 1994, *Astronomische Nachrichten* 315, 157
Kitchatinov LL, Rüdiger G, Küker M, 1994, *Astronomy & Astrophysics* 292, 125
Kitchatinov LL, Rüdiger G, 1995, *Astronomy & Astrophysics* 299, 446
Kitchatinov LL, Mazur MV, 1997, *Astronomy & Astrophysics* 324, 821

Kitchatinov LL, Rüdiger G, 1997, *Month. Not. Roy. Astr. Soc.* 286, 757

Kitchatinov LL, Rüdiger G, 1999, *Astronomy & Astrophysics* 344, 911

Kitchatinov LL, Mazur MV, 2000, *Solar Physics* 191, 325

Kitchatinov LL, 2002, *Astronomy & Astrophysics* 394, 1135

Kitchatinov LL, Rüdiger G, 2004, *Astronomy & Astrophysics*, acc

Kivelson MG, Khurana KK, Russell CT, et al., 2000, *Science* 289, 1340

Klahr H, Henning T, 1997, *Icarus* 128, 213

Klahr H, Bodenheimer P, 2003, *The Astrophysical Journal* 582, 869

Kleeorin N, Rogachevskii I, Ruzmaikin A, et al., 1997, *J. Fluid Mechanics* 344, 213

Kleeorin N, Kuzanyan K, Moss D, et al., 2003, *Astronomy & Astrophysics* 409, 1097

Klessen R, Heitsch F, Mac Low MM, 2000, *The Astrophysical Journal* 535, 887

Klessen R, 2003, *The Relation between Interstellar Turbulence and Star Formation*, Habilitationsschrift, Universität Potsdam

Kley W, Papaloizou JCB, Lin DNC, 1993, *The Astrophysical Journal* 416, 679

Knobloch E, 1992, *Month. Not. Roy. Astr. Soc.* 255, 25

Knobloch E, Landsberg AS, 1996, *Month. Not. Roy. Astr. Soc.* 278, 294

Knobloch E, Tobias SM, Weiss NO, 1998, *Month. Not. Roy. Astr. Soc.* 297, 1123

Ko CM, Parker EN, 1989, *The Astrophysical Journal* 341, 828

Ko CM, 1993, *The Astrophysical Journal* 410, 128

Köhler H, 1969, *Differentielle Rotation als Folge anisotroper turbulenter Viskosität*, Thesis, Universität Göttingen

Köhler H, 1970, *Solar Physics* 13, 3

Köhler H, 1973, *Astronomy & Astrophysics* 25, 467

König B, Neuhäuser R, Stelzer B, 2001, *Astronomy & Astrophysics* 369, 971

Kono M, Sakuraba A, Ishida M, 2000, *Phil. Trans. Royal Society London A* 358, 1123

Kono M, Roberts PH, 2001, *Physics Earth Planetary Interiors* 128, 13

Kono M, Roberts PH, 2002, *Rev. Geophysics* 40, art. no. 1013

Korhonen H, Berdyugina SV, Strassmeier KG, Tuominen I, 2001, *Astronomy & Astrophysics* 379, L30

Korhonen H, Berdyugina SV, Tuominen I, 2002, *Astronomy & Astrophysics* 390, 179

Korpi MJ, 1999, *Interstellar Turbulence and Magnetic Fields: The Role of Supernova Explosions*, Thesis, University of Oulu

Korpi MJ, Brandenburg A, Shukurov A, Tuominen I, 1999, *Astronomy & Astrophysics* 350, 230

Korpi MJ, Brandenburg A, Shukurov A, et al., 1999, *The Astrophysical Journal* 514, L99

Koschmieder EL, 1993, *Benard Cells and Taylor Vortices*, Cambridge University Press

Kosovichev AG, 2002, *Astronomische Nachrichten* 323, 186

Krause F, 1967, *Eine Lösung des Dynamoproblems auf der Grundlage einer linearen Theorie der magnetohydrodynamischen Turbulenz*, Thesis, Universität Jena

Krause F, Rüdiger G, 1975, *Solar Physics* 42, 107

Krause F, Oetken L, 1976, in *Physics of Ap Stars*, W. W. Weiss et al. (Eds.), Universitätssternwarte Wien, p. 29

Krause F, Rädler KH, 1980, *Mean-field Magnetohydrodynamics and Dynamo Theory*, Pergamon Press

Krause F, Meinel R, 1988, *Geophysical Astrophysical Fluid Dyn.* 43, 95

Krause F, Meinel R, Elstner D, Rüdiger G, 1990, in *Galactic and Intergalactic Magnetic Fields*, R. Beck et al. (Eds.), Kluwer, p. 97

Krause F, Beck R, 1998, *Astronomy & Astrophysics* 335, 789

Krause M, Camenzind M, 2001, *Astronomy & Astrophysics* 380, 789

Krivodubskij VN, Schultz M, 1993, in *The Cosmic Dynamo*, F. Krause et al. (Eds.), Kluwer, p. 25

Krivova NA, Solanki SK, 2002, *Astronomy & Astrophysics* 394, 701

Kronberg PP, Perry JJ, Zukowski ELH, 1992, *The Astrophysical Journal* 387, 528

Kronberg PP, Dufton QW, Li H, Colgate SA, 2001, *The Astrophysical Journal* 560, 178

Kuang WJ, Bloxham J, 1997, *Nature* 389, 371

Kuang WJ, 1999, *Physics Earth Planetary Interiors* 116, 65

Kuang WJ, Bloxham J, 1999, *J. Computational Physics* 153, 51

Küker M, Rüdiger G, Kitchatinov LL, 1993, *Astronomy & Astrophysics* 279, L1

Küker M, Rüdiger G, 1997, *Astronomy & Astrophysics* 328, 253

Küker M, Arlt R, Rüdiger G, 1999, *Astronomy & Astrophysics* 343, 977

Küker M, Rüdiger G, Schultz M, 2001, *Astronomy & Astrophysics* 374, 301

Küker M., Stix M, 2001, *Astronomy & Astrophysics* 366, 668

Küker M, Henning T, Rüdiger G, 2003, *Astrophys. Space Sci.* 287, 83

Kulkarni SR, Fich M, 1985, *The Astrophysical Journal* 289, 792

Kulsrud RM, Andersen SW, 1992, *The Astrophysical Journal* 396, 606

Kulsrud RM, 1999, *Ann. Rev. Astronomy Astrophysics* 37, 37

Kurths J, Brandenburg A, Feudel U, Jansen W, 1993, in *The Cosmic Dynamo*, F. Krause et al. (Eds.), Kluwer, p. 83

Kurths J, Feudel U, Jansen W, et al., 1997, in *Past and Present Variability of the Solar-Terrestrial System: Measurement, Data Analysis and Theoretical Models*, G. Castagnoli et al. (Eds.), IOS Press, p. 247

Kurzweg UH, 1963, *J. Fluid Mechanics* 17, 52

Labrosse S, Macouin M, 2003, *CR Geosci* 335, 37

Lamm M, 2003, *Angular Momentum Evolution Of Low Mass Stars*, Thesis, Universität Heidelberg

Landau LD, Lifschitz EM, 1969, *Hydrodynamik*, Akademieverlag

Landstreet JD, 1987, *Month. Not. Roy. Astr. Soc.* 225, 437

Landstreet JD, Mathys G, 2000, *Astronomy & Astrophysics* 359, 213

Lantz SR, Fan Y, 1999, *Astrophys. J. Suppl.* 121, 247

Lanza AF, Rodonò M, Rosner R, 1998, *Month. Not. Roy. Astr. Soc.* 296, 893

Lanza AF, Rodonò M, 1999, *Astronomy & Astrophysics* 349, 887

Lanza AF, Rodonò M, 2004, *Astronomische Nachrichten* 325, 392

Larmor J, 1919, *Rep. Brit. Assoc. Adv. Sci. A* 159

Laske G, Masters G, 1999, *Nature* 402, 66

Latour J, Spiegel EA, Toomre J, Zahn JP, 1976, *The Astrophysical Journal* 207, 233

Laure P, Chossat P, Daviaud F, 2000, in *Dynamo and Dynamics, a Mathematical Challenge*, P. Chossat et al. (Eds.), Kluwer, p. 17

Lean J, 1994, in *The Solar Engine and Its Influence on Terrestrial Atmosphere and Climate*, E. Nesme-Ribes (Ed.), Springer, p. 163

van Leer B, 1977, *J. Computational Physics* 23, 276

Le Huy M, Alexandrescu M, Hulot G, Le Mouel JL, 1998, *Earth Planets Space* 50, 723

Leighton RB, 1969, *The Astrophysical Journal* 156, 1

Lesch H, Hanasz M, 2003, *Astronomy & Astrophysics* 401, 809

Levy EH, Sonett CP, 1978, in *Protostars and Planets*, T. Gehrels (Ed.), University of Arizona Press, p. 516

Levy EH, 1992, in *Cool Stars, Stellar Systems, and the Sun*, M. S. Giampapa et al. (Eds.), ASP, p. 223

Li Y, Wilson PR, 1998, *The Astrophysical Journal* 499, 504

Libbrecht KG, Woodard MF, 1990, *Nature* 345, 779

Linsky JL, 1988, in *Multiwavelength Astrophysics*, F. A. Cordova (Ed.), Cambridge University Press, p. 49

Linsky JL, 1994, in *Solar Coronal Structures*, V. Rušin et al. (Eds.), VEDA Publishing Company, p. 641

Linsky JL, 1999, in *Solar and Stellar Activity: Similarities and Differences*, C. J. Butler et al. (Eds.), ASP, p. 3

Lissauer JJ, 1993, *Ann. Rev. Astronomy Astrophysics* 31, 129

Livio M, Shaviv G, 1977, *Astronomy & Astrophysics* 55, 95

Livio M, 1997, in *Accretion Phenomena and Related Outflows*, D. T. Wickramasinghe et al. (Eds.), ASP, p. 845

Lockman FJ, 1984, *The Astrophysical Journal* 283, 90

Love JJ, 2000, *Geophysical Research Letters* 27, 2889

Low BC, 1996, *Solar Physics* 167, 217

Lubow SH, Papaloizou JCB, Pringle JE, 1994, *Month. Not. Roy. Astr. Soc.* 267, 235

Lüst R, 1952, *Z. Naturforschung* 7a, 87

Lufkin G, Quinn T, Wadsley J, et al., 2004, *Month. Not. Roy. Astr. Soc.* 347, 421

Lynden-Bell D, Ostriker JP, 1967, *Month. Not. Roy. Astr. Soc.* 136, 293

Lyne AG, 2000, *Phil. Trans. Royal Society London A* 358, 831

Ma CY, Biermann PL, 1998, *Astronomy & Astrophysics* 334, 736

MacGregor KB, Charbonneau P, 1999, *The Astrophysical Journal* 519, 911

Mac Low MM, 1999, *The Astrophysical Journal* 524, 169

Maeder A, 1975, *Astronomy & Astrophysics* 40, 303

Malhotra S, 1994, *The Astrophysical Journal* 433, 687

Malkus WVR, 1959, *The Astrophysical Journal* 130, 259

Malkus WVR, 1967, *J. Fluid Mechanics* 28, 793

Malkus WVR, Proctor MRE, 1975, *J. Fluid Mechanics* 67, 417

Markiel JA, Thomas JH, 1999, *The Astrophysical Journal* 523, 827

Martínez Pillet V, Moreno-Insertis F, Vázquez M, 1993, *Astronomy & Astrophysics* 274, 521

Martínez Pillet V, 2002, *Astronomische Nachrichten* 323, 342

Matese JJ, Whitmire DP, 1983, *Astronomy & Astrophysics* 117, L7

Matsumoto R, Tajima T, 1995, *The Astrophysical Journal* 445, 767

Mayer L, Quinn T, Wadsley J, Stadel J, 2003, in *Scientific Frontiers in Research on Extrasolar Planets*, D. Deming et al. (Eds.), ASP, p. 281

McElhinny MW, Senanayake WE, 1980, *J. Geophysical Research* 85, 3523

McIntyre M, 1994, in *The Solar Engine and Its Influence on the Terrestrial Atmosphere and Climate*, E. Nesme-Ribes (Ed.), Springer, p. 293

Meinel R, 1983, *Astronomische Nachrichten* 304, 65

Meinel R, Elstner D, Rüdiger G, 1990, *Astronomy & Astrophysics* 236, L33

Meinel R, 1991, in *The Sun and Cool Stars. Activity, Magnetism, Dynamos*, I. Tuominen et al. (Eds.), Springer, p. 103

Ménard F, Duchêne, 2003, in *Magnetism and Activity of the Sun and Stars*, J. Arnaud et al. (Eds.), EAS, EDP Sciences, p. 279

Meneguzzi M, Pouquet A, 1989, *J. Fluid Mechanics* 205, 297

Merrill RT, McElhinny MW, McFadden PL, 1998, *The Magnetic Field of the Earth*, Academic Press

Merrill RT, McFadden PL, 1999, *Rev. Geophysics* 37, 201

Messina S, Guinan EF, 2003, *Astronomy & Astrophysics* 409, 1017

Mestel L, Spitzer L, 1956, *Month. Not. Roy. Astr. Soc.* 116, 503

Mestel L, Weiss NO, 1987, *Month. Not. Roy. Astr. Soc.* 226, 123

Mestel L, Subramanian K, 1991, *Month. Not. Roy. Astr. Soc.* 248, 677

Mestel L, 1999, *Stellar Magnetism*, Clarendon

Michaud G, Zahn JP, 1998, *Theoretical Computational Fluid Dyn.* 11, 183

Miesch MS, Elliott JR, Toomre J, et al., 2000, *The Astrophysical Journal* 532, 593

Miesch MS, 2003, *The Astrophysical Journal* 586, 663

Miller K, Stone JM, 1997, *The Astrophysical Journal* 489, 890

Mininni PD, Gómez DO, 2002, *The Astrophysical Journal* 573, 454

Mininni PD, Gómez DO, Mahajan SM, 2002, *The Astrophysical Journal* 567, L81

Moffatt KH, 1966, *J. Fluid Mechanics* 28, 571

Moffatt KH, 1970, *J. Fluid Mechanics* 44, 705

Moffatt KH, 1972, *J. Fluid Mechanics* 53, 385

Moffatt KH, 1978, *Magnetic Field Generation in Electrically Conducting Fluids*, Cambridge University Press

Molemaker MJ, McWilliams JC, Yavneh I, 2001, *Physical Review Letters* 86, 5270

Montero I, Galán L, Najmi O, Albella JM, 1994, *Physical Review B* 50, 4881

Mordvinov AV, Willson RC, 2003, *Solar Physics* 215, 5

Moreno-Insertis F, 1983, *Astronomy & Astrophysics* 122, 241

Morfill G, Spruit H, Levy EH, 1993, in *Protostars and Planets III*, E. H. Levy et al. (Eds.), University of Arizona Press, p. 939

Morrison G, Fearn DR, 2000, *Physics Earth Planetary Interiors* 117, 237

Moss D, 1990, *Month. Not. Roy. Astr. Soc.* 243, 537

Moss D, Tuominen I, Brandenburg A, 1990, *Astronomy & Astrophysics* 228, 284

Moss D, Tuominen I, Brandenburg A, 1991, *Astronomy & Astrophysics* 245, 129

Moss D, Brandenburg A, 1992, *Astronomy & Astrophysics* 256, 371

Moss D, Brandenburg A, Tavakol R, Tuominen I, 1992, *Astronomy & Astrophysics* 265, 843

Moss D, Brandenburg A, Donner KJ, Thomasson M, 1993, *The Astrophysical Journal* 409, 179

Moss D, Barker DM, Brandenburg A, Tuominen I, 1995, *Astronomy & Astrophysics* 294, 155

Moss D, Brandenburg A, 1995, *Geophysical Astrophysical Fluid Dyn.* 80, 229

Moss D, 1996, *Astronomy & Astrophysics* 315, 63

Moss D, 1997, *Month. Not. Roy. Astr. Soc.* 289, 554

Moss D, Shukurov A, Sokoloff DD, et al., 1998, *Astronomy & Astrophysics* 335, 500

Moss D, 1999, *Month. Not. Roy. Astr. Soc.* 306, 300

Moss D, Rautiainen P, Salo H, 1999, *Month. Not. Roy. Astr. Soc.* 303, 125

Moss D, Brooke J, 2000, *Month. Not. Roy. Astr. Soc.* 315, 521

Moss D, 2003a, *Astronomy & Astrophysics* 403, 693

Moss D, 2003b, in *Magnetism and Activity of the Sun and Stars*, J. Arnaud et al. (Eds.), EAS,
    EDP Sciences, p. 21

Mouschovias TC, 1976a, *The Astrophysical Journal* 206, 753

Mouschovias TC, 1976b, *The Astrophysical Journal* 207, 141

Mouschovias TC, Spitzer L, 1976, *The Astrophysical Journal* 210, 326

Mouschovias TC, 1996, in *The Role of Dust in the Formation of Stars*, H. U. Käufl et al.
    (Eds.), Springer, p. 382

Münch G, Zirin H, 1961, *The Astrophysical Journal* 133, 11

Murawski K, Roberts B, 1993, *Astronomy & Astrophysics* 272, 601

Muslimov AG, van Horn HM, 1995, *The Astrophysical Journal* 442, 758

Nakagawa Y, 1959, *Proc. Royal Society London A* 249, 138

Nakamura F, Hanawa T, 1997, *The Astrophysical Journal* 480, 701

Nakano T, 1998, *The Astrophysical Journal* 494, 587

Nandy D, Choudhuri AR, 2002, *Science* 296, 1671

Nellis WJ, Holmes NC, Mitchell AC, et al., 1997, *J. Chem. Phys.* 107, 9096

Nellis WJ, 2000, *Planetary Space Science* 48, 671

Nelson RW, Bildsten L, Chakrabarty D, et al, 1997, *The Astrophysical Journal* 488, L117

Nesme-Ribes E, Ferreira EN, Mein P, 1993, *Astronomy & Astrophysics* 274, 563

Nesme-Ribes E, Ferreira EN, Vince L, 1993, *Astronomy & Astrophysics* 276, 211

Ness NF, 1994, *Phil. Trans. Royal Society London A* 349, 249

Neuhäuser R, Sterzik MF, Schmitt JHMM, et al., 1995, *Astronomy & Astrophysics* 295, L5

Nimmo F, Stevenson DJ, 2000, *J. Geophysical Research* 105, 11969

Noguchi K, Pariev VI, Colgate SA, et al., 2002, *The Astrophysical Journal* 575, 1151

Nordlund Å, Stein RF, 1989, in *Solar and Stellar Granulation*, R. J. Rutten et al. (Eds.),
    Kluwer, p. 453

Nordlund Å, Brandenburg A, Jennings RJ, et al., 1992, *The Astrophysical Journal* 392, 647

Nordlund Å, Galsgaard K, Stein R, 1994, in *Solar Surface Magnetism*, R. J. Rutten et al.
    (Eds.), Kluwer, p. 471

Noyes RW, Weiss NO, Vaughan AH, 1984, *The Astrophysical Journal* 287, 769

Oetken L, 1979, *Astronomische Nachrichten* 300, 1

Ogilvie GI, Pringle JE, 1996, *Month. Not. Roy. Astr. Soc.* 279, 152

Ogilvie GI, Livio M, 1998, *The Astrophysical Journal* 499, 329

Oláh K, Kolláth Z, Strassmeier KG, 2000, *Astronomy & Astrophysics* 356, 643

Oláh K, Jurczik J, Strassmeier KG, 2003, *Astronomy & Astrophysics* 410, 685

Olson P, Hagee VL, 1990, *J. Geophysical Research* 95, 4609

Oren AL, Wolfe AM, 1995, *The Astrophysical Journal* 445, 624

Orszag SA, 1970, *J. Atmospheric Science* 27, 890

Ossendrijver M, Hoyng P, Schmitt D, 1996, *Astronomy & Astrophysics* 313, 938

Ossendrijver M, 1997, *Astronomy & Astrophysics* 323, 151

Ossendrijver M, Stix M, Brandenburg A, 2001, *Astronomy & Astrophysics* 376, 713

Ossendrijver M, Stix M, Brandenburg A, Rüdiger G, 2002, *Astronomy & Astrophysics* 394, 735

Ossendrijver M, 2003, *Astronomy & Astrophysics Rev.* 11, 287

Ostriker EC, Stone JM, Gammie CF, 2001, *The Astrophysical Journal* 546, 980

Otmianowska-Mazur K, Urbanik M, Terech A, 1992, *Geophysical Astrophysical Fluid Dyn.* 66, 209

Otmianowska-Mazur K, Chiba M, 1995, *Astronomy & Astrophysics* 301, 41

Otmianowska-Mazur K, Rüdiger G, Elstner D, Arlt R, 1997, *Geophysical Astrophysical Fluid Dyn.* 86, 229

Otmianowska-Mazur K, Elstner D, Soida M, Urbanik M, 2002, *Astronomy & Astrophysics* 384, 48

Panesar JS, Nelson AH, 1992, *Astronomy & Astrophysics* 264, 77

Papaloizou JCB, Szuszkiewicz E, 1992, *Geophysical Astrophysical Fluid Dyn.* 66, 223

Papaloizou JCB, Terquem C, 1997, *Month. Not. Roy. Astr. Soc.* 287, 771

Parker EN, 1955, *The Astrophysical Journal* 122, 293

Parker EN, 1971, *The Astrophysical Journal* 164, 491

Parker EN, 1979, *Cosmical Magnetic Fields: Their Origin and Their Activity*, Oxford University Press

Parker EN, 1987, *Solar Physics* 110, 11

Parker EN, 1992, *The Astrophysical Journal* 401, 137

Parker EN, 1993, *The Astrophysical Journal* 408, 707

Paternò L, 1991, in *The Sun and Cool Stars. Activity, Magnetism, Dynamos*, I. Tuominen et al. (Eds.), Springer, p. 182

Pelletier G, Pudritz RE, 1992, *The Astrophysical Journal* 394, 117

Pelt J, Brooke J, Pulkkinen PJ, Tuominen I, 2000, *Astronomy & Astrophysics* 362, 1143

Petrovay K, Moreno-Insertis F, 1997, *The Astrophysical Journal* 485, 398

Petrovay K, van Driel-Gesztelyi L, 1997, *Solar Physics* 166, 249

Pevtsov AA, Canfield RC, Latushko SM, 2001, *The Astrophysical Journal* 549, L261

Piau L, Randich S, Palla F, 2003, *Astronomy & Astrophysics* 408, 1037

Pidatella RM, Stix M, Belvedere G, Paternò L, 1986, *Astronomy & Astrophysics* 156, 22

Pipin VV, 1999, *Astronomy & Astrophysics* 346, 295

Pipin VV, 2003, *Geophysical Astrophysical Fluid Dyn.* 97, 25

Poezd A, Shukurov A, Sokoloff DD, 1993, *Month. Not. Roy. Astr. Soc.* 264, 285

Poirier JP, 1994, *CR Acad. Sci. II A* 318, 341

Poppe T, Blum J, Henning T, 2000, *The Astrophysical Journal* 533, 472

Potekhin AY, Yakovlev DG, 2001, *Astronomy & Astrophysics* 374, 213

Pouquet A, Frisch U, Léorat J, 1976, *J. Fluid Mechanics* 77, 321

Priest ER, 1999, in *Solar and Stellar Activity: Similarities and Differences*, C. J. Butler et al. (Eds.), ASP, p. 321

Priklonsky V, Shukurov A, Sokoloff DD, Soward A, 2000, *Geophysical Astrophysical Fluid Dyn.* 93, 97

Pringle JE, 1981, *Ann. Rev. Astronomy Astrophysics* 19, 137

Pringle JE, 1996, *Month. Not. Roy. Astr. Soc.* 281, 357

Proctor MRE, 1977, *Astronomische Nachrichten* 298, 19

Pudritz RE, 1981, *Month. Not. Roy. Astr. Soc.* 195, 897

Pudritz RE, Norman CA, 1986, *The Astrophysical Journal* 301, 571

Pulkkinen PJ, Tuominen I, Brandenburg A, et al., 1993, *Astronomy & Astrophysics* 267, 265

Pulkkinen PJ, Brooke J, Pelt J, Tuominen I, 1999, *Astronomy & Astrophysics* 341, L43

Rädler KH, 1969, *Monatsbericht der Deutschen Akademie der Wissenschaften zu Berlin* 11, 272

Rädler KH, 1974, *Astronomische Nachrichten* 295, 265

Rädler KH, 1986a, *Astronomische Nachrichten* 307, 89

Rädler KH, 1986b, in *Plasma Astrophysics*, T. D. Guyenne et al. (Eds.), ESA Publications Division, p. 569

Rädler KH, Bräuer HJ, 1987, *Astronomische Nachrichten* 308, 101

Rädler KH, Seehafer N, 1990, in *Topological Fluid Mechanics*, H. K. Moffatt et al. (Eds.), Cambridge University Press, p. 157

Rädler KH, Wiedemann E, 1990, in *Galctic and Intergalactic Magnetic Fields*, R. Beck et al. (Eds.), Kluwer, p. 107

Rädler KH, Wiedemann E, Meinel R, et al., 1990, *Astronomy & Astrophysics* 239, 413

Rädler KH, Rheinhardt M, Apstein E, Fuchs H, 2002, *Nonlinear Processes in Geophysics* 9, 171

Rand RJ, Kulkarni SR, Hester JJ, 1992, *The Astrophysical Journal* 396, 97

Ray TP, Muxlow TWB, Axon DJ, et al., 1997, *Nature* 385, 415

Régulo C, Jiménez A, Pallé PL, et al., 1994, *The Astrophysical Journal* 434, 384

Reighard AB, Brown MR, 2001, *Physical Review Letters* 86, 2794

Reiners A, Schmitt JHMM, 2003a, *Astronomy & Astrophysics* 398, 647

Reiners A, Schmitt JHMM, 2003b, *Astronomy & Astrophysics* 412, 813

Reisenegger A, 2001, in *Magnetic Fields Across the Hertzsprung-Russell Diagram*, G. Mathys et al. (Eds.), ASP, p. 469

v. Rekowski B, Brandenburg A, Dobler W, Shukurov A, 2003, *Astronomy & Astrophysics* 398, 825

v. Rekowski M, Rüdiger G, Elstner D, 2000, *Astronomy & Astrophysics* 353, 813

Rempel M, Schüssler M, Tóth G, 2000, *Astronomy & Astrophysics* 363, 789

Reyes-Ruiz M, Stepinski TF, 1996, *The Astrophysical Journal* 459, 653

Reynolds RJ, 1989, *The Astrophysical Journal* 339, L29

Ribes E, 1986, *Advances Space Research* 6, 221

Ribes JC, Nesme-Ribes E, 1993, *Astronomy & Astrophysics* 276, 549

Rice JB, Strassmeier KG, 1996, *Astronomy & Astrophysics* 316, 164

Richard D, Zahn JP, 1999, *Astronomy & Astrophysics* 347, 734

Richard D, Dauchot O, Daviaud F, Zahn JP, 2001, in *International Couette-Taylor Workshop*, R. Lueptow (Ed.), Evanston, IL

Richard D, 2003, *Astronomy & Astrophysics* 408, 409

Rieutord M, Brandenburg A, Mangeney A, Drossart P, 1994, *Astronomy & Astrophysics* 286, 471

Roberts PH, 1964, *Proc. Cambridge Phil. Soc.* 60, 635

Roberts PH, 1967, *An Introduction to Magnetohydrodynamics*, Elsevier

Roberts PH, 1968, *Phil. Trans. Royal Society London A* 263, 93

Roberts PH, 1972, *Phil. Trans. Royal Society London A* 272, 663

Roberts PH, Stix M, 1972, *Astronomy & Astrophysics* 18, 453

Roberts PH, Soward A, 1975, *Astronomische Nachrichten* 296, 49

Roberts PH, 1978, in *Rotating Fluids in Geophysics*, P. H. Roberts et al. (Eds.), Academic Press, p. 421

Roberts PH, Glatzmaier GA, 2000, *Rev. Modern Physics* 72, 1081

Roberts PH, Glatzmaier GA, 2001, *Geophysical Astrophysical Fluid Dyn.* 94, 47

Rodonò M., 1992, in *Evolutionary Processes in Interacting Binary Stars*, Y. Kondo et al. (Eds.), Kluwer, p. 71

Rodonò M, Lanza AF, Catalano S, 1995, *Astronomy & Astrophysics* 301, 75

Rogachevskii I, Kleeorin N, 2001, *Physical Review E* 64, 056307

Rohde R, Elstner D, 1998, *Astronomy & Astrophysics* 333, 27

Rohde R, Elstner D, Rüdiger G, 1998, *Astronomy & Astrophysics* 329, 911

Rohde R, Rüdiger G, Elstner D, 1999, *Astronomy & Astrophysics* 347, 860

Rohlfs K, 1977, *Lectures on Density Wave Theory*, Springer

Romani RW, 1990, *Nature* 347, 741

Romanova MM, Ustyugova GV, Koldoba AV, Lovelace RVE, 2002, *The Astrophysical Journal* 578, 420

Rosenthal CS, 1998, in *Solar Convection and Oscillations and Their Relationship*, F. P. Pijpers et al. (Eds.), Kluwer, p. 145

Rosenthal CS, Bogdan TJ, Carlsson M, et al., 2002, *The Astrophysical Journal* 564, 508

Rosner R, 2003, *Nature* 425, 672

Roxburgh IW, 1978, *Astronomy & Astrophysics* 65, 281

Rozelot JP, 1995, *Astronomy & Astrophysics* 297, L45

Ruden SP, Papaloizou JCB, Lin DNC, 1988, *The Astrophysical Journal* 329, 739

Rüdiger G, 1974, *Astronomische Nachrichten* 295, 275

Rüdiger G, 1977, *Solar Physics* 51, 257

Rüdiger G, 1978, *Astronomische Nachrichten* 299, 217

Rüdiger G, 1980, *Geophysical Astrophysical Fluid Dyn.* 16, 239

Rüdiger G, Tuominen I, Krause F, Virtanen H, 1986, *Astronomy & Astrophysics* 166, 306

Rüdiger G, Scholz G, 1988, *Astronomische Nachrichten* 309, 181

Rüdiger G, 1989, *Differential Rotation and Stellar Convection: Sun and Solar-Type Stars*, Gordon and Breach Science Publishers

Rüdiger G, Kitchatinov LL, 1990, *Astronomy & Astrophysics* 236, 503

Rüdiger G, Tuominen I, 1990, in *Solar Photosphere: Structure, Convection, and Magnetic Fields*, J. O. Stenflo (Ed.), Kluwer, p. 315

Rüdiger G, Elstner D, Schultz M, 1993, *Astronomy & Astrophysics* 270, 53

Rüdiger G, Kitchatinov LL, 1993, *Astronomy & Astrophysics* 269, 581

Rüdiger G, Elstner D, 1994, *Astronomy & Astrophysics* 281, 46

Rüdiger G, Kitchatinov LL, Küker M, Schultz M, 1994, *Geophysical Astrophysical Fluid Dyn.* 78, 247

Rüdiger G, Brandenburg A, 1995, *Astronomy & Astrophysics* 296, 557

Rüdiger G, Elstner D, Stepinski TF, 1995, *Astronomy & Astrophysics* 298, 934

Rüdiger G, Arlt R, 1996, *Astronomy & Astrophysics* 316, L17

Rüdiger G, Kitchatinov LL, 1996, *The Astrophysical Journal* 466, 1078

Rüdiger G, Kitchatinov LL, 1997, *Astronomische Nachrichten* 318, 273

Rüdiger G, Schultz M, 1997, *Astronomy & Astrophysics* 319, 781

Rüdiger G, Brandenburg A, Pipin VV, 1999, *Astronomische Nachrichten* 320, 135
Rüdiger G, Primavera L, Arlt R, Elstner D, 1999, *Month. Not. Roy. Astr. Soc.* 306, 913
Rüdiger G, Kitchatinov LL, 2000a, *Astronomische Nachrichten* 321, 75
Rüdiger G, Kitchatinov LL, 2000b, *Astronomische Nachrichten* 321, 181
Rüdiger G, Pipin VV, 2000, *Astronomy & Astrophysics* 362, 756
Rüdiger G, Drecker A, 2001, *Astronomische Nachrichten* 322, 179
Rüdiger G, Pipin VV, 2001, *Astronomy & Astrophysics* 375, 149
Rüdiger G, Shalybkov D, 2002, *Physical Review E* 66, 016307
Rüdiger G, Arlt R, Shalybkov D, 2002, *Astronomy & Astrophysics* 391, 781
Rüdiger G, Tschäpe R, Kitchatinov LL, 2002, *Month. Not. Roy. Astr. Soc.* 332, 435
Rüdiger G, Elstner D, Lanza A, Granzer T, 2002, *Astronomy & Astrophysics* 392, 605
Rüdiger G, Arlt R, 2003, in *Advances in Nonlinear Dynamos, The Fluid Mechanics of Astrophysics and Geophysics*, A. Ferriz-Mas et al. (Eds.), vol. 9, Taylor & Francis, p. 147
Rüdiger G, Elstner D, Ossendrijver M, 2003, *Astronomy & Astrophysics* 406, 15
Rüdiger G, Küker M, Chan KL, 2003, *Astronomy & Astrophysics* 399, 743
Rüdiger G, Schultz M, Shalybkov D, 2003, *Physical Review E* 67, 046312
Rüdiger G, Shalybkov D, 2004a, in *Magnetic Fields and Star Formation: Theory vs. Observations*, Madrid
Rüdiger G, Shalybkov D, 2004b, *Physical Review E* 69, 016303
Runcorn SK, 1994, *Phil. Trans. Royal Society London A* 349, 181
Russell CT, 1993, *J. Geophysical Research* 98, 18681
Russell CT, 2000, *Advances Space Research* 26, 1653
Rust DM, Kumar A, 1994, *Solar Physics* 155, 69
Ruzmaikin AA, Shukurov AM, Sokoloff DD, 1988, *Magnetic Fields of Galaxies*, Kluwer
Ryu D, Goodman J, 1992, *The Astrophysical Journal* 388, 438
Saar SH, Baliunas SL, 1992, in *The Solar Cycle*, K. L. Harvey (Ed.), ASP, p. 150
Saar SH, 1998, in *Cool Stars, Stellar Systems and the Sun*, R. A. Donahue et al. (Eds.), ASP, p. 211
Saar SH, Brandenburg A, 1999, *The Astrophysical Journal* 524, 295
Saar SH, Brandenburg A, 2001, in *Magnetic Fields Across the Hertzsprung-Russell Diagram*, G. Mathys et al. (Eds.), ASP, p. 231
Safronov VS, 1969, *Evolution of the Protoplanetary Cloud and Formation of the Earth and the Planets*, Nauka
Saikia E, Singh HP, Chan KL, et al., 2000, *The Astrophysical Journal* 529, 402
Sakuraba A, Kono M, 1999, *Physics Earth Planetary Interiors* 111, 105
Sakurai T, 1985, *Astronomy & Astrophysics* 152, 121
Sano T, Inutsuka SI, Miyama SM, 1998, *The Astrophysical Journal* 506, L57
Sano T, Stone JM, 2002a, *The Astrophysical Journal* 570, 314
Sano T, Stone JM, 2002b, *The Astrophysical Journal* 577, 534
Sarson GR, Jones CA, Zhang K, Schubert G, 1997, *Science* 276, 1106
Schaaf ME, 1990, *Astronomy & Astrophysics* 227, 61
Schekochihin AA, Cowley SC, Maron JL, McWilliams JC, 2004, *Physical Review Letters* 92, 054502
Schlattl H, Weiss A, 1999, *Astronomy & Astrophysics* 347, 272
Schlichenmaier R, Stix M, 1995, *Astronomy & Astrophysics* 302, 264

Schmidt GD, 2001, in *Magnetic Fields Across the Hertzsprung-Russell Diagram*, G. Mathys et al. (Eds.), ASP, p. 443

Schmitt D, Schüssler M, 1989, *Astronomy & Astrophysics* 223, 343

Schmitt D, Rüdiger G, 1992, *Astronomy & Astrophysics* 264, 319

Schmitt D, 1993, in *The Cosmic Dynamo*, F. Krause et al. (Eds.), Kluwer, p. 1

Schou J, 2001, in *Recent Insights Into the Physics of the Sun and Heliosphere: Highlights From SOHO and Other Space Missions*, P. Brekke et al. (Eds.), ASP, p. 21

Schrijver CJ, Zwaan C, 2000, *Solar and Stellar Magnetic Activity*, Cambridge University Press

Schubert G, Russell CT, Moore WB, 2000, *Nature* 408, 666

Schubert G, Zhang K, 2000, *The Astrophysical Journal* 532, L149

Schubert G, Turcotte DL, Olson P, 2001, *Mantle Convection in the Earth and Planets*, Cambridge University Press

Schultz M, Elstner D, Rüdiger G, 1994, *Astronomy & Astrophysics* 286, 72

Schultz-Grunow F, 1959, *Z. Angewandte Mathematik und Mechanik* 39, 101

Schüssler M, 1975, *Astronomy & Astrophysics* 38, 263

Schüssler M, 1979, *Astronomy & Astrophysics* 72, 348

Schüssler M, 1981, *Astronomy & Astrophysics* 94, L17

Schüssler M, 1987, in *The Internal Solar Angular Velocity*, R. Durney (Ed.), Reidel, p. 303

Schüssler M, Caligari P, Ferriz-Mas A, Moreno-Insertis F, 1994, *Astronomy & Astrophysics* 281, L69

Schüssler M, 1996, in *Solar and Astrophysical Magnetohydrodynamic Flows*, K. Tsinganos (Ed.), Kluwer, p. 17

Schüssler M, 1996, in *Stellar Surface Structure*, K.G. Strassmeier et al. (Eds.), Kluwer, p. 269

Schüssler M, 2002, *Astronomische Nachrichten* 323, 377

Schwarz U, 1994, *Zeitreihenanalyse astrophysikalischer Aktivitätsphänomene*, Thesis, Universität Potsdam

Seehafer N, 1990, *Solar Physics* 125, 219

Seehafer N, 1996, *Physical Review E* 53, 1283

Sellwood JA, 1985, *Month. Not. Roy. Astr. Soc.* 217, 127

Sellwood JA, Balbus SA, 1999, *The Astrophysical Journal* 511, 660

Shakura NI, Sunyaev RA, 1973, *Astronomy & Astrophysics* 24, 337

Shalybkov D, Urpin VA, 1995, *Month. Not. Roy. Astr. Soc.* 273, 643

Shalybkov D, Urpin VA, 1997, *Astronomy & Astrophysics* 321, 685

Shalybkov D, Rüdiger G, 2000, *Month. Not. Roy. Astr. Soc.* 315, 762

Shalybkov D, Rüdiger G, Schultz M, 2002, *Astronomy & Astrophysics* 395, 339

Shapiro S, Teukolsky S, 1983, *Black Holes, White Dwarfs and Neutron Stars*, Wiley

Sheeley NR, 1992, in *The Solar Cycle*, K. L. Harvey (Ed.), ASP, p. 1

Showman AP, Malhotra R, 1999, *Science* 286, 77

Shu FH, 1992, *Gas Dynamics*, University Science Books

Sigl G, Olinto AV, Jedamzik K, 1997, *Physical Review D* 55, 4582

Simon GW, Title AM, Topka KP, et al., 1988, *The Astrophysical Journal* 327, 964

Simon GW, Brandt PN, November LJ, et al., 1994, in *Solar Surface Magnetism*, R. J. Rutten et al. (Eds.), Kluwer, p. 261

Simon GW, Weiss NO, 1997, *The Astrophysical Journal* 489, 960

Singh HP, Roxburgh IW, Chan KL, 1998, *Astronomy & Astrophysics* 340, 178

Skumanich A, 1972, *The Astrophysical Journal* 171, 565

Skumanich A, Lites BW, Martínez Pillet, 1994, in *Solar Surface Magnetism*, R. J. Rutten et al. (Eds.), Kluwer, p. 99

Soida M, Beck R, Urbanik M, Braine J, 2002, *Astronomy & Astrophysics* 394, 47

Sokoloff DD, Nesme-Ribes E, 1994, *Astronomy & Astrophysics* 288, 293

Solanki SK, Fligge M, 1998, *Geophysical Research Letters* 25, 341

Solanki SK, Fligge M, 1999, *Geophysical Research Letters* 26, 2465

Solanki SK, 2002, *Astronomische Nachrichten* 323, 165

Solanki SK, Krivova NA, Schüssler M, Fligge M, 2002, *Astronomy & Astrophysics* 396, 1029

Solanki SK, 2003, *Astronomy & Astrophysics Rev.* 11, 153

Solanki SK, Lagg A, Woch J, et al., 2003, *Nature* 425, 692

Song XD, Richards PG, 1996, *Nature* 382, 221

Souriau A, Garcia R, Poupinet G, 2003, *CR Geosci.* 335, 51

Soward AM, 1974, *Phil. Trans. Royal Society London A* 275, 611

Soward AM, 1978, *Astronomische Nachrichten* 299, 25

Soward AM, Jones CA, 1983, *Geophysical Astrophysical Fluid Dyn.* 27, 87

Soward AM, Hollerbach R, 2000, *J. Fluid Mechanics* 408, 239

Spencer SJ, 1994, *Global Magnetic Fields in Spiral Galaxies*, Thesis, University of Sydney

Spiegel EA, 1963, *The Astrophysical Journal* 138, 216

Spiegel EA, Weiss NO, 1980, *Nature* 287, 616

Spiegel EA, Zahn JP, 1992, *Astronomy & Astrophysics* 265, 106

Spitzer L, 1978, *Physical Processes in the Interstellar Medium*, Wiley

Spörer G, 1887, *Vierteljahresschrift der Astronomischen Gesellschaft* 22, 323

Spörer G, 1894, *Publicationen des Astrophysikalischen Observatoriums zu Potsdam* 32

Spruit HC, 1987, *Astronomy & Astrophysics* 184, 173

Spruit HC, Foglizzo T, Stehle R, 1997, *Month. Not. Roy. Astr. Soc.* 288, 333

Spruit HC, 1999, *Astronomy & Astrophysics* 349, 189

Staude J, 2002, *Astronomische Nachrichten* 323, 317

Stauffer JR, Soderblom DR, 1991, in *The Sun in Time*, C.P. Sonett et al. (Eds.), University of Arizona Press, p. 832

Steenbeck M, Krause F, 1966, *Z. Naturforsch.* 21a, 1285

Steenbeck M, Krause F, Rädler KH, 1966, *Z. Naturforsch.* 21a, 369

Steenbeck M, Krause F, 1969, *Astronomische Nachrichten* 291, 49

Stefani F, Gerbeth G, 2003, *Physical Review E* 67, 7302

Stegman DR, Jellinek AM, Zatman SA, et al., 2003, *Nature* 421, 143

Stein RF, Nordlund Å, Kuhn JR, 1988, in *Seismology of the Sun and Sun-Like Stars*, E.J. Rolfe (Ed.), ESA, p. 529

Stein RF, Nordlund Å, 1989, *The Astrophysical Journal* 342, L95

Steinacker A, Papaloizou JCB, 2002, *The Astrophysical Journal* 571, 413

Steiner O, Grossmann-Doerth U, Knölker M, Schüssler M, 1998, *The Astrophysical Journal* 495, 468

Stelzer B, Fernández M, Costa VM, et al., 2003, *Astronomy & Astrophysics* 411, 517

Stenflo JO, Vogel M, 1986, *Solar Physics* 319, 285

Stępień K, 1978, *Astronomy & Astrophysics* 70, 509

Stępień K, 1991, in *The Sun and Cool Stars. Activity, Magnetism, Dynamos*, I. Tuominen et al. (Eds.), Springer, p. 288

Stępień K, 2000, *Astronomy & Astrophysics* 353, 227

Stępień K, Landstreet JD, 2002, *Astronomy & Astrophysics* 384, 554

Stepinski TF, Levy EH, 1988, *The Astrophysical Journal* 331, 416

Stepinski TF, Levy EH, 1991, *The Astrophysical Journal* 379, 343

Stevenson DJ, 1982, *Geophysical Astrophysical Fluid Dyn.* 21, 113

Stevenson DJ, 1983, *Reports on Progress in Physics* 46, 555

Stevenson DJ, 2001, *Nature* 412, 214

Stevenson DJ, 2003, *Earth Planetary Science Letters* 208, 1

Stewartson K, 1957, *J. Fluid Mechanics* 3, 17

Stewartson K, 1966, *J. Fluid Mechanics* 26, 131

Stieglitz R, Müller U, 2001, *Physics of Fluids* 13, 561

Stix M, 1975, *Astronomy & Astrophysics* 42, 85

Stix M, 1976, *Astronomy & Astrophysics* 47, 243

Stix M, 1981, *Astronomy & Astrophysics* 93, 339

Stix M, 1989, *Rev. Modern Astronomy* 2, 248

Stix M, Skaley D, 1990, *Astronomy & Astrophysics* 232, 234

Stix M, 1991, *Geophysical Astrophysical Fluid Dyn.* 62, 211

Stix M, Rüdiger G, Knölker M, Grabowski U, 1993, *Astronomy & Astrophysics* 272, 340

Stix M, 2002, *The Sun: An Introduction*, Springer

Stix M, Zhugzhda YD, 2004, *Astronomy & Astrophysics* 418, 305

Stone JM, Mihalas D, Norman ML, 1992, *Astrophys. J. Suppl.* 80, 819

Stone JM, Norman ML, 1992a, *Astrophys. J. Suppl.* 80, 753

Stone JM, Norman ML, 1992b, *Astrophys. J. Suppl.* 80, 791

Stone JM, Balbus SA, 1996, *The Astrophysical Journal* 464, 364

Stone JM, Hawley JF, Gammie CF, Balbus SA, 1996, *The Astrophysical Journal* 463, 656

St. Pierre MG, 1993, in *Solar and Planetary Dynamos*, M. R. E. Proctor et al. (Eds.), Cambridge University Press, p. 295

St. Pierre MG, 1996, *Geophysical Astrophysical Fluid Dyn.* 83, 293

Strassmeier KG, Hall DS, Fekel FC, Scheck M, 1993, *Astronomy & Astrophysics Suppl.* 100, 173

Strassmeier KG, Rice JB, 1998, *Astronomy & Astrophysics* 330, 685

Strassmeier KG, 1999, *Astronomy & Astrophysics* 347, 225

Strassmeier KG, Serkowitsch E, Granzer T, 1999, *Astronomy & Astrophysics Suppl.* 140, 29

Strassmeier KG, 2002, *Astronomische Nachrichten* 323, 309

Strassmeier KG, 2004, in *Stars as Suns: Activity, Evolution and Planets*, A. K. Dupree et al. (Eds.), ASP

Strittmatter PA, 1966, *Month. Not. Roy. Astr. Soc.* 132, 359

Su WJ, Dziewonsky AM, Jeanloz R, 1996, *Science* 274, 1883

Sweet PA, 1950, *Month. Not. Roy. Astr. Soc.* 110, 548

Tassoul JL, 1978, *Theory of Rotating Stars*, Princeton University Press

Taylor GI, 1923, *Phil. Trans. Royal Society London A* 233, 289

Taylor JB, 1963, *Proc. Royal Society London A* 274, 274

Ternullo M, 2004, *private communication*

Terquem C, Papaloizou JCB, 1996, *Month. Not. Roy. Astr. Soc.* 279, 767

Thierbach M, Klein U, Wielebinski R, 2003, *Astronomy & Astrophysics* 397, 53

Thomas JH, Weiss NO, Tobias SM, Brummell NH, 2002, *Astronomische Nachrichten* 323, 383

Thompson C, Duncan RC, 1993, *The Astrophysical Journal* 408, 194

Thompson C, 2001, in *The Neutron Star - Black Hole Connection*, C. Kouveliotou et al. (Eds.), Kluwer, p. 369

Thompson MJ, Toomre J, Anderson E, et al., 1996, *Science* 272, 1300

Thompson MJ, Christensen-Dalsgaard J, Miesch MS, Toomre J, 2003, *Ann. Rev. Astronomy Astrophysics* 41, 599

Tilgner A, Busse FH, 1997, *J. Fluid Mechanics* 332, 359

Tilgner A, 1999, *Int. J. Numer. Meth. Fluids* 30, 713

Tobias SM, 1996, *The Astrophysical Journal* 467, 870

Tobias SM, 1997, *Astronomy & Astrophysics* 322, 1007

Tobias SM, 1998, *Month. Not. Roy. Astr. Soc.* 296, 653

Toomre A, 1964, *The Astrophysical Journal* 139, 1217

Torkelsson U, Brandenburg A, 1994, *Astronomy & Astrophysics* 283, 677

Torkelsson U, Brandenburg A, Nordlund Å, Stein RF, 1996, *Astrophysical Letters and Communications* 34, 383

Torkelsson U, 1998, *Month. Not. Roy. Astr. Soc.* 298, L55

Tout CA, Pringle JE, 1992, *Month. Not. Roy. Astr. Soc.* 259, 604

Tscharnuter WM, Winkler KH, 1979, *Computer Physics Communications* 18, 171

Tscharnuter WM, 1985, in *Birth and Infancy of Stars*, R. Lucas et al. (Eds.), Elsevier, p. 601

Tüllmann R, Dettmar RJ, Soida M, et al., 2000, *Astronomy & Astrophysics* 364, L36

Tuominen I, Rüdiger G, Brandenburg A, 1988, in *Activity in Cool Star Envelopes*, O. Havnes et al. (Eds.), Kluwer, p. 13

Tuominen I, Berdyugina SV, Korpi MJ, 2002, *Astronomische Nachrichten* 323, 367

Tuominen J, 1941, *Z. Astrophysik* 21, 96

Tuominen J, 1961, *Z. Astrophysik* 51, 91

Ulrich RK, 1976, *The Astrophysical Journal* 207, 564

Ulrich RK, Hawkins GW, 1981, *The Astrophysical Journal* 246, 985

Urpin VA, Geppert U, 1995, *Month. Not. Roy. Astr. Soc.* 275, 1117

Urpin VA, 1996, *Month. Not. Roy. Astr. Soc.* 280, 149

Urpin VA, Brandenburg A, 1998, *Month. Not. Roy. Astr. Soc.* 294, 399

Urpin VA, Rüdiger G, 2004, *Astronomy & Astrophysics*, subm.

Vainshtein SI, Kitchatinov LL, 1983, *Geophysical Astrophysical Fluid Dyn.* 24, 273

Vainshtein SI, Cattaneo F, 1992, *The Astrophysical Journal* 393, 165

Vainshtein SI, Chitre SM, Olinto AV, 2000, *Physical Review E* 61, 4422

Valenti JA, Basri G, Johns CM, 1993, *Astronomical Journal* 106, 2024

Velikhov EP, 1959, *Soviet Physics JETP* 9, 995

Verma VK, 1993, *The Astrophysical Journal* 403, 797

Vidale JE, Dodge DA, Earle PS, 2000, *Nature* 405, 445

Vishniac ET, Brandenburg A, 1997, *The Astrophysical Journal* 475, 263

Völk HJ, Jones FC, Morfill GE, Röser S, 1980, *Astronomy & Astrophysics* 85, 316

Vorontsov SV, Christensen-Dalsgaard J, Schou J, et al., 2002, *Science* 296, 101

Voss H, Kurths J, Schwarz U, 1996, *J. Geophysical Research* 101, 15637
Voss H, Sanchez A, Zolitschka B, et al., 1997, *Surveys Geophysics* 18, 163
Vršnak B, Brajša R, Wöhl H, et al., 2003, *Astronomy & Astrophysics* 404, 1117
Wälder M, Deinzer W, Stix M, 1980, *J. Fluid Mechanics* 96, 207
Walker MR, Barenghi CF, Jones CA, 1998, *Geophysical Astrophysical Fluid Dyn.* 88, 261
Walker MR, Hollerbach R, 1999, *Physics Earth Planetary Interiors* 114, 181
Wang Y, Sheeley NR, Nash AG, 1991, *The Astrophysical Journal* 383, 431
Wang Y, Noyes RW, Tarbell TD, Title AM, 1995, *The Astrophysical Journal* 447, 419
Ward F, 1965, *The Astrophysical Journal* 141, 534
Ward F, 1973, *Solar Physics* 31, 131
Wardle M, 1999, *Month. Not. Roy. Astr. Soc.* 307, 849
Wasiutynski J, 1946, *Astrophysica Norvegica* 4, 1
Watson M, 1981, *Geophysical Astrophysical Fluid Dyn.* 16, 285
Weiss NO, 1966, *Proc. Royal Society London A* 293, 310
Weiss NO, Cattaneo F, Jones CA, 1984, *Geophysical Astrophysical Fluid Dyn.* 30, 305
Weiss NO, 1994, in *Lectures on Solar and Planetary Dynamos*, M. R. E. Proctor et al. (Eds.), Cambridge University Press, p. 59
Weiss NO, Brownjohn DP, Matthews PC, Proctor MRE, 1996, *Month. Not. Roy. Astr. Soc.* 283, 1153
Weiss NO, 1997, in *Past and Present Variability of the Solar-Terrestrial System: Measurement, Data Analysis and Theoretical Models*, G. C. Castagnoli et al. (Eds.), IOS Press, p. 325
Weisshaar E, 1982, *Geophysical Astrophysical Fluid Dyn.* 21, 285
v. Weizsäcker CF, 1943, *Z. Astrophysik* 22, 319
v. Weizsäcker CF, 1948, *Z. Naturforschung* 3a, 524
Wendt F, 1933, *Ingenieur-Archiv* 4, 577
White M, 1978, *Astronomische Nachrichten* 299, 209
Wicht J, 2002, *Physics Earth Planetary Interiors* 132, 281
Wickramasinghe DT, Ferrario L, 2000, *Publications of the Astronomical Society of the Pacific* 112, 873
Wickramasinghe DT, 2001, in *Magnetic Fields Across the Hertzsprung-Russell Diagram*, G. Mathys et al. (Eds.), ASP, p. 453
Widrow LM, 2002, *Rev. Modern Physics* 74, 775
Wiebicke HJ, Geppert U, 1996, *Astronomy & Astrophysics* 309, 203
Wielebinski R, Krause F, 1993, *Astronomy & Astrophysics Rev.* 4, 449
Willis AP, Barenghi CF, 2002, *Astronomy & Astrophysics* 393, 339
Witt A, 1996, *Komplexitätsmaße und ihre Anwendungen*, Thesis, Universität Potsdam
Wittmann A, 1978, *Astronomy & Astrophysics* 66, 93
Wolfe AM, Lanzetta KM, Oren AL, 1992, *The Astrophysical Journal* 388, 17
Worster MG, 2000, in *Perspectives in Fluid Dynamics*, G. K. Batchelor et al. (Eds.), Cambridge University Press, p. 393
Wrubel MH, 1952, *The Astrophysical Journal* 116, 291
Xie X, Toomre J, 1993, *The Astrophysical Journal* 405, 747
Yoshimura H, 1981, *The Astrophysical Journal* 247, 1102
Youdin AN, Shu FH, 2002, *The Astrophysical Journal* 580, 494

Yousef T, Brandenburg A, Rüdiger G, 2003, *Astronomy & Astrophysics* 411, 321

Zahn JP, 1979, *Lecture Notes in Physics* 114, 1

Zahn JP, 1989, *Astronomy & Astrophysics* 220, 112

Zatman S, Bloxham J, 1997, *Nature* 388, 760

Zatman S, Bloxham J, 1999, *Geophysical J. International* 138, 679

Zhang K, Fearn DR, 1993, *Geophysical Research Letters* 20, 2083

Zhang K, 1994, *J. Fluid Mechanics* 268, 211

Zhang K, 1995, *Proc. Royal Society London A* 448, 245

Zhang K, Jones CA, 1996, *Proc. Royal Society London A* 452, 981

Zhang K, Jones CA, 1997, *Geophysical Research Letters* 24, 2869

Zhang K, Gubbins D, 2000, *Geophysical J. International* 140, 1

Zhao J, Kosovichev AG, 2004, *The Astrophysical Journal* 603, 776

Zhugzhda YD, Stix M, 1994, *Astronomy & Astrophysics* 291, 310

Ziegler U, 1996, *Astronomy & Astrophysics* 313, 448

Ziegler U, Yorke HW, Kaisig M, 1996, *Astronomy & Astrophysics* 305, 114

Ziegler U, 1998, *Computer Physics Communications* 109, 111

Ziegler U, Rüdiger G, 2000, *Astronomy & Astrophysics* 356, 1141

Ziegler U, Rüdiger G, 2001, *Astronomy & Astrophysics* 378, 668

Ziegler U, 2002, *Astronomy & Astrophysics* 386, 331

Ziegler U, Rüdiger G, 2003, *Astronomy & Astrophysics* 401, 433

Zöllner F, 1881, in *Wissenschaftliche Abhandlungen*, Bd. 4, Commissionsverlag von L. Staackmann

Zwaan C, 1992, in *Sunspots: Theory and Observations*, J. H. Thomas et al. (Eds.), Kluwer, p. 75

Index